Kalman Filtering

C.K. Chui · G. Chen

Kalman Filtering
with Real-Time Applications

Fourth Edition

 Springer

Professor Charles K. Chui
Texas A&M University
Department Mathematics
608K Blocker Hall
College Station, TX, 77843
USA

Professor Guanrong Chen
City University Hong Kong
Department of Electronic Engineering
83 Tat Chee Avenue
Kowloon
Hong Kong/PR China

Second printing of the third edition with ISBN 3-540-64611-6, published as softcover edition in *Springer Series in Information Sciences*.

ISBN 978-3-642-09966-3   e-ISBN 978-3-540-87849-0

DOI 10.1007/978-3-540-87849-0

Cover design: eStudioCalamar S.L., F. Steinen-Broo, Girona, Spain

Printed on acid-free paper

9 8 7 6 5 4 3 2 1

springer.com

Preface to the Third Edition

Two modern topics in Kalman filtering are new additions to this Third Edition of **Kalman Filtering with Real-Time Applications**. Interval Kalman Filtering (Chapter 10) is added to expand the capability of Kalman filtering to uncertain systems, and Wavelet Kalman Filtering (Chapter 11) is introduced to incorporate efficient techniques from wavelets and splines with Kalman filtering to give more effective computational schemes for treating problems in such areas as signal estimation and signal decomposition. It is hoped that with the addition of these two new chapters, the current edition gives a more complete and up-to-date treatment of Kalman filtering for real-time applications.

College Station and Houston   Charles K. Chui
August 1998   Guanrong Chen

Preface to the Second Edition

In addition to making a number of minor corrections and updating the list of references, we have expanded the section on "real-time system identification" in Chapter 10 of the first edition into two sections and combined it with Chapter 8. In its place, a very brief introduction to wavelet analysis is included in Chapter 10. Although the pyramid algorithms for wavelet decompositions and reconstructions are quite different from the Kalman filtering algorithms, they can also be applied to time-domain filtering, and it is hoped that splines and wavelets can be incorporated with Kalman filtering in the near future.

College Station and Houston   Charles K. Chui
September 1990   Guanrong Chen

Preface to the First Edition

Kalman filtering is an optimal state estimation process applied to a dynamic system that involves random perturbations. More precisely, the Kalman filter gives a linear, unbiased, and minimum error variance recursive algorithm to optimally estimate the unknown state of a dynamic system from noisy data taken at discrete real-time. It has been widely used in many areas of industrial and government applications such as video and laser tracking systems, satellite navigation, ballistic missile trajectory estimation, radar, and fire control. With the recent development of high-speed computers, the Kalman filter has become more useful even for very complicated real-time applications.

In spite of its importance, the mathematical theory of Kalman filtering and its implications are not well understood even among many applied mathematicians and engineers. In fact, most practitioners are just told what the filtering algorithms are without knowing why they work so well. One of the main objectives of this text is to disclose this mystery by presenting a fairly thorough discussion of its mathematical theory and applications to various elementary real-time problems.

A very elementary derivation of the filtering equations is first presented. By assuming that certain matrices are nonsingular, the advantage of this approach is that the optimality of the Kalman filter can be easily understood. Of course these assumptions can be dropped by using the more well known method of orthogonal projection usually known as the innovations approach. This is done next, again rigorously. This approach is extended first to take care of correlated system and measurement noises, and then colored noise processes. Kalman filtering for nonlinear systems with an application to adaptive system identification is also discussed in this text. In addition, the limiting or steady-state Kalman filtering theory and efficient computational schemes such as the sequential and square-root algorithms are included for real-time application purposes. One such application is the design of a digital tracking filter such as the $\alpha - \beta - \gamma$ and $\alpha - \beta - \gamma - \theta$

trackers. Using the limit of Kalman gains to define the α, β, γ parameters for white noise and the $\alpha, \beta, \gamma, \theta$ values for colored noise processes, it is now possible to characterize this tracking filter as a limiting or steady-state Kalman filter. The state estimation obtained by these much more efficient prediction-correction equations is proved to be near-optimal, in the sense that its error from the optimal estimate decays exponentially with time. Our study of this topic includes a decoupling method that yields the filtering equations for each component of the state vector.

The style of writing in this book is intended to be informal, the mathematical argument throughout elementary and rigorous, and in addition, easily readable by anyone, student or professional, with a minimal knowledge of linear algebra and system theory. In this regard, a preliminary chapter on matrix theory, determinants, probability, and least-squares is included in an attempt to ensure that this text be self-contained. Each chapter contains a variety of exercises for the purpose of illustrating certain related view-points, improving the understanding of the material, or filling in the gaps of some proofs in the text. Answers and hints are given at the end of the text, and a collection of notes and references is included for the reader who might be interested in further study.

This book is designed to serve three purposes. It is written not only for self-study but also for use in a one-quarter or one-semester introductory course on Kalman filtering theory for upper-division undergraduate or first-year graduate applied mathematics or engineering students. In addition, it is hoped that it will become a valuable reference to any industrial or government engineer.

The first author would like to thank the U.S. Army Research Office for continuous support and is especially indebted to Robert Green of the White Sands Missile Range for his encouragement and many stimulating discussions. To his wife, Margaret, he would like to express his appreciation for her understanding and constant support. The second author is very grateful to Professor Mingjun Chen of Zhongshan University for introducing him to this important research area, and to his wife Qiyun Xian for her patience and encouragement.

Among the colleagues who have made valuable suggestions, the authors would especially like to thank Professors Andrew Chan (Texas A&M), Thomas Huang (Illinois), and Thomas Kailath (Stanford). Finally, the friendly cooperation and kind assistance from Dr. Helmut Lotsch, Dr. Angela Lahee, and their editorial staff at Springer-Verlag are greatly appreciated.

College Station   Charles K. Chui
Texas, January, 1987   Guanrong Chen

Contents

Notation

A, A_k	system matrices	
A^c	"square-root" of A in Cholesky factorization	103
A^u	"square-root" of A in upper triangular decomposition	107
B, B_k	control input matrices	
C, C_k	measurement matrices	
$\text{Cov}(X, Y)$	covariance of random variables X and Y	13
$E(X)$	expectation of random variable X	9
$E(X|Y = \mathbf{y})$	conditional expectation	14
$\mathbf{e}_j, \hat{\mathbf{e}}_j$		36,37
$f(x)$	probability density function	9
$f(x_1, x_2)$	joint probability density function	10
$f(x_1|x_2)$	conditional probability density function	12
$\mathbf{f}_k(\mathbf{x}_k)$	vector-valued nonlinear functions	108
G	limiting Kalman gain matrix	78
G_k	Kalman gain matrix	23
$H_k(\mathbf{x}_k)$	matrix-valued nonlinear function	108
H^*		50
I_n	$n \times n$ identity matrix	
J	Jordan canonical form of a matrix	7
K_k		56
$L(\mathbf{x}, \mathbf{v})$		51
$M_{A\Gamma}$	controllability matrix	85
N_{CA}	observability matrix	79
$O_{n \times m}$	$n \times m$ zero matrix	
P	limiting (error) covariance matrix	79
$P_{k,k}$	estimate (error) covariance matrix	26
$P[i, j]$	(i, j)th entry of matrix P	
$P(X)$	probability of random variable X	8
Q_k	variance matrix of random vector $\underline{\xi}_k$	
R_k	variance matrix of random vector $\underline{\eta}_k$	

XIV   Notation

$\mathbf{R}^n$	space of column vectors $\mathbf{x} = [x_1 \ \cdots \ x_n]^{\top}$	
S_k	covariance matrix of $\underline{\xi}_k$ and $\underline{\eta}_k$	55
tr	trace	5
$\mathbf{u}_k$	deterministic control input (at the kth time instant)	
$\mathrm{Var}(X)$	variance of random variable X	10
$\mathrm{Var}(X|Y=\mathbf{y})$	conditional variance	14
$\mathbf{v}_k$	observation (or measurement) data (at the kth time instant)	
$\mathbf{v}^{2\#}$		53
W_k	weight matrix	15
$\mathbf{w}_j$		34
$(W_\psi f)(b,a)$	integral wavelet transform	164
$\mathbf{x}_k$	state vector (at the kth time instant)	
$\hat{\mathbf{x}}_k, \hat{\mathbf{x}}_{k|k}$	optimal filtering estimate of $\mathbf{x}_k$	
$\hat{\mathbf{x}}_{k|k-1}$	optimal prediction of $\mathbf{x}_k$	
$\check{\mathbf{x}}_k$	suboptimal estimate of $\mathbf{x}_k$	
$\vec{\mathbf{x}}_k$	near-optimal estimate of $\mathbf{x}_k$	
$\mathbf{x}^*$		52
$\mathbf{x}^{\#}, \mathbf{x}_k^{\#}$		53,57
$\|\mathbf{w}\|$	"norm" of $\mathbf{w}$	34
$\langle \mathbf{x}, \mathbf{w} \rangle$	"inner product" of $\mathbf{x}$ and $\mathbf{w}$	33
$Y(\mathbf{w}_0, \cdots, \mathbf{w}_r)$	"linear span" of vectors $\mathbf{w}_0, \cdots, \mathbf{w}_r$	34
$\{\mathbf{z}_j\}$	innovations sequence of data	35
$\alpha, \beta, \gamma, \theta$	tracker parameters	136,141,180
$\{\underline{\beta}_k\}, \{\underline{\gamma}_k\}$	white noise sequences	67
Γ, Γ_k	system noise matrices	
δ_{ij}	Kronecker delta	15
$\underline{\xi}_{k,\ell}, \overline{\underline{\xi}}_{k,\ell}$	random (noise) vectors	22
$\underline{\eta}_k$	measurement noise (at the kth time instant)	
$\underline{\xi}_k$	system noise (at the kth time instant)	
$\Phi_{k\ell}$	transition matrix	22
df/dA	Jacobian matrix	65
$\partial\mathbf{h}/\partial\mathbf{x}$	Jacobian matrix	110

1. Preliminaries

The importance of Kalman filtering in engineering applications is well known, and its mathematical theory has been rigorously established. The main objective of this treatise is to present a thorough discussion of the mathematical theory, computational algorithms, and application to real-time tracking problems of the Kalman filter.

In explaining how the Kalman filtering algorithm is obtained and how well it performs, it is necessary to use some formulas and inequalities in matrix algebra. In addition, since only the statistical properties of both the system and measurement noise processes in real-time applications are being considered, some knowledge of certain basic concepts in probability theory will be helpful. This chapter is devoted to the study of these topics.

1.1 Matrix and Determinant Preliminaries

Let $\mathbf{R}^n$ denote the space of all column vectors $\mathbf{x} = [x_1 \cdots x_n]^\top$, where $x_1, \cdots, x_n$ are real numbers. An $n \times n$ real matrix A is said to be *positive definite* if $\mathbf{x}^\top A\mathbf{x}$ is a positive number for all nonzero vectors $\mathbf{x}$ in $\mathbf{R}^n$. It is said to be *non-negative definite* if $\mathbf{x}^\top A\mathbf{x}$ is non-negative for any $\mathbf{x}$ in $\mathbf{R}^n$. If A and B are any two $n \times n$ matrices of real numbers, we will use the notation

$$A > B$$

when the matrix $A - B$ is positive definite, and

$$A \geq B$$

when $A - B$ is non-negative definite.

We first recall the so-called *Schwarz inequality*:

$$|\mathbf{x}^\mathsf{T}\mathbf{y}| \le |\mathbf{x}|\,|\mathbf{y}|, \quad \mathbf{x}, \mathbf{y} \in \mathbf{R}^n,$$

where, as usual, the notation

$$|\mathbf{x}| = (\mathbf{x}^\mathsf{T}\mathbf{x})^{1/2}$$

is used. In addition, we recall that the above inequality becomes equality if and only if $\mathbf{x}$ and $\mathbf{y}$ are parallel, and this, in turn, means that

$$\mathbf{x} = \lambda\mathbf{y} \quad or \quad \mathbf{y} = \lambda\mathbf{x}$$

for some scalar λ. Note, in particular, that if $\mathbf{y} \neq 0$, then the Schwarz inequality may be written as

$$\mathbf{x}^\mathsf{T}\mathbf{x} \ge (\mathbf{y}^\mathsf{T}\mathbf{x})^\mathsf{T}(\mathbf{y}^\mathsf{T}\mathbf{y})^{-1}(\mathbf{y}^\mathsf{T}\mathbf{x}).$$

This formulation allows us to generalize the Schwarz inequality to the matrix setting.

Lemma 1.1. (*Matrix Schwarz inequality*) *Let P and Q be $m \times n$ and $m \times \ell$ matrices, respectively, such that $P^\mathsf{T}P$ is nonsingular. Then*

$$Q^\mathsf{T}Q \ge (P^\mathsf{T}Q)^\mathsf{T}(P^\mathsf{T}P)^{-1}(P^\mathsf{T}Q). \tag{1.1}$$

Furthermore, equality in (1.1) holds if and only if $Q = PS$ for some $n \times \ell$ matrix S.

The proof of the (*vector*) *Schwarz inequality* is simple. It amounts to observing that the minimum of the quadratic polynomial

$$(\mathbf{x} - \lambda\mathbf{y})^\mathsf{T}(\mathbf{x} - \lambda\mathbf{y}), \quad \mathbf{y} \neq 0,$$

of λ is attained at

$$\lambda = (\mathbf{y}^\mathsf{T}\mathbf{y})^{-1}(\mathbf{y}^\mathsf{T}\mathbf{x})$$

and using this λ value in the above inequality. Hence, in the matrix setting, we consider

$$(Q - PS)^\mathsf{T}(Q - PS) \ge 0$$

and choose

$$S = (P^\mathsf{T}P)^{-1}(P^\mathsf{T}Q),$$

so that

$$Q^\mathsf{T}Q \ge S^\mathsf{T}(P^\mathsf{T}Q) + (P^\mathsf{T}Q)^\mathsf{T}S - S^\mathsf{T}(P^\mathsf{T}P)S = (P^\mathsf{T}Q)^\mathsf{T}(P^\mathsf{T}P)^{-1}(P^\mathsf{T}Q)$$

as stated in (1.1). Furthermore, this inequality becomes equality if and only if

$$(Q - PS)^\mathsf{T}(Q - PS) = 0,$$

or equivalently, $Q = PS$ for some $n \times \ell$ matrix S. This completes the proof of the lemma.

We now turn to the following so-called *matrix inversion lemma*.

Lemma 1.2. *(Matrix inversion lemma) Let*

$$A = \begin{bmatrix} A_{11} & A_{12} \\ A_{21} & A_{22} \end{bmatrix},$$

where A_{11} and A_{22} are $n \times n$ and $m \times m$ nonsingular submatrices, respectively, such that

$$(A_{11} - A_{12}A_{22}^{-1}A_{21}) \quad and \quad (A_{22} - A_{21}A_{11}^{-1}A_{12})$$

are also nonsingular. Then A is nonsingular with

$$A^{-1}$$

$$= \begin{bmatrix} A_{11}^{-1} + A_{11}^{-1}A_{12}(A_{22} & -A_{11}^{-1}A_{12}(A_{22} - A_{21}A_{11}^{-1}A_{12})^{-1} \\ \quad -A_{21}A_{11}^{-1}A_{12})^{-1}A_{21}A_{11}^{-1} & \\ -(A_{22} - A_{21}A_{11}^{-1}A_{12})^{-1}A_{21}A_{11}^{-1} & (A_{22} - A_{21}A_{11}^{-1}A_{12})^{-1} \end{bmatrix}$$

$$= \begin{bmatrix} (A_{11} - A_{12}A_{22}^{-1}A_{21})^{-1} & -(A_{11} - A_{12}A_{22}^{-1}A_{21})^{-1}A_{12}A_{22}^{-1} \\ -A_{22}^{-1}A_{21}(A_{11} - A_{12}A_{22}^{-1}A_{21})^{-1} & A_{22}^{-1} + A_{22}^{-1}A_{21}(A_{11} \\ & \quad -A_{12}A_{22}^{-1}A_{21})^{-1}A_{12}A_{22}^{-1} \end{bmatrix}.$$

$$(1.2)$$

In particular,

$$(A_{11} - A_{12}A_{22}^{-1}A_{21})^{-1}$$
$$= A_{11}^{-1} + A_{11}^{-1}A_{12}(A_{22} - A_{21}A_{11}^{-1}A_{12})^{-1}A_{21}A_{11}^{-1} \qquad (1.3)$$

and

$$A_{11}^{-1}A_{12}(A_{22} - A_{21}A_{11}^{-1}A_{12})^{-1}$$
$$= (A_{11} - A_{12}A_{22}^{-1}A_{21})^{-1}A_{12}A_{22}^{-1}. \qquad (1.4)$$

Furthermore,

$$det A = (det A_{11}) det(A_{22} - A_{21}A_{11}^{-1}A_{12})$$
$$= (det A_{22}) det(A_{11} - A_{12}A_{22}^{-1}A_{21}). \qquad (1.5)$$

To prove this lemma, we write

$$A = \begin{bmatrix} I_n & 0 \\ A_{21}A_{11}^{-1} & I_m \end{bmatrix} \begin{bmatrix} A_{11} & A_{12} \\ 0 & A_{22} - A_{21}A_{11}^{-1}A_{12} \end{bmatrix}$$

and

$$A = \begin{bmatrix} I_n & A_{12}A_{22}^{-1} \\ 0 & I_m \end{bmatrix} \begin{bmatrix} A_{11} - A_{12}A_{22}^{-1}A_{21} & 0 \\ A_{21} & A_{22} \end{bmatrix}.$$

Taking determinants, we obtain (1.5). In particular, we have

$$det\, A \neq 0\,,$$

or A is nonsingular. Now observe that

$$\begin{bmatrix} A_{11} & A_{12} \\ 0 & A_{22} - A_{21}A_{11}^{-1}A_{12} \end{bmatrix}^{-1}$$
$$= \begin{bmatrix} A_{11}^{-1} & -A_{11}^{-1}A_{12}(A_{22} - A_{21}A_{11}^{-1}A_{12})^{-1} \\ 0 & (A_{22} - A_{21}A_{11}^{-1}A_{12})^{-1} \end{bmatrix}$$

and

$$\begin{bmatrix} I_n & 0 \\ A_{21}A_{11}^{-1} & I_m \end{bmatrix}^{-1} = \begin{bmatrix} I_n & 0 \\ -A_{21}A_{11}^{-1} & I_m \end{bmatrix}.$$

Hence, we have

$$A^{-1} = \begin{bmatrix} A_{11} & A_{12} \\ 0 & A_{22} - A_{21}A_{11}^{-1}A_{12} \end{bmatrix}^{-1} \begin{bmatrix} I_n & 0 \\ A_{21}A_{11}^{-1} & I_m \end{bmatrix}^{-1}$$

which gives the first part of (1.2). A similar proof also gives the second part of (1.2). Finally, (1.3) and (1.4) follow by equating the appropriate blocks in (1.2).

An immediate application of Lemma 1.2 yields the following result.

Lemma 1.3. If $P \geq Q > 0$, then $Q^{-1} \geq P^{-1} > 0$.

Let $P(\epsilon) = P + \epsilon I$ where $\epsilon > 0$. Then $P(\epsilon) - Q > 0$. By Lemma 1.2, we have

$$P^{-1}(\epsilon) = [Q + (P(\epsilon) - Q)]^{-1}$$
$$= Q^{-1} - Q^{-1}[(P(\epsilon) - Q)^{-1} + Q^{-1}]^{-1}Q^{-1}\,,$$

so that

$$Q^{-1} - P^{-1}(\epsilon) = Q^{-1}[(P(\epsilon) - Q)^{-1} + Q^{-1}]^{-1}Q^{-1} > 0\,.$$

Letting $\epsilon \to 0$ gives $Q^{-1} - P^{-1} \geq 0$, so that

$$Q^{-1} \geq P^{-1} > 0.$$

Now let us turn to discussing the trace of an $n \times n$ matrix A. The *trace* of A, denoted by $tr A$, is defined as the sum of its diagonal elements, namely:

$$tr A = \sum_{i=1}^{n} a_{ii},$$

where $A = [a_{ij}]$. We first state some elementary properties.

Lemma 1.4. *If A and B are $n \times n$ matrices, then*

$$tr A^{\top} = tr A, \tag{1.6}$$

$$tr(A + B) = tr A + tr B, \tag{1.7}$$

and

$$tr(\lambda A) = \lambda \, tr A. \tag{1.8}$$

If A is an $n \times m$ matrix and B is an $m \times n$ matrix, then

$$tr AB = tr B^{\top} A^{\top} = tr BA = tr A^{\top} B^{\top} \tag{1.9}$$

and

$$tr A^{\top} A = \sum_{i=1}^{n} \sum_{j=1}^{m} a_{ij}^2. \tag{1.10}$$

The proof of the above identities is immediate from the definition and we leave it to the reader (cf. Exercise 1.1). The following result is important.

Lemma 1.5. *Let A be an $n \times n$ matrix with eigenvalues $\lambda_1, \cdots, \lambda_n$, multiplicities being listed. Then*

$$tr A = \sum_{i=1}^{n} \lambda_i. \tag{1.11}$$

To prove the lemma, we simply write $A = UJU^{-1}$ where J is the Jordan canonical form of A and U is some nonsingular matrix. Then an application of (1.9) yields

$$tr A = tr(AU)U^{-1} = tr U^{-1}(AU) = tr J = \sum_{i=1}^{n} \lambda_i.$$

It follows from this lemma that if $A > 0$ then $tr A > 0$, and if $A \geq 0$ then $tr A \geq 0$.

Next, we state some useful inequalities on the trace.

Lemma 1.6. *Let A be an $n \times n$ matrix. Then*

$$tr A \leq (n \, tr AA^\top)^{1/2}. \tag{1.12}$$

We leave the proof of this inequality to the reader (cf. Exercise 1.2).

Lemma 1.7. *If A and B are $n \times m$ and $m \times \ell$ matrices respectively, then*

$$tr(AB)(AB)^\top \leq (tr AA^\top)(tr BB^\top).$$

Consequently, for any matrices $A_1, \cdots, A_p$ with appropriate dimensions,

$$tr(A_1 \cdots A_p)(A_1 \cdots A_p)^\top \leq (tr A_1 A_1^\top) \cdots (tr A_p A_p^\top). \tag{1.13}$$

If $A = [a_{ij}]$ and $B = [b_{ij}]$, then

$$tr(AB)(AB)^\top = tr \left[\sum_{k=1}^m a_{ik} b_{kj} \right] \left[\sum_{k=1}^m a_{jk} b_{ki} \right]$$

$$= tr \begin{bmatrix} \sum_{p=1}^\ell \left(\sum_{k=1}^m a_{1k} b_{kp} \right)^2 & & * \\ & \ddots & \\ * & & \sum_{p=1}^\ell \left(\sum_{k=1}^m a_{nk} b_{kp} \right)^2 \end{bmatrix}$$

$$= \sum_{i=1}^n \sum_{p=1}^\ell \left(\sum_{k=1}^m a_{ik} b_{kp} \right)^2 \leq \sum_{i=1}^n \sum_{p=1}^\ell \sum_{k=1}^m a_{ik}^2 \sum_{k=1}^m b_{kp}^2$$

$$= \left(\sum_{i=1}^n \sum_{k=1}^m a_{ik}^2 \right) \left(\sum_{p=1}^\ell \sum_{k=1}^m b_{kp}^2 \right) = (tr AA^\top)(tr BB^\top),$$

where the Schwarz inequality has been used. This completes the proof of the lemma.

It should be remarked that $A \geq B > 0$ does not necessarily imply $tr AA^\top \geq tr BB^\top$. An example is

$$A = \begin{bmatrix} \frac{12}{5} & 0 \\ 0 & 1 \end{bmatrix} \quad and \quad B = \begin{bmatrix} 2 & -1 \\ 1 & 1 \end{bmatrix}.$$

Here, it is clear that $A - B \geq 0$ and $B > 0$, but

$$tr AA^\top = \frac{169}{25} < 7 = tr BB^\top$$

(cf. Exercise 1.3).

For a symmetric matrix, however, we can draw the expected conclusion as follows.

Lemma 1.8. *Let A and B be non-negative definite symmetric matrices with $A \geq B$. Then $tr AA^\top \geq tr BB^\top$, or $tr A^2 \geq tr B^2$.*

We leave the proof of this lemma as an exercise (cf. Exercise 1.4).

Lemma 1.9. *Let B be an $n \times n$ non-negative definite symmetric matrix. Then*

$$tr B^2 \leq (tr B)^2. \tag{1.14}$$

Consequently, if A is another $n \times n$ non-negative definite symmetric matrix such that $B \leq A$, then

$$tr B^2 \leq (tr A)^2. \tag{1.15}$$

To prove (1.14), let $\lambda_1, \cdots, \lambda_n$ be the eigenvalues of B. Then $\lambda_1^2, \cdots, \lambda_n^2$ are the eigenvalues of B^2. Now, since $\lambda_1, \cdots, \lambda_n$ are non-negative, Lemma 1.5 gives

$$tr B^2 = \sum_{i=1}^n \lambda_i^2 \leq \left(\sum_{i=1}^n \lambda_i\right)^2 = (tr B)^2.$$

(1.15) follows from the fact that $B \leq A$ implies $tr B \leq tr A$.

We also have the following result which will be useful later.

Lemma 1.10. *Let F be an $n \times n$ matrix with eigenvalues $\lambda_1, \cdots, \lambda_n$ such that*

$$\lambda := max(|\lambda_1|, \cdots, |\lambda_n|) < 1.$$

Then there exist a real number r satisfying $0 < r < 1$ and a constant C such that

$$\left|tr F^k (F^k)^\top\right| \leq C r^k$$

for all $k = 1, 2, \cdots$.

Let J be the Jordan canonical form for F. Then $F = UJU^{-1}$ for some nonsingular matrix U. Hence, using (1.13), we have

$$\left|tr F^k (F^k)^\top\right| = \left|tr U J^k U^{-1} (U^{-1})^\top (J^k)^\top U^\top\right|$$

$$\leq \left|tr U U^\top\right|\left|tr J^k (J^k)^\top\right|\left|tr U^{-1}(U^{-1})^\top\right|$$

$$\leq p(k)\lambda^{2k},$$

where $p(k)$ is a polynomial in k. Now, any choice of r satisfying $\lambda^2 < r < 1$ yields the desired result, by choosing a positive constant C that satisfies

$$p(k)\left(\frac{\lambda^2}{r}\right)^k \leq C$$

for all k.

1.2 Probability Preliminaries

Consider an experiment in which a fair coin is tossed such that on each toss either the head (denoted by H) or the tail (denoted by T) occurs. The actual result that occurs when the experiment is performed is called an *outcome* of the experiment and the set of all possible outcomes is called the *sample space* (denoted by S) of the experiment. For instance, if a fair coin is tossed twice, then each result of two tosses is an outcome, the possibilities are HH, TT, HT, TH, and the set $\{HH, TT, HT, TH\}$ is the sample space S. Furthermore, any subset of the sample space is called an *event* and an event consisting of a single outcome is called a *simple event*.

Since there is no way to predict the outcomes, we have to assign a real number P, between 0 and 1, to each event to indicate the probability that a certain outcome occurs. This is specified by a real-valued function, called a *random variable*, defined on the sample space. In the above example, if the random variable $X = X(s)$, $s \in S$, denotes the number of H's in the outcome s, then the number $P = P(X(s))$ gives the probability in percentage in the number of H's of the outcome s. More generally, let S be a sample space and $X : S \to \mathbf{R}^1$ be a random variable. For each measurable set $A \subset \mathbf{R}^1$ (and in the above example, $A = \{0\}$, $\{1\}$, or $\{2\}$ indicating no H, one H, or two H's, respectively) define $P : \{\text{events}\} \to [0,1]$, where each event is a set $\{s \in S : X(s) \in A \subset \mathbf{R}^1\} := \{X \in A\}$, subject to the following conditions:

(1) $P(X \in A) \geq 0$ for any measurable set $A \subset \mathbf{R}^1$,
(2) $P(X \in \mathbf{R}^1) = 1$, and
(3) for any countable sequence of pairwise disjoint measurable sets A_i in $\mathbf{R}^1$,

$$P\left(X \in \bigcup_{i=1}^{\infty} A_i\right) = \sum_{i=1}^{\infty} P(X \in A_i).$$

P is called the *probability distribution* (or *probability distribution function*) of the random variable X.

If there exists an integrable function f such that

$$P(X \in A) = \int_A f(x)dx \tag{1.16}$$

for all measurable sets A, we say that P is a *continuous* probability distribution and f is called the *probability density function* of the random variable X. Note that actually we could have defined $f(x)dx = d\lambda$ where λ is a *measure* (for example, step functions) so that the discrete case such as the example of "tossing coins" can be included.

If the probability density function f is given by

$$f(x) = \frac{1}{\sqrt{2\pi}\,\sigma} e^{-\frac{1}{2\sigma^2}(x-\mu)^2}, \quad \sigma > 0 \quad \text{and} \quad \mu \in \mathbf{R}, \tag{1.17}$$

called the *Gaussian* (or *normal*) *probability density function*, then P is called a *normal distribution* of the random variable X, and we use the notation: $X \sim N(\mu, \sigma^2)$. It can be easily verified that the normal distribution P is a probability distribution. Indeed, (1) since $f(x) > 0$, $P(X \in A) = \int_A f(x)dx \geq 0$ for any measurable set $A \subset \mathbf{R}$, (2) by substituting $y = (x - \mu)/(\sqrt{2}\,\sigma)$,

$$P(X \in \mathbf{R}^1) = \int_{-\infty}^{\infty} f(x)dx = \frac{1}{\sqrt{\pi}} \int_{-\infty}^{\infty} e^{-y^2} dy = 1,$$

(cf. Exercise 1.5), and (3) since

$$\int_{\cup_i A_i} f(x)dx = \sum_i \int_{A_i} f(x)dx$$

for any countable sequence of pairwise disjoint measurable sets $A_i \subset \mathbf{R}^1$, we have

$$P\left(X \in \bigcup_i A_i\right) = \sum_i P(X \in A_i).$$

Let X be a random variable. The *expectation* of X indicates the mean of the values of X, and is defined by

$$E(X) = \int_{-\infty}^{\infty} x f(x)dx. \tag{1.18}$$

Note that $E(X)$ is a real number for any random variable X with probability density function f. For the normal distribution, using the substitution $y = (x - \mu)/(\sqrt{2}\,\sigma)$ again, we have

$$\begin{aligned}
E(X) &= \int_{-\infty}^{\infty} x f(x)dx \\
&= \frac{1}{\sqrt{2\pi}\,\sigma} \int_{-\infty}^{\infty} x e^{-\frac{1}{2\sigma^2}(x-\mu)^2} dx \\
&= \frac{1}{\sqrt{\pi}} \int_{-\infty}^{\infty} (\sqrt{2}\,\sigma y + \mu) e^{-y^2} dy \\
&= \mu \frac{1}{\sqrt{\pi}} \int_{-\infty}^{\infty} e^{-y^2} dy \\
&= \mu.
\end{aligned} \tag{1.19}$$

Note also that $E(X)$ is the *first moment* of the probability density function f. The *second moment* gives the *variance* of X defined by

$$Var\ (X) = E(X - E(X))^2 = \int_{-\infty}^{\infty} (x - E(X))^2 f(x) dx. \qquad (1.20)$$

This number indicates the dispersion of the values of X from its mean $E(X)$. For the normal distribution, using the substitution $y = (x - \mu)/(\sqrt{2}\,\sigma)$ again, we have

$$
\begin{aligned}
Var(X) &= \int_{-\infty}^{\infty} (x - \mu)^2 f(x) dx \\
&= \frac{1}{\sqrt{2\pi}\,\sigma} \int_{-\infty}^{\infty} (x - \mu)^2 e^{-\frac{1}{2\sigma^2}(x-\mu)^2} dx \\
&= \frac{2\sigma^2}{\sqrt{\pi}} \int_{-\infty}^{\infty} y^2 e^{-y^2} dy \\
&= \sigma^2, \qquad\qquad\qquad\qquad\qquad\qquad (1.21)
\end{aligned}
$$

where we have used the equality $\int_{-\infty}^{\infty} y^2 e^{-y^2} dy = \sqrt{\pi}/2$ (cf. Exercise 1.6).

We now turn to *random vectors* whose components are random variables. We denote a random n-vector $X = [X_1 \cdots X_n]^\mathsf{T}$ where each $X_i(s) \in \mathbf{R}^1$, $s \in S$.

Let P be a continuous probability distribution function of X. That is,

$$
\begin{aligned}
&P(X_1 \in A_1, \cdots, X_n \in A_n) \\
&= \int_{A_1} \cdots \int_{A_n} f(x_1, \cdots, x_n) dx_1 \cdots dx_n, \qquad (1.22)
\end{aligned}
$$

where $A_1, \cdots, A_n$ are measurable sets in $\mathbf{R}^1$ and f an integrable function. f is called a *joint probability density function* of X and P is called a *joint probability distribution (function)* of X. For each i, $i = 1, \cdots, n$, define

$$
f_i(x) =
$$
$$
\int_{-\infty}^{\infty} \cdots \int_{-\infty}^{\infty} f(x_1, \cdots, x_{i-1}, x, x_{i+1}, \cdots, x_n) dx_1 \cdots dx_{i-1} dx_{i+1} \cdots dx_n.
$$
$$(1.23)$$

Then it is clear that $\int_{-\infty}^{\infty} f_i(x) dx = 1$. f_i is called the *ith marginal probability density function* of X corresponding to the joint probability density function $f(x_1, \cdots, x_n)$. Similarly, we define f_{ij} and

f_{ijk} by deleting the integrals with respect to x_i, x_j and x_i, x_j, x_k, respectively, etc., as in the definition of f_i. If

$$f(\mathbf{x}) = \frac{1}{(2\pi)^{n/2}(det\,R)^{1/2}}\, exp\left\{-\frac{1}{2}(\mathbf{x} - \underline{\mu})^\top R^{-1}(\mathbf{x} - \underline{\mu})\right\}, \qquad (1.24)$$

where $\underline{\mu}$ is a constant n-vector and R is a symmetric positive definite matrix, we say that $f(\mathbf{x})$ is a *Gaussian* (or *normal*) *probability density function* of X. It can be verified that

$$\int_{-\infty}^{\infty} f(\mathbf{x})d\mathbf{x} := \int_{-\infty}^{\infty} \cdots \int_{-\infty}^{\infty} f(\mathbf{x})dx_1 \cdots dx_n = 1, \qquad (1.25)$$

$$E(X) = \int_{-\infty}^{\infty} \mathbf{x}f(\mathbf{x})d\mathbf{x}$$

$$:= \int_{-\infty}^{\infty} \cdots \int_{-\infty}^{\infty} \begin{bmatrix} x_1 \\ \vdots \\ x_n \end{bmatrix} f(\mathbf{x})dx_1 \cdots dx_n$$

$$= \underline{\mu}, \qquad (1.26)$$

and

$$Var(X) = E(X - \underline{\mu})(X - \underline{\mu})^\top = R. \qquad (1.27)$$

Indeed, since R is symmetric and positive definite, there is a unitary matrix U such that $R = U^\top J U$ where $J = diag[\lambda_1, \cdots, \lambda_n]$ and $\lambda_1, \cdots, \lambda_n > 0$. Let $\mathbf{y} = \frac{1}{\sqrt{2}}diag[\sqrt{\lambda_1}, \cdots, \sqrt{\lambda_n}]U(\mathbf{x} - \underline{\mu})$. Then

$$\int_{-\infty}^{\infty} f(\mathbf{x})d\mathbf{x}$$

$$= \frac{2^{n/2}\sqrt{\lambda_1} \cdots \sqrt{\lambda_n}}{(2\pi)^{n/2}(\lambda_1 \cdots \lambda_n)^{1/2}} \int_{-\infty}^{\infty} e^{-y_1^2}dy_1 \cdots \int_{-\infty}^{\infty} e^{-y_n^2}dy_n$$

$$= 1.$$

Equations (1.26) and (1.27) can be verified by using the same substitution as that used for the scalar case (cf. (1.21) and Exercise 1.7).

Next, we introduce the concept of *conditional probability*. Consider an experiment in which balls are drawn one at a time from an urn containing M_1 white balls and M_2 black balls. What is the probability that the second ball drawn from the urn is also black (event A_2) under the condition that the first one is black (event A_1)? Here, we sample without replacement; that is, the first ball is not returned to the urn after being drawn.

To solve this simple problem, we reason as follows: since the first ball drawn from the urn is black, there remain M_1 white balls and $M_2 - 1$ black balls in the urn before the second drawing. Hence, the probability that a black ball is drawn is now

$$\frac{M_2 - 1}{M_1 + M_2 - 1}.$$

Note that

$$\frac{M_2 - 1}{M_1 + M_2 - 1} = \frac{M_2}{M_1 + M_2} \cdot \frac{M_2 - 1}{M_1 + M_2 - 1} \Big/ \frac{M_2}{M_1 + M_2},$$

where $M_2/(M_1 + M_2)$ is the probability that a black ball is picked at the first drawing, and $\frac{M_2}{(M_1 + M_2)} \cdot \frac{M_2 - 1}{M_1 + M_2 - 1}$ is the probability that black balls are picked at both the first and second drawings. This example motivates the following definition of conditional probability: The *conditional probability of* $X_1 \in A_1$ given $X_2 \in A_2$ is defined by

$$P(X_1 \in A_1 | X_2 \in A_2) = \frac{P(X_1 \in A_1, X_2 \in A_2)}{P(X_2 \in A_2)}. \qquad (1.28)$$

Suppose that P is a continuous probability distribution function with joint probability density function f. Then (1.28) becomes

$$P(X_1 \in A_1 | X_2 \in A_2) = \frac{\int_{A_1} \int_{A_2} f(x_1, x_2) dx_1 dx_2}{\int_{A_2} f_2(x_2) dx_2},$$

where f_2 defined by

$$f_2(x_2) = \int_{-\infty}^{\infty} f(x_1, x_2) dx_1$$

is the second marginal probability density function of f. Let $f(x_1|x_2)$ denote the probability density function corresponding to $P(X_1 \in A_1 | X_2 \in A_2)$. $f(x_1|x_2)$ is called the *conditional probability density function* corresponding to the conditional probability distribution function $P(X_1 \in A_1 | X_2 \in A_2)$. It is known that

$$f(x_1|x_2) = \frac{f(x_1, x_2)}{f_2(x_2)} \qquad (1.29)$$

which is called the *Bayes formula* (see, for example, *Probability* by A. N. Shiryayev (1984)). By symmetry, the Bayes formula can be written as

$$f(x_1, x_2) = f(x_1|x_2) f_2(x_2) = f(x_2|x_1) f_1(x_1). \qquad (1.30)$$

We remark that this formula also holds for random vectors X_1 and X_2.

Let X and Y be random n- and m-vectors, respectively. The *covariance* of X and Y is defined by the $n \times m$ matrix

$$Cov(X,Y) = E[(X - E(X))(Y - E(Y))^\top]. \tag{1.31}$$

When $Y = X$, we have the variance matrix, which is sometimes called a *covariance* matrix of X, $Var(X) = Cov(X,X)$.

It can be verified that the expectation, variance, and covariance have the following properties:

$$E(AX + BY) = AE(X) + BE(Y) \tag{1.32a}$$

$$E((AX)(BY)^\top) = A(E(XY^\top))B^\top \tag{1.32b}$$

$$Var(X) \geq 0, \tag{1.32c}$$

$$Cov(X,Y) = (Cov(Y,X))^\top, \tag{1.32d}$$

and

$$Cov(X,Y) = E(XY^\top) - E(X)(E(Y))^\top, \tag{1.32e}$$

where A and B are constant matrices (cf. Exercise 1.8). X and Y are said to be *independent* if $f(\mathbf{x}|\mathbf{y}) = f_1(\mathbf{x})$ and $f(\mathbf{y}|\mathbf{x}) = f_2(\mathbf{y})$, and X and Y are said to be *uncorrelated* if $Cov(X,Y) = 0$. It is easy to see that if X and Y are independent then they are uncorrelated. Indeed, if X and Y are independent then $f(\mathbf{x},\mathbf{y}) = f_1(\mathbf{x})f_2(\mathbf{y})$. Hence,

$$\begin{aligned}
E(XY^\top) &= \int_{-\infty}^{\infty} \int_{-\infty}^{\infty} \mathbf{x}\mathbf{y}^\top f(\mathbf{x},\mathbf{y})d\mathbf{x}d\mathbf{y} \\
&= \int_{-\infty}^{\infty} \mathbf{x}f_1(\mathbf{x})d\mathbf{x} \int_{-\infty}^{\infty} \mathbf{y}^\top f_2(\mathbf{y})d\mathbf{y} \\
&= E(X)(E(Y))^\top,
\end{aligned}$$

so that by property (1.32e) $Cov(X,Y) = 0$. But the converse does not necessarily hold, unless the probability distribution is normal. Let

$$X = \begin{bmatrix} X_1 \\ X_2 \end{bmatrix} \sim N\left(\begin{bmatrix} \mu_1 \\ \mu_2 \end{bmatrix}, R\right),$$

where

$$R = \begin{bmatrix} R_{11} & R_{12} \\ R_{21} & R_{22} \end{bmatrix}, \qquad R_{12} = R_{21}^\top,$$

R_{11} and R_{22} are symmetric, and R is positive definite. Then it can be shown that X_1 and X_2 are independent if and only if $R_{12} = Cov(X_1, X_2) = 0$ (cf. Exercise 1.9).

Let X and Y be two random vectors. Similar to the definitions of expectation and variance, the *conditional expectation* of X *under the condition that* $Y = \mathbf{y}$ is defined to be

$$E(X|Y = \mathbf{y}) = \int_{-\infty}^{\infty} \mathbf{x} f(\mathbf{x}|\mathbf{y}) d\mathbf{x} \tag{1.33}$$

and the *conditional variance*, which is sometimes called the *conditional covariance of X, under the condition that* $Y = \mathbf{y}$ to be

$$Var(X|Y = \mathbf{y})$$
$$= \int_{-\infty}^{\infty} [\mathbf{x} - E(X|Y = \mathbf{y})][\mathbf{x} - E(X|Y = \mathbf{y})]^{\mathsf{T}} f(\mathbf{x}|\mathbf{y}) d\mathbf{x}. \tag{1.34}$$

Next, suppose that

$$E\left(\begin{bmatrix} X \\ Y \end{bmatrix}\right) = \begin{bmatrix} \underline{\mu}_x \\ \underline{\mu}_y \end{bmatrix} \quad and \quad Var\left(\begin{bmatrix} X \\ Y \end{bmatrix}\right) = \begin{bmatrix} R_{xx} & R_{xy} \\ R_{yx} & R_{yy} \end{bmatrix}.$$

Then it follows from (1.24) that

$$f(\mathbf{x}, \mathbf{y}) = f\left(\begin{bmatrix} \mathbf{x} \\ \mathbf{y} \end{bmatrix}\right)$$

$$= \frac{1}{(2\pi)^{n/2}\left(det\begin{bmatrix} R_{xx} & R_{xy} \\ R_{yx} & R_{yy} \end{bmatrix}\right)^{1/2}}$$

$$\cdot exp\left\{-\frac{1}{2}\left(\begin{bmatrix} \mathbf{x} \\ \mathbf{y} \end{bmatrix} - \begin{bmatrix} \underline{\mu}_x \\ \underline{\mu}_y \end{bmatrix}\right)^{\mathsf{T}} \begin{bmatrix} R_{xx} & R_{xy} \\ R_{yx} & R_{yy} \end{bmatrix}^{-1} \left(\begin{bmatrix} \mathbf{x} \\ \mathbf{y} \end{bmatrix} - \begin{bmatrix} \underline{\mu}_x \\ \underline{\mu}_y \end{bmatrix}\right)\right\}.$$

It can be verified that

$$f(\mathbf{x}|\mathbf{y}) = \frac{f(\mathbf{x}, \mathbf{y})}{f(\mathbf{y})}$$

$$= \frac{1}{(2\pi)^{n/2}(det\,\tilde{R})^{1/2}} \; exp\left\{-\frac{1}{2}(\mathbf{x} - \underline{\tilde{\mu}})^{\mathsf{T}} \tilde{R}^{-1}(\mathbf{x} - \underline{\tilde{\mu}})\right\}, \tag{1.35}$$

where

$$\underline{\tilde{\mu}} = \underline{\mu}_x + R_{xy} R_{yy}^{-1}(\mathbf{y} - \underline{\mu}_y)$$

and

$$\tilde{R} = R_{xx} - R_{xy} R_{yy}^{-1} R_{yx}$$

(cf. Exercise 1.10). Hence, by rewriting $\underline{\tilde{\mu}}$ and $\tilde{R}$, we have

$$E(X|Y = \mathbf{y}) = E(X) + Cov(X, Y)[Var(Y)]^{-1}(\mathbf{y} - E(Y)) \tag{1.36}$$

and

$$Var(X|Y = \mathbf{y}) = Var(X) - Cov(X, Y)[Var(Y)]^{-1}Cov(Y, X). \quad (1.37)$$

1.3 Least-Squares Preliminaries

Let $\{\underline{\xi}_k\}$ be a sequence of random vectors, called a *random sequence*. Denote $E(\underline{\xi}_k) = \underline{\mu}_k$, $Cov(\underline{\xi}_k, \underline{\xi}_j) = R_{kj}$ so that $Var(\underline{\xi}_k) = R_{kk} := R_k$. A random sequence $\{\underline{\xi}_k\}$ is called a *white noise sequence* if $Cov(\underline{\xi}_k, \underline{\xi}_j) = R_{kj} = R_k\delta_{kj}$ where $\delta_{kj} = 1$ if $k = j$ and 0 if $k \neq j$. $\{\underline{\xi}_k\}$ is called a *sequence of Gaussian* (or *normal*) *white noise* if it is white and each $\underline{\xi}_k$ is normal.

Consider the observation equation of a linear system where the observed data is contaminated with noise, namely:

$$\mathbf{v}_k = C_k\mathbf{x}_k + D_k\mathbf{u}_k + \underline{\xi}_k,$$

where, as usual, $\{\mathbf{x}_k\}$ is the state sequence, $\{\mathbf{u}_k\}$ the control sequence, and $\{\mathbf{v}_k\}$ the data sequence. We assume, for each k, that the $q \times n$ constant matrix C_k, $q \times p$ constant matrix D_k, and the deterministic control p-vector $\mathbf{u}_k$ are given. Usually, $\{\underline{\xi}_k\}$ is not known but will be assumed to be a sequence of zero-mean Gaussian white noise, namely: $E(\underline{\xi}_k) = 0$ and $E(\underline{\xi}_k\underline{\xi}_j^\top) = R_{kj}\delta_{kj}$ with R_k being symmetric and positive definite, $k, j = 1, 2, \cdots$.

Our goal is to obtain an optimal estimate $\hat{\mathbf{y}}_k$ of the state vector $\mathbf{x}_k$ from the information $\{\mathbf{v}_k\}$. If there were no noise, then it is clear that $\mathbf{z}_k - C_k\hat{\mathbf{y}}_k = 0$, where $\mathbf{z}_k := \mathbf{v}_k - D_k\mathbf{u}_k$, whenever this linear system has a solution; otherwise, some measurement of the error $\mathbf{z}_k - C_k\mathbf{y}_k$ must be minimized over all $\mathbf{y}_k$. In general, when the data is contaminated with noise, we will minimize the quantity:

$$F(\mathbf{y}_k, W_k) = E(\mathbf{z}_k - C_k\mathbf{y}_k)^\top W_k(\mathbf{z}_k - C_k\mathbf{y}_k)$$

over all n-vectors $\mathbf{y}_k$ where W_k is a positive definite and symmetric $q \times q$ matrix, called a *weight matrix*. That is, we wish to find a $\hat{\mathbf{y}}_k = \hat{\mathbf{y}}_k(W_k)$ such that

$$F(\hat{\mathbf{y}}_k, W_k) = \min_{\mathbf{y}_k} F(\mathbf{y}_k, W_k).$$

In addition, we wish to determine the *optimal weight* $\hat{W}_k$. To find $\hat{\mathbf{y}}_k = \hat{\mathbf{y}}_k(W_k)$, assuming that $(C_k^\top W_k C_k)$ is nonsingular, we rewrite

$$F(\mathbf{y}_k, W_k)$$
$$=E(\mathbf{z}_k - C_k\mathbf{y}_k)^\top W_k(\mathbf{z}_k - C_k\mathbf{y}_k)$$
$$=E[(C_k^\top W_k C_k)\mathbf{y}_k - C_k^\top W_k\mathbf{z}_k]^\top (C_k^\top W_k C_k)^{-1}[(C_k^\top W_k C_k)\mathbf{y}_k - C_k^\top W_k\mathbf{z}_k]$$
$$+ E(\mathbf{z}_k^\top [I - W_k C_k(C_k^\top W_k C_k)^{-1}C_k^\top]W_k\mathbf{z}_k),$$

where the first term on the right hand side is non-negative definite. To minimize $F(\mathbf{y}_k, W_k)$, the first term on the right must vanish, so that

$$\hat{\mathbf{y}}_k = (C_k^\top W_k C_k)^{-1}C_k^\top W_k\mathbf{z}_k.$$

Note that if $(C_k^\top W_k C_k)$ is singular, then $\hat{\mathbf{y}}_k$ is not unique. To find the optimal weight $\hat{W}_k$, let us consider

$$F(\hat{\mathbf{y}}_k, W_k) = E(\mathbf{z}_k - C_k\hat{\mathbf{y}}_k)^\top W_k(\mathbf{z}_k - C_k\hat{\mathbf{y}}_k).$$

It is clear that this quantity does not attain a minimum value at a positive definite weight W_k since such a minimum would result from $W_k = 0$. Hence, we need another measurement to determine an optimal $\hat{W}_k$. Noting that the original problem is to estimate the state vector $\mathbf{x}_k$ by $\hat{\mathbf{y}}_k(W_k)$, it is natural to consider a measurement of the error $(\mathbf{x}_k - \hat{\mathbf{y}}_k(W_k))$. But since not much about $\mathbf{x}_k$ is known and only the noisy data can be measured, this measurement should be determined by the variance of the error. That is, we will minimize $Var(\mathbf{x}_k - \hat{\mathbf{y}}_k(W_k))$ over all positive definite symmetric matrices W_k. We write $\hat{\mathbf{y}}_k = \hat{\mathbf{y}}_k(W_k)$ and

$$\mathbf{x}_k - \hat{\mathbf{y}}_k = (C_k^\top W_k C_k)^{-1}(C_k^\top W_k C_k)\mathbf{x}_k - (C_k^\top W_k C_k)^{-1}C_k^\top W_k\mathbf{z}_k$$
$$= (C_k^\top W_k C_k)^{-1}C_k^\top W_k(C_k\mathbf{x}_k - \mathbf{z}_k)$$
$$= -(C_k^\top W_k C_k)^{-1}C_k^\top W_k\underline{\xi}_k.$$

Therefore, by the linearity of the expectation operation, we have

$$Var(\mathbf{x}_k - \hat{\mathbf{y}}_k) = (C_k^\top W_k C_k)^{-1}C_k^\top W_k E(\underline{\xi}_k\underline{\xi}_k^\top)W_k C_k(C_k^\top W_k C_k)^{-1}$$
$$= (C_k^\top W_k C_k)^{-1}C_k^\top W_k R_k W_k C_k(C_k^\top W_k C_k)^{-1}.$$

This is the quantity to be minimized. To write this as a perfect square, we need the *positive square root* of the positive definite symmetric matrix R_k defined as follows: Let the eigenvalues of R_k be $\lambda_1, \cdots, \lambda_n$, which are all positive, and write $R_k = U^\top \operatorname{diag}[\lambda_1, \cdots, \lambda_n]U$ where U is a unitary matrix (formed by the normalized eigenvectors of $\lambda_i, i = 1, \cdots, n$). Then we define

$R_k^{1/2} = U^\top \operatorname{diag}[\sqrt{\lambda_1}, \cdots, \sqrt{\lambda_n}]U$ which gives $(R_k^{1/2})(R_k^{1/2})^\top = R_k$. It follows that

$$Var(\mathbf{x}_k - \hat{\mathbf{y}}_k) = Q^\top Q,$$

where $Q = (R_k^{1/2})^\top W_k C_k (C_k^\top W_k C_k)^{-1}$. By Lemma 1.1 (the *matrix Schwarz inequality*), under the assumption that P is a $q \times n$ matrix with nonsingular $P^\top P$, we have

$$Q^\top Q \geq (P^\top Q)^\top (P^\top P)^{-1}(P^\top Q).$$

Hence, if $(C_k^\top R_k^{-1} C_k)$ is nonsingular, we may choose $P = (R_k^{1/2})^{-1} C_k$, so that

$$P^\top P = C_k^\top ((R_k^{1/2})^\top)^{-1}(R_k^{1/2})C_k = C_k^\top R_k^{-1} C_k$$

is nonsingular, and

$$
\begin{aligned}
&(P^\top Q)^\top (P^\top P)^{-1}(P^\top Q) \\
&= [C_k^\top ((R_k^{1/2})^{-1})^\top (R_k^{1/2})^\top W_k C_k (C_k^\top W_k C_k)^{-1}]^\top (C_k^\top R_k^{-1} C_k)^{-1} \\
&\quad \cdot [C_k^\top ((R_k^{1/2})^{-1})^\top (R_k^{1/2})^\top W_k C_k (C_k^\top W_k C_k)^{-1}] \\
&= (C_k^\top R_k^{-1} C_k)^{-1} \\
&= Var(\mathbf{x}_k - \hat{\mathbf{y}}_k(R_k^{-1})).
\end{aligned}
$$

Hence, $Var(\mathbf{x}_k - \hat{\mathbf{y}}_k(W_k)) \geq Var(\mathbf{x}_k - \hat{\mathbf{y}}_k(R_k^{-1}))$ for all positive definite symmetric weight matrices W_k. Therefore, the optimal weight matrix is $\hat{W}_k = R_k^{-1}$, and the optimal estimate of $\mathbf{x}_k$ using this optimal weight is

$$\hat{\mathbf{x}}_k := \hat{\mathbf{y}}_k(R_k^{-1}) = (C_k^\top R_k^{-1} C_k)^{-1} C_k^\top R_k^{-1} (\mathbf{v}_k - D_k \mathbf{u}_k). \tag{1.38}$$

We call $\hat{\mathbf{x}}_k$ the *least-squares optimal estimate* of $\mathbf{x}_k$. Note that $\hat{\mathbf{x}}_k$ is a *linear estimate* of $\mathbf{x}_k$. Being the image of a linear transformation of the data $\mathbf{v}_k - D_k \mathbf{u}_k$, it gives an *unbiased estimate* of $\mathbf{x}_k$, in the sense that $E\hat{\mathbf{x}}_k = E\mathbf{x}_k$ (cf. Exercise 1.12), and it also gives a *minimum variance estimate* of $\mathbf{x}_k$, since

$$Var(\mathbf{x}_k - \hat{\mathbf{x}}_k) \leq Var(\mathbf{x}_k - \hat{\mathbf{y}}_k(W_k))$$

for all positive definite symmetric weight matrices W_k.

Exercises

1.1. Prove Lemma 1.4.
1.2. Prove Lemma 1.6.
1.3. Give an example of two matrices A and B such that $A \geq B > 0$ but for which the inequality $AA^{\top} \geq BB^{\top}$ is not satisfied.
1.4. Prove Lemma 1.8.
1.5. Show that $\int_{-\infty}^{\infty} e^{-y^2} dy = \sqrt{\pi}$.
1.6. Verify that $\int_{-\infty}^{\infty} y^2 e^{-y^2} dy = \frac{1}{2}\sqrt{\pi}$. (Hint: Differentiate the integral $- \int_{-\infty}^{\infty} e^{-xy^2} dy$ with respect to x and then let $x \to 1$.)
1.7. Let

$$f(\mathbf{x}) = \frac{1}{(2\pi)^{n/2}(\det R)^{1/2}} exp\left\{ -\frac{1}{2}(\mathbf{x} - \underline{\mu})^{\top} R^{-1} (\mathbf{x} - \underline{\mu}) \right\}.$$

Show that
(a)

$$E(X) = \int_{-\infty}^{\infty} \mathbf{x} f(\mathbf{x}) d\mathbf{x}$$

$$:= \int_{-\infty}^{\infty} \cdots \int_{-\infty}^{\infty} \begin{bmatrix} x_1 \\ \vdots \\ x_n \end{bmatrix} f(\mathbf{x}) dx_1 \cdots dx_n$$

$$= \underline{\mu},$$

and
(b)
$$Var(X) = E(X - \underline{\mu})(X - \underline{\mu})^{\top} = R.$$

1.8. Verify the properties (1.32a-e) of the expectation, variance, and covariance.
1.9. Prove that two random vectors X_1 and X_2 with normal distributions are independent if and only if $Cov(X_1, X_2) = 0$.
1.10. Verify (1.35).
1.11. Consider the minimization of the quantity

$$F(\mathbf{y}_k) = E(\mathbf{z}_k - C_k \mathbf{y}_k)^{\top} W_k (\mathbf{z}_k - C_k \mathbf{y}_k)$$

over all n-vectors $\mathbf{y}_k$, where $\mathbf{z}_k$ is a $q \times 1$ vector, C_k, a $q \times n$ matrix, and W_k, a $q \times q$ weight matrix, such that the matrix $(C_k^{\top} W_k C_k)$ is nonsingular. By letting $dF(\mathbf{y}_k)/d\mathbf{y}_k = 0$, show that the optimal solution $\hat{\mathbf{y}}_k$ is given by

$$\hat{\mathbf{y}}_k = (C_k^{\top} W_k C_k)^{-1} C_k^{\top} W_k \mathbf{z}_k.$$

(Hint: The differentiation of a scalar-valued function $F(\mathbf{y})$ with respect to the n-vector $\mathbf{y} = [y_1 \cdots y_n]^\top$ is defined to be

$$\frac{dF(\mathbf{y})}{d\mathbf{y}} = \left[\frac{\partial F}{\partial y_1} \cdots \frac{\partial F}{\partial y_n} \right]^\top .)$$

1.12. Verify that the estimate $\hat{\mathbf{x}}_k$ given by (1.38) is an unbiased estimate of $\mathbf{x}_k$ in the sense that $E\hat{\mathbf{x}}_k = E\mathbf{x}_k$.

2. Kalman Filter: An Elementary Approach

This chapter is devoted to a most elementary introduction to the Kalman filtering algorithm. By assuming invertibility of certain matrices, the Kalman filtering "prediction-correction" algorithm will be derived based on the optimality criterion of least-squares unbiased estimation of the state vector with the optimal weight, using all available data information. The filtering algorithm is first obtained for a system with no deterministic (control) input. By superimposing the deterministic solution, we then arrive at the general Kalman filtering algorithm.

2.1 The Model

Consider a linear system with state-space description

$$\begin{cases} \mathbf{y}_{k+1} = A_k \mathbf{y}_k + B_k \mathbf{u}_k + \Gamma_k \underline{\xi}_k \\ \mathbf{w}_k = C_k \mathbf{y}_k + D_k \mathbf{u}_k + \underline{\eta}_k \, , \end{cases}$$

where $A_k, B_k, \Gamma_k, C_k, D_k$ are $n \times n, n \times m, n \times p, q \times n, q \times m$ (known) constant matrices, respectively, with $1 \le m, p, q \le n$, $\{\mathbf{u}_k\}$ a (known) sequence of m-vectors (called a *deterministic input sequence*), and $\{\underline{\xi}_k\}$ and $\{\underline{\eta}_k\}$ are, respectively, (unknown) system and observation noise sequences, with known statistical information such as mean, variance, and covariance. Since both the deterministic input $\{\mathbf{u}_k\}$ and noise sequences $\{\underline{\xi}_k\}$ and $\{\underline{\eta}_k\}$ are present, the system is usually called a *linear deterministic/stochastic system*. This system can be decomposed into the sum of a linear deterministic system:

$$\begin{cases} \mathbf{z}_{k+1} = A_k \mathbf{z}_k + B_k \mathbf{u}_k \\ \mathbf{s}_k = C_k \mathbf{z}_k + D_k \mathbf{u}_k \, , \end{cases}$$

and a linear (purely) stochastic system:

$$\begin{cases} \mathbf{x}_{k+1} = A_k\mathbf{x}_k + \Gamma_k\underline{\xi}_k \\ \mathbf{v}_k = C_k\mathbf{x}_k + \underline{\eta}_k, \end{cases} \tag{2.1}$$

with $\mathbf{w}_k = \mathbf{s}_k + \mathbf{v}_k$ and $\mathbf{y}_k = \mathbf{z}_k + \mathbf{x}_k$. The advantage of the decomposition is that the solution of $\mathbf{z}_k$ in the linear deterministic system is well known and is given by the so-called *transition equation*

$$\mathbf{z}_k = (A_{k-1}\cdots A_0)\mathbf{z}_0 + \sum_{i=1}^{k}(A_{k-1}\cdots A_{i-1})B_{i-1}\mathbf{u}_{i-1}.$$

Hence, it is sufficient to derive the optimal estimate $\hat{\mathbf{x}}_k$ of $\mathbf{x}_k$ in the stochastic state-space description (2.1), so that

$$\hat{\mathbf{y}}_k = \mathbf{z}_k + \hat{\mathbf{x}}_k$$

becomes the optimal estimate of the state vector $\mathbf{y}_k$ in the original linear system. Of course, the estimate has to depend on the statistical information of the noise sequences. In this chapter, we will only consider zero-mean Gaussian white noise processes.

Assumption 2.1. Let $\{\underline{\xi}_k\}$ and $\{\underline{\eta}_k\}$ be sequences of zero-mean Gaussian white noise such that $Var(\underline{\xi}_k) = Q_k$ and $Var(\underline{\eta}_k) = R_k$ are positive definite matrices and $E(\underline{\xi}_k\underline{\eta}_\ell^\top) = 0$ for all k and ℓ. The initial state $\mathbf{x}_0$ is also assumed to be independent of $\underline{\xi}_k$ and $\underline{\eta}_k$ in the sense that $E(\mathbf{x}_0\underline{\xi}_k^\top) = 0$ and $E(\mathbf{x}_0\underline{\eta}_k^\top) = 0$ for all k.

2.2 Optimality Criterion

In determining the optimal estimate $\hat{\mathbf{x}}_k$ of $\mathbf{x}_k$, it will be seen that the optimality is in the sense of least-squares followed by choosing the optimal weight matrix that gives a minimum variance estimate as discussed in Section 1.3. However, we will incorporate the information of all data $\mathbf{v}_j$, $j = 0, 1, \cdots, k$, in determining the estimate $\hat{\mathbf{x}}_k$ of $\mathbf{x}_k$ (instead of just using $\mathbf{v}_k$ as discussed in Section 1.3). To accomplish this, we introduce the vectors

$$\overline{\mathbf{v}}_j = \begin{bmatrix} \mathbf{v}_0 \\ \vdots \\ \mathbf{v}_j \end{bmatrix}, \qquad j = 0, 1, \cdots,$$

and obtain $\hat{\mathbf{x}}_k$ from the data vector $\bar{\mathbf{v}}_k$. For this approach, we assume for the time being that all the system matrices A_j are nonsingular. Then it can be shown that the state-space description of the linear stochastic system can be written as

$$\bar{\mathbf{v}}_j = H_{k,j}\mathbf{x}_k + \bar{\underline{\xi}}_{k,j}, \tag{2.2}$$

where

$$H_{k,j} = \begin{bmatrix} C_0\Phi_{0k} \\ \vdots \\ C_j\Phi_{jk} \end{bmatrix} \quad \text{and} \quad \bar{\underline{\xi}}_{k,j} = \begin{bmatrix} \underline{\xi}_{k,0} \\ \vdots \\ \underline{\xi}_{k,j} \end{bmatrix},$$

with $\Phi_{\ell k}$ being the transition matrices defined by

$$\Phi_{\ell k} = \begin{cases} A_{\ell-1}\cdots A_k & \text{if } \ell > k, \\ I & \text{if } \ell = k, \end{cases}$$

$\Phi_{\ell k} = \Phi_{k\ell}^{-1}$ if $\ell < k$, and

$$\underline{\xi}_{k,\ell} = \underline{\eta}_\ell - C_\ell \sum_{i=\ell+1}^{k} \Phi_{\ell i}\Gamma_{i-1}\underline{\xi}_{i-1}.$$

Indeed, by applying the inverse transition property of Φ_{ki} described above and the transition equation

$$\mathbf{x}_k = \Phi_{k\ell}\mathbf{x}_\ell + \sum_{i=\ell+1}^{k} \Phi_{ki}\Gamma_{i-1}\underline{\xi}_{i-1},$$

which can be easily obtained from the first recursive equation in (2.1), we have

$$\mathbf{x}_\ell = \Phi_{\ell k}\mathbf{x}_k - \sum_{i=\ell+1}^{k} \Phi_{\ell i}\Gamma_{i-1}\underline{\xi}_{i-1};$$

and this yields

$$\begin{aligned} &H_{k,j}\mathbf{x}_k + \bar{\underline{\xi}}_{k,j} \\ &= \begin{bmatrix} C_0\Phi_{0k} \\ \vdots \\ C_j\Phi_{jk} \end{bmatrix} \mathbf{x}_k + \begin{bmatrix} \underline{\eta}_0 - C_0\sum_{i=1}^{k}\Phi_{0i}\Gamma_{i-1}\underline{\xi}_{i-1} \\ \vdots \\ \underline{\eta}_j - C_j\sum_{i=j+1}^{k}\Phi_{ji}\Gamma_{i-1}\underline{\xi}_{i-1} \end{bmatrix} \\ &= \begin{bmatrix} C_0\mathbf{x}_0 + \underline{\eta}_0 \\ \vdots \\ C_j\mathbf{x}_j + \underline{\eta}_j \end{bmatrix} = \begin{bmatrix} \mathbf{v}_0 \\ \vdots \\ \mathbf{v}_j \end{bmatrix} = \bar{\mathbf{v}}_j \end{aligned}$$

which is (2.2).

Now, using the least-squares estimate discussed in Chapter 1, Section 1.3, with weight $W_{kj} = (Var(\underline{\varepsilon}_{k,j}))^{-1}$, where the inverse is assumed only for the purpose of illustrating the optimality criterion, we arrive at the linear, unbiased, minimum variance least-squares estimate $\hat{\mathbf{x}}_{k|j}$ of $\mathbf{x}_k$ using the data $\mathbf{v}_0, \cdots, \mathbf{v}_j$.

Definition 2.1. (1) For $j = k$, we denote $\hat{\mathbf{x}}_k = \hat{\mathbf{x}}_{k|k}$ and call the estimation process a *digital filtering process*. (2) For $j < k$, we call $\hat{\mathbf{x}}_{k|j}$ an *optimal prediction* of $\mathbf{x}_k$ and the process a *digital prediction process*. (3) For $j > k$, we call $\hat{\mathbf{x}}_{k|j}$ a *smoothing estimate* of $\mathbf{x}_k$ and the process a *digital smoothing process*.

We will only discuss digital filtering. However, since $\hat{\mathbf{x}}_k = \hat{\mathbf{x}}_{k|k}$ is determined by using *all* data $\mathbf{v}_0, \cdots, \mathbf{v}_k$, the process is not applicable to real-time problems for very large values of k, since the need for storage of the data and the computational requirement grow with time. Hence, we will derive a recursive formula that gives $\hat{\mathbf{x}}_k = \hat{\mathbf{x}}_{k|k}$ from the "prediction" $\hat{\mathbf{x}}_{k|k-1}$ and $\hat{\mathbf{x}}_{k|k-1}$ from the estimate $\hat{\mathbf{x}}_{k-1} = \hat{\mathbf{x}}_{k-1|k-1}$. At each step, we only use the incoming bit of the data information so that very little storage of the data is necessary. This is what is usually called the *Kalman filtering algorithm*.

2.3 Prediction-Correction Formulation

To compute $\hat{\mathbf{x}}_k$ in real-time, we will derive the recursive formula

$$\begin{cases} \hat{\mathbf{x}}_{k|k} = \hat{\mathbf{x}}_{k|k-1} + G_k(\mathbf{v}_k - C_k\hat{x}_{k|k-1}) \\ \hat{\mathbf{x}}_{k|k-1} = A_{k-1}\hat{\mathbf{x}}_{k-1|k-1}, \end{cases}$$

where G_k will be called the *Kalman gain* matrices. The starting point is the initial estimate $\hat{\mathbf{x}}_0 = \hat{\mathbf{x}}_{0|0}$. Since $\hat{\mathbf{x}}_0$ is an unbiased estimate of the initial state $\mathbf{x}_0$, we could use $\hat{\mathbf{x}}_0 = E(\mathbf{x}_0)$, which is a constant vector. In the actual Kalman filtering, G_k must also be computed recursively. The two recursive processes together will be called the *Kalman filtering process*.

Let $\hat{\mathbf{x}}_{k|j}$ be the (optimal) least-squares estimate of $\mathbf{x}_k$ with minimum variance by choosing the weight matrix to be

$$W_{k,j} = (Var(\underline{\varepsilon}_{k,j}))^{-1}$$

using $\overline{v}_j$ in (2.2) (see Section 1.3 for details). It is easy to verify that

$$W_{k,k-1}^{-1} = \begin{bmatrix} R_0 & & 0 \\ & \ddots & \\ 0 & & R_{k-1} \end{bmatrix} + Var \begin{bmatrix} C_0 \sum_{i=1}^{k} \Phi_{0i}\Gamma_{i-1}\underline{\xi}_{i-1} \\ \vdots \\ C_{k-1}\Phi_{k-1,k}\Gamma_{k-1}\underline{\xi}_{k-1} \end{bmatrix} \quad (2.3)$$

and

$$W_{k,k}^{-1} = \begin{bmatrix} W_{k,k-1}^{-1} & 0 \\ 0 & R_k \end{bmatrix} \quad (2.4)$$

(cf. Exercise 2.1). Hence, $W_{k,k-1}$ and $W_{k,k}$ are positive definite (cf. Exercise 2.2).

In this chapter, we also assume that the matrices

$$(H_{k,j}^{\top} W_{k,j} H_{k,j}), \quad j = k - 1 \quad and \quad k,$$

are nonsingular. Then it follows from Chapter 1, Section 1.3, that

$$\hat{\mathbf{x}}_{k|j} = (H_{k,j}^{\top} W_{k,j} H_{k,j})^{-1} H_{k,j}^{\top} W_{k,j} \overline{\mathbf{v}}_j. \quad (2.5)$$

Our first goal is to relate $\hat{\mathbf{x}}_{k|k-1}$ with $\hat{\mathbf{x}}_{k|k}$. To do so, we observe that

$$H_{k,k}^{\top} W_{k,k} H_{k,k} = [H_{k,k-1}^{\top} C_k^{\top}] \begin{bmatrix} W_{k,k-1} & 0 \\ 0 & R_k^{-1} \end{bmatrix} \begin{bmatrix} H_{k,k-1} \\ C_k \end{bmatrix}$$
$$= H_{k,k-1}^{\top} W_{k,k-1} H_{k,k-1} + C_k^{\top} R_k^{-1} C_k$$

and

$$H_{k,k}^{\top} W_{k,k} \overline{\mathbf{v}}_k = H_{k,k-1}^{\top} W_{k,k-1} \overline{\mathbf{v}}_{k-1} + C_k^{\top} R_k^{-1} \mathbf{v}_k.$$

Using (2.5) and the above two equalities, we have

$$(H_{k,k-1}^{\top} W_{k,k-1} H_{k,k-1} + C_k^{\top} R_k^{-1} C_k)\hat{\mathbf{x}}_{k|k-1}$$
$$= H_{k,k-1}^{\top} W_{k,k-1} \overline{\mathbf{v}}_{k-1} + C_k^{\top} R_k^{-1} C_k \hat{\mathbf{x}}_{k|k-1}$$

and

$$(H_{k,k-1}^{\top} W_{k,k-1} H_{k,k-1} + C_k^{\top} R_k^{-1} C_k)\hat{\mathbf{x}}_{k|k}$$
$$= (H_{k,k}^{\top} W_{k,k} H_{k,k})\hat{\mathbf{x}}_{k|k}$$
$$= H_{k,k-1}^{\top} W_{k,k-1} \overline{\mathbf{v}}_{k-1} + C_k^{\top} R_k^{-1} \mathbf{v}_k \ .$$

A simple subtraction gives

$$(H_{k,k-1}^{\top} W_{k,k-1} H_{k,k-1} + C_k^{\top} R_k^{-1} C_k)(\hat{\mathbf{x}}_{k|k} - \hat{\mathbf{x}}_{k|k-1})$$
$$= C_k^{\top} R_k^{-1}(\mathbf{v}_k - C_k \hat{\mathbf{x}}_{k|k-1}).$$

Now define

$$G_k = (H_{k,k-1}^\top W_{k,k-1} H_{k,k-1} + C_k^\top R_k^{-1} C_k)^{-1} C_k^\top R_k^{-1}$$
$$= (H_{k,k}^\top W_{k,k} H_{k,k})^{-1} C_k^\top R_k^{-1}.$$

Then we have

$$\hat{\mathbf{x}}_{k|k} = \hat{\mathbf{x}}_{k|k-1} + G_k(\mathbf{v}_k - C_k \hat{\mathbf{x}}_{k|k-1}). \tag{2.6}$$

Since $\hat{\mathbf{x}}_{k|k-1}$ is a one-step prediction and $(\mathbf{v}_k - C_k\hat{\mathbf{x}}_{k|k-1})$ is the error between the real data and the prediction, (2.6) is in fact a "prediction-correction" formula with the Kalman gain matrix G_k as a weight matrix. To complete the recursive process, we need an equation that gives $\hat{\mathbf{x}}_{k|k-1}$ from $\hat{\mathbf{x}}_{k-1|k-1}$. This is simply the equation

$$\hat{\mathbf{x}}_{k|k-1} = A_{k-1}\hat{\mathbf{x}}_{k-1|k-1}. \tag{2.7}$$

To prove this, we first note that

$$\bar{\underline{\epsilon}}_{k,k-1} = \bar{\underline{\epsilon}}_{k-1,k-1} - H_{k,k-1}\Gamma_{k-1}\underline{\xi}_{k-1}$$

so that

$$W_{k,k-1}^{-1} = W_{k-1,k-1}^{-1} + H_{k-1,k-1}\Phi_{k-1,k}\Gamma_{k-1}Q_{k-1}\Gamma_{k-1}^\top\Phi_{k-1,k}^\top H_{k-1,k-1}^\top \tag{2.8}$$

(cf. Exercise 2.3). Hence, by Lemma 1.2, we have

$$W_{k,k-1} = W_{k-1,k-1} - W_{k-1,k-1}H_{k-1,k-1}\Phi_{k-1,k}\Gamma_{k-1}(Q_{k-1}^{-1}$$
$$+ \Gamma_{k-1}^\top\Phi_{k-1,k}^\top H_{k-1,k-1}^\top W_{k-1,k-1}H_{k-1,k-1}\Phi_{k-1,k}\Gamma_{k-1})^{-1}$$
$$\cdot \Gamma_{k-1}^\top\Phi_{k-1,k}^\top H_{k-1,k-1}^\top W_{k-1,k-1} \tag{2.9}$$

(cf. Exercise 2.4). Then by the transition relation

$$H_{k,k-1} = H_{k-1,k-1}\Phi_{k-1,k}$$

we have

$$H_{k,k-1}^\top W_{k,k-1}$$
$$= \Phi_{k-1,k}^\top\{I - H_{k-1,k-1}^\top W_{k-1,k-1}H_{k-1,k-1}\Phi_{k-1,k}\Gamma_{k-1}(Q_{k-1}^{-1}$$
$$+ \Gamma_{k-1}^\top\Phi_{k-1,k}^\top H_{k-1,k-1}^\top W_{k-1,k-1}H_{k-1,k-1}\Phi_{k-1,k}\Gamma_{k-1})^{-1}$$
$$\cdot \Gamma_{k-1}^\top\Phi_{k-1,k}^\top\}H_{k-1,k-1}^\top W_{k-1,k-1} \tag{2.10}$$

(cf. Exercise 2.5). It follows that

$$(H_{k,k-1}^\top W_{k,k-1} H_{k,k-1}) \Phi_{k,k-1} (H_{k-1,k-1}^\top W_{k-1,k-1} H_{k-1,k-1})^{-1}$$
$$\cdot H_{k-1,k-1}^\top W_{k-1,k-1} = H_{k,k-1}^\top W_{k,k-1} \tag{2.11}$$

(cf. Exercise 2.6). This, together with (2.5) with $j = k - 1$ and k, gives (2.7).

Our next goal is to derive a recursive scheme for calculating the Kalman gain matrices G_k. Write

$$G_k = P_{k,k} C_k^\top R_k^{-1}$$

where

$$P_{k,k} = (H_{k,k}^\top W_{k,k} H_{k,k})^{-1}$$

and set

$$P_{k,k-1} = (H_{k,k-1}^\top W_{k,k-1} H_{k,k-1})^{-1} .$$

Then, since

$$P_{k,k}^{-1} = P_{k,k-1}^{-1} + C_k^\top R_k^{-1} C_k ,$$

we obtain, using Lemma 1.2,

$$P_{k,k} = P_{k,k-1} - P_{k,k-1} C_k^\top (C_k P_{k,k-1} C_k^\top + R_k)^{-1} C_k P_{k,k-1} .$$

It can be proved that

$$G_k = P_{k,k-1} C_k^\top (C_k P_{k,k-1} C_k^\top + R_k)^{-1} \tag{2.12}$$

(cf. Exercise 2.7), so that

$$P_{k,k} = (I - G_k C_k) P_{k,k-1} . \tag{2.13}$$

Furthermore, we can show that

$$P_{k,k-1} = A_{k-1} P_{k-1,k-1} A_{k-1}^\top + \Gamma_{k-1} Q_{k-1} \Gamma_{k-1}^\top \tag{2.14}$$

(cf. Exercise 2.8). Hence, using (2.13) and (2.14) with the initial matrix $P_{0,0}$, we obtain a recursive scheme to compute $P_{k-1,k-1}$, $P_{k,k-1}$, G_k and $P_{k,k}$ for $k = 1, 2, \cdots$. Moreover, it can be shown that

$$P_{k,k-1} = E(\mathbf{x}_k - \hat{\mathbf{x}}_{k|k-1})(\mathbf{x}_k - \hat{\mathbf{x}}_{k|k-1})^\top$$
$$= Var(\mathbf{x}_k - \hat{\mathbf{x}}_{k|k-1}) \tag{2.15}$$

(cf. Exercise 2.9) and that

$$P_{k,k} = E(\mathbf{x}_k - \hat{\mathbf{x}}_{k|k})(\mathbf{x}_k - \hat{\mathbf{x}}_{k|k})^\top = Var(\mathbf{x}_k - \hat{\mathbf{x}}_{k|k}). \qquad (2.16)$$

In particular, we have

$$P_{0,0} = E(\mathbf{x}_0 - E\mathbf{x}_0)(\mathbf{x}_0 - E\mathbf{x}_0)^\top = Var(\mathbf{x}_0).$$

Finally, combining all the results obtained above, we arrive at the following Kalman filtering process for the linear stochastic system with state-space description (2.1):

$$\left\{ \begin{array}{l} P_{0,0} = Var(\mathbf{x}_0) \\ P_{k,k-1} = A_{k-1}P_{k-1,k-1}A_{k-1}^\top + \Gamma_{k-1}Q_{k-1}\Gamma_{k-1}^\top \\ G_k = P_{k,k-1}C_k^\top(C_k P_{k,k-1}C_k^\top + R_k)^{-1} \\ P_{k,k} = (I - G_k C_k)P_{k,k-1} \\ \hat{\mathbf{x}}_{0|0} = E(\mathbf{x}_0) \\ \hat{\mathbf{x}}_{k|k-1} = A_{k-1}\hat{\mathbf{x}}_{k-1|k-1} \\ \hat{\mathbf{x}}_{k|k} = \hat{\mathbf{x}}_{k|k-1} + G_k(\mathbf{v}_k - C_k\hat{\mathbf{x}}_{k|k-1}) \\ k = 1, 2, \cdots . \end{array} \right. \qquad (2.17)$$

This algorithm may be realized as shown in Fig. 2.1.

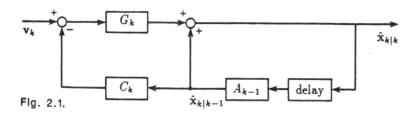

Fig. 2.1.

2.4 Kalman Filtering Process

Let us now consider the general linear deterministic/stochastic system where the deterministic control input $\{\mathbf{u}_k\}$ is present. More precisely, let us consider the state-space description

$$\left\{ \begin{array}{l} \mathbf{x}_{k+1} = A_k\mathbf{x}_k + B_k\mathbf{u}_k + \Gamma_k\underline{\xi}_k \\ \mathbf{v}_k = C_k\mathbf{x}_k + D_k\mathbf{u}_k + \underline{\eta}_k, \end{array} \right.$$

where $\{\mathbf{u}_k\}$ is a sequence of m-vectors with $1 \le m \le n$. Then by superimposing the deterministic solution with (2.17), the Kalman filtering process for this system is given by

$$
\left\{
\begin{aligned}
&P_{0,0} = Var(\mathbf{x}_0) \\
&P_{k,k-1} = A_{k-1}P_{k-1,k-1}A_{k-1}^\top + \Gamma_{k-1}Q_{k-1}\Gamma_{k-1}^\top \\
&G_k = P_{k,k-1}C_k^\top (C_k P_{k,k-1}C_k^\top + R_k)^{-1} \\
&P_{k,k} = (I - G_k C_k)P_{k,k-1} \\
&\hat{\mathbf{x}}_{0|0} = E(\mathbf{x}_0) \\
&\hat{\mathbf{x}}_{k|k-1} = A_{k-1}\hat{\mathbf{x}}_{k-1|k-1} + B_{k-1}\mathbf{u}_{k-1} \\
&\hat{\mathbf{x}}_{k|k} = \hat{\mathbf{x}}_{k|k-1} + G_k(\mathbf{v}_k - D_k\mathbf{u}_k - C_k\hat{\mathbf{x}}_{k|k-1}) \\
&k = 1, 2, \cdots,
\end{aligned}
\right.
\tag{2.18}
$$

(cf. Exercise 2.13). This algorithm may be implemented as shown in Fig.2.2.

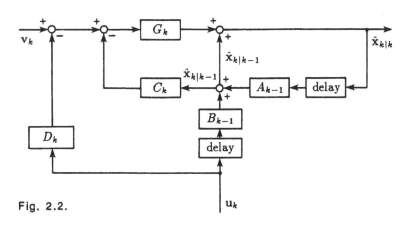

Fig. 2.2.

Exercises

2.1. Let

$$
\bar{\varepsilon}_{k,j} = \begin{bmatrix} \varepsilon_{k,0} \\ \vdots \\ \varepsilon_{k,j} \end{bmatrix} \qquad and \qquad \varepsilon_{k,\ell} = \underline{\eta}_\ell - C_\ell \sum_{i=\ell+1}^{k} \Phi_{\ell i} \Gamma_{i-1} \underline{\xi}_{i-1} ,
$$

where $\{\underline{\xi}_k\}$ and $\{\underline{\eta}_k\}$ are both zero-mean Gaussian white noise sequences with $Var(\underline{\xi}_k) = Q_k$ and $Var(\underline{\eta}_k) = R_k$. Define $W_{k,j} = (Var(\bar{\varepsilon}_{k,j}))^{-1}$. Show that

$$
W_{k,k-1}^{-1} = \begin{bmatrix} R_0 & & 0 \\ & \ddots & \\ 0 & & R_{k-1} \end{bmatrix} + Var \begin{bmatrix} C_0 \sum_{i=1}^{k} \Phi_{0i} \Gamma_{i-1} \underline{\xi}_{i-1} \\ \vdots \\ C_{k-1} \Phi_{k-1,k} \Gamma_{k-1} \underline{\xi}_{k-1} \end{bmatrix}
$$

and

$$
W_{k,k}^{-1} = \begin{bmatrix} W_{k,k-1}^{-1} & 0 \\ 0 & R_k \end{bmatrix} .
$$

2.2. Show that the sum of a positive definite matrix A and a non-negative definite matrix B is positive definite.

2.3. Let $\bar{\varepsilon}_{k,j}$ and $W_{k,j}$ be defined as in Exercise 2.1. Verify the relation

$$
\bar{\varepsilon}_{k,k-1} = \bar{\varepsilon}_{k-1,k-1} - H_{k,k-1} \Gamma_{k-1} \underline{\xi}_{k-1}
$$

where

$$
H_{k,j} = \begin{bmatrix} C_0 \Phi_{0k} \\ \vdots \\ C_j \Phi_{jk} \end{bmatrix} ,
$$

and then show that

$$
W_{k,k-1}^{-1} = W_{k-1,k-1}^{-1} + H_{k-1,k-1} \Phi_{k-1,k} \Gamma_{k-1} Q_{k-1} \Gamma_{k-1}^\top \Phi_{k-1,k}^\top H_{k-1,k-1}^\top .
$$

2.4. Use Exercise 2.3 and Lemma 1.2 to show that

$$
\begin{aligned}
W_{k,k-1} = W_{k-1,k-1} &- W_{k-1,k-1} H_{k-1,k-1} \Phi_{k-1,k} \Gamma_{k-1} (Q_{k-1}^{-1} \\
&+ \Gamma_{k-1}^\top \Phi_{k-1,k}^\top H_{k-1,k-1}^\top W_{k-1,k-1} H_{k-1,k-1} \Phi_{k-1,k} \Gamma_{k-1})^{-1} \\
&\cdot \Gamma_{k-1}^\top \Phi_{k-1,k}^\top H_{k-1,k-1}^\top W_{k-1,k-1} .
\end{aligned}
$$

2.5. Use Exercise 2.4 and the relation $H_{k,k-1} = H_{k-1,k-1} \Phi_{k-1,k}$ to show that

$$
\begin{aligned}
H_{k,k-1}^\top & W_{k,k-1} \\
= \Phi_{k-1,k}^\top & \{ I - H_{k-1,k-1}^\top W_{k-1,k-1} H_{k-1,k-1} \Phi_{k-1,k} \Gamma_{k-1} (Q_{k-1}^{-1} \\
&+ \Gamma_{k-1}^\top \Phi_{k-1,k}^\top H_{k-1,k-1}^\top W_{k-1,k-1} H_{k-1,k-1} \Phi_{k-1,k} \Gamma_{k-1})^{-1} \\
&\cdot \Gamma_{k-1}^\top \Phi_{k-1,k}^\top \} H_{k-1,k-1}^\top W_{k-1,k-1} .
\end{aligned}
$$

2.6.   Use Exercise 2.5 to derive the identity:

$$(H_{k,k-1}^\top W_{k,k-1} H_{k,k-1}) \Phi_{k,k-1} (H_{k-1,k-1}^\top W_{k-1,k-1} H_{k-1,k-1})^{-1}$$
$$\cdot H_{k-1,k-1}^\top W_{k-1,k-1} = H_{k,k-1}^\top W_{k,k-1}.$$

2.7.   Use Lemma 1.2 to show that

$$P_{k,k-1} C_k^\top (C_k P_{k,k-1} C_k^\top + R_k)^{-1} = P_{k,k} C_k^\top R_k^{-1} = G_k.$$

2.8.   Start with $P_{k,k-1} = (H_{k,k-1}^\top W_{k,k-1} H_{k,k-1})^{-1}$. Use Lemma 1.2,
       (2.8), and the definition of $P_{k,k} = (H_{k,k}^\top W_{k,k} H_{k,k})^{-1}$ to show
       that

$$P_{k,k-1} = A_{k-1} P_{k-1,k-1} A_{k-1}^\top + \Gamma_{k-1} Q_{k-1} \Gamma_{k-1}^\top.$$

2.9.   Use (2.5) and (2.2) to prove that

$$E(\mathbf{x}_k - \hat{\mathbf{x}}_{k|k-1})(\mathbf{x}_k - \hat{\mathbf{x}}_{k|k-1})^\top = P_{k,k-1}$$

and

$$E(\mathbf{x}_k - \hat{\mathbf{x}}_{k|k})(\mathbf{x}_k - \hat{\mathbf{x}}_{k|k})^\top = P_{k,k}.$$

2.10. Consider the one-dimensional linear stochastic dynamic
       system

$$x_{k+1} = a x_k + \xi_k, \qquad x_0 = 0,$$

where $E(x_k) = 0, Var(x_k) = \sigma^2, E(x_k \xi_j) = 0, E(\xi_k) = 0$, and
$E(\xi_k \xi_j) = \mu^2 \delta_{kj}$. Prove that $\sigma^2 = \mu^2/(1 - a^2)$ and $E(x_k x_{k+j}) = a^{|j|} \sigma^2$ for all integers j.

2.11. Consider the one-dimensional stochastic linear system

$$\begin{cases} x_{k+1} = x_k \\ v_k = x_k + \eta_k \end{cases}$$

with $E(\eta_k) = 0, Var(\eta_k) = \sigma^2, E(x_0) = 0$ and $Var(x_0) = \mu^2$.
Show that

$$\begin{cases} \hat{x}_{k|k} = \hat{x}_{k-1|k-1} + \dfrac{\mu^2}{\sigma^2 + k\mu^2}(v_k - \hat{x}_{k-1|k-1}) \\ \hat{x}_{0|0} = 0 \end{cases}$$

and that $\hat{x}_{k|k} \to c$ for some constant c as $k \to \infty$.

2.12. Let $\{\mathbf{v}_k\}$ be a sequence of data obtained from the observa-
       tion of a zero-mean random vector $\mathbf{y}$ with unknown variance
       Q. The variance of $\mathbf{y}$ can be estimated by

$$\hat{Q}_N = \frac{1}{N} \sum_{k=1}^{N} (\mathbf{v}_k \mathbf{v}_k^\top).$$

Derive a prediction-correction recursive formula for this estimation.

2.13. Consider the linear deterministic/stochastic system

$$\begin{cases} \mathbf{x}_{k+1} = A_k \mathbf{x}_k + B_k \mathbf{u}_k + \Gamma_k \underline{\xi}_k \\ \mathbf{v}_k = C_k \mathbf{x}_k + D_k \mathbf{u}_k + \underline{\eta}_k \,, \end{cases}$$

where $\{\mathbf{u}_k\}$ is a given sequence of deterministic control input m-vectors, $1 \le m \le n$. Suppose that Assumption 2.1 is satisfied and the matrix $Var(\underline{\xi}_{k,j})$ is nonsingular (cf. (2.2) for the definition of $\underline{\xi}_{k,j}$). Derive the Kalman filtering equations for this model.

2.14. In digital signal processing, a widely used mathematical model is the following so-called *ARMA* (*autoregressive moving-average*) process:

$$\mathbf{v}_k = \sum_{i=1}^{N} B_i \mathbf{v}_{k-i} + \sum_{i=0}^{M} A_i \mathbf{u}_{k-i} \,,$$

where the $n \times n$ matrices $B_1, \cdots, B_N$ and the $n \times q$ matrices $A_0, A_1, \cdots, A_M$ are independent of the time variable k, and $\{\mathbf{u}_k\}$ and $\{\mathbf{v}_k\}$ are input and output digital signal sequences, respectively (cf. Fig. 2.3). Assuming that $M \le N$, show that the input-output relationship can be described as a state-space model

$$\begin{cases} \mathbf{x}_{k+1} = A\mathbf{x}_k + B\mathbf{u}_k \\ \mathbf{v}_k = C\mathbf{x}_k + D\mathbf{u}_k \end{cases}$$

with $\mathbf{x}_0 = 0$, where

$$A = \begin{bmatrix} B_1 & I & 0 & \cdots & 0 \\ B_2 & 0 & I & & \vdots \\ \vdots & \vdots & & \ddots & \vdots \\ B_{N-1} & 0 & \cdots & \cdots & I \\ B_N & 0 & \cdots & \cdots & 0 \end{bmatrix}, \quad B = \begin{bmatrix} A_1 + B_1 A_0 \\ A_2 + B_2 A_0 \\ \vdots \\ A_M + B_M A_0 \\ B_{M+1} A_0 \\ \vdots \\ B_N A_0 \end{bmatrix},$$

$$C = \begin{bmatrix} I & 0 & \cdots & 0 \end{bmatrix} \quad and \quad D = \begin{bmatrix} A_0 \end{bmatrix}.$$

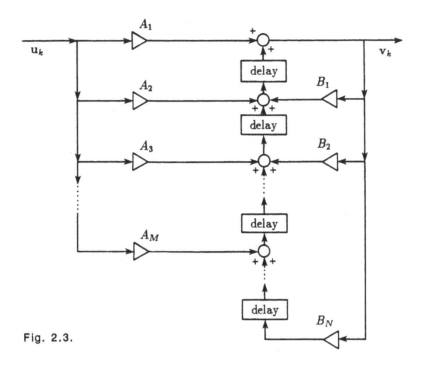

Fig. 2.3.

3. Orthogonal Projection and Kalman Filter

The elementary approach to the derivation of the optimal Kalman filtering process discussed in Chapter 2 has the advantage that the optimal estimate $\hat{\mathbf{x}}_k = \hat{\mathbf{x}}_{k|k}$ of the state vector $\mathbf{x}_k$ is easily understood to be a least-squares estimate of $\mathbf{x}_k$ with the properties that (i) the transformation that yields $\hat{\mathbf{x}}_k$ from the data $\bar{\mathbf{v}}_k = [\mathbf{v}_0^\top \cdots \mathbf{v}_k^\top]^\top$ is linear, (ii) $\hat{\mathbf{x}}_k$ is unbiased in the sense that $E(\hat{\mathbf{x}}_k) = E(\mathbf{x}_k)$, and (iii) it yields a minimum variance estimate with $(Var(\bar{\epsilon}_{k,k}))^{-1}$ as the optimal weight. The disadvantage of this elementary approach is that certain matrices must be assumed to be nonsingular. In this chapter, we will drop the nonsingularity assumptions and give a rigorous derivation of the Kalman filtering algorithm.

3.1 Orthogonality Characterization of Optimal Estimates

Consider the linear stochastic system described by (2.1) such that Assumption 2.1 is satisfied. That is, consider the state-space description

$$\begin{cases} \mathbf{x}_{k+1} = A_k\mathbf{x}_k + \Gamma_k\underline{\xi}_k \\ \mathbf{v}_k = C_k\mathbf{x}_k + \underline{\eta}_k \,, \end{cases} \tag{3.1}$$

where A_k, Γ_k and C_k are known $n \times n$, $n \times p$ and $q \times n$ constant matrices, respectively, with $1 \leq p, q \leq n$, and

$$E(\underline{\xi}_k) = 0\,, \qquad E(\underline{\xi}_k\underline{\xi}_\ell^\top) = Q_k\delta_{k\ell}\,,$$

$$E(\underline{\eta}_k) = 0\,, \qquad E(\underline{\eta}_k\underline{\eta}_\ell^\top) = R_k\delta_{k\ell}\,,$$

$$E(\underline{\xi}_k\underline{\eta}_\ell^\top) = 0\,, \quad E(\mathbf{x}_0\underline{\xi}_k^\top) = 0\,, \quad E(\mathbf{x}_0\underline{\eta}_k^\top) = 0\,,$$

for all $k, \ell = 0, 1, \cdots$, with Q_k and R_k being positive definite and symmetric matrices.

Let $\mathbf{x}$ be a random n-vector and $\mathbf{w}$ a random q-vector. We define the "inner product" $\langle \mathbf{x}, \mathbf{w} \rangle$ to be the $n \times q$ matrix

$$\langle \mathbf{x}, \mathbf{w} \rangle = Cov(\mathbf{x}, \mathbf{w}) = E(\mathbf{x} - E(\mathbf{x}))(\mathbf{w} - E(\mathbf{w}))^\top .$$

Let $\|\mathbf{w}\|_q$ be the positive square root of $\langle \mathbf{w}, \mathbf{w} \rangle$. That is, $\|\mathbf{w}\|_q$ is a non-negative definite $q \times q$ matrix with

$$\|\mathbf{w}\|_q^2 = \|\mathbf{w}\|_q \|\mathbf{w}\|_q^\top = \langle \mathbf{w}, \mathbf{w} \rangle .$$

Similarly, let $\|\mathbf{x}\|_n$ be the positive square root of $\langle \mathbf{x}, \mathbf{x} \rangle$. Now, let $\mathbf{w}_0, \cdots, \mathbf{w}_r$ be random q-vectors and consider the "linear span":

$$Y(\mathbf{w}_0, \cdots, \mathbf{w}_r)$$

$$= \{ \mathbf{y} : \ \mathbf{y} = \sum_{i=0}^{r} P_i \mathbf{w}_i, \ P_0, \cdots, P_r, \ n \times q \ \text{constant matrices} \} .$$

The first minimization problem we will study is to determine a $\hat{\mathbf{y}}$ in $Y(\mathbf{w}_0, \cdots, \mathbf{w}_r)$ such that $tr\|\mathbf{x}_k - \hat{\mathbf{y}}\|_n^2 = F_k$, where

$$F_k := min\{ tr\|\mathbf{x}_k - \mathbf{y}\|_n^2 : \ \mathbf{y} \in Y(\mathbf{w}_0, \cdots, \mathbf{w}_r) \} . \tag{3.2}$$

The following result characterizes $\hat{\mathbf{y}}$.

Lemma 3.1. $\hat{\mathbf{y}} \in Y(\mathbf{w}_0, \cdots, \mathbf{w}_r)$ *satisfies* $tr\|\mathbf{x}_k - \hat{\mathbf{y}}\|_n^2 = F_k$ *if and only if*

$$\langle \mathbf{x}_k - \hat{\mathbf{y}}, \mathbf{w}_j \rangle = O_{n \times q}$$

for all $j = 0, 1, \cdots, r$. *Furthermore,* $\hat{\mathbf{y}}$ *is unique in the sense that*

$$tr\|\mathbf{x}_k - \hat{\mathbf{y}}\|_n^2 = tr\|\mathbf{x}_k - \tilde{\mathbf{y}}\|_n^2$$

only if $\hat{\mathbf{y}} = \tilde{\mathbf{y}}$.

To prove this lemma, we first suppose that $tr\|\mathbf{x}_k - \hat{\mathbf{y}}\|_n^2 = F_k$ but $\langle \mathbf{x}_k - \hat{\mathbf{y}}, \mathbf{w}_{j_0} \rangle = C \neq O_{n \times q}$ for some j_0 where $0 \leq j_0 \leq r$. Then $\mathbf{w}_{j_0} \neq 0$ so that $\|\mathbf{w}_{j_0}\|_q^2$ is a positive definite symmetric matrix and so is its inverse $\|\mathbf{w}_{j_0}\|_q^{-2}$. Hence, $C\|\mathbf{w}_{j_0}\|_q^{-2}C^\top \neq O_{n \times n}$ and is a non-negative definite and symmetric matrix. It can be shown that

$$tr\{ C\|\mathbf{w}_{j_0}\|_q^{-2}C^\top \} > 0 \tag{3.3}$$

(cf. Exercise 3.1). Now, the vector $\hat{\mathbf{y}} + C\|\mathbf{w}_{j_0}\|_q^{-2}\mathbf{w}_{j_0}$ is in $Y(\mathbf{w}_0, \cdots, \mathbf{w}_r)$ and

$$tr\|\mathbf{x}_k - (\hat{\mathbf{y}} + C\|\mathbf{w}_{j_0}\|_q^{-2}\mathbf{w}_{j_0})\|_n^2$$

$$= tr\{ \|\mathbf{x}_k - \hat{\mathbf{y}}\|_n^2 - \langle \mathbf{x}_k - \hat{\mathbf{y}}, \mathbf{w}_{j_0} \rangle (C\|\mathbf{w}_{j_0}\|_q^{-2})^\top - C\|\mathbf{w}_{j_0}\|_q^{-2}\langle \mathbf{w}_{j_0}, \mathbf{x}_k - \hat{\mathbf{y}} \rangle$$

$$+ C\|\mathbf{w}_{j_0}\|_q^{-2}\|\mathbf{w}_{j_0}\|_q^2(C\|\mathbf{w}_{j_0}\|_q^{-2})^\top \}$$

$$= tr\{ \|\mathbf{x}_k - \hat{\mathbf{y}}\|_n^2 - C\|\mathbf{w}_{j_0}\|_q^{-2}C^\top \}$$

$$< tr\|\mathbf{x}_k - \hat{\mathbf{y}}\|_n^2 = F_k$$

by using (3.3). This contradicts the definition of F_k in (3.2).

Conversely, let $\langle \mathbf{x}_k - \hat{\mathbf{y}}, \mathbf{w}_j \rangle = O_{n \times q}$ for all $j = 0, 1, \cdots, r$. Let $\mathbf{y}$ be an arbitrary random n-vector in $Y(\mathbf{w}_0, \cdots, \mathbf{w}_r)$ and write $\mathbf{y}_0 = \mathbf{y} - \hat{\mathbf{y}} = \sum_{j=0}^r P_{0j} \mathbf{w}_j$ where P_{0j} are constant $n \times q$ matrices, $j = 0, 1, \cdots, r$. Then

$$
tr\|\mathbf{x}_k - \mathbf{y}\|_n^2
$$
$$
= tr\|(\mathbf{x}_k - \hat{\mathbf{y}}) - \mathbf{y}_0\|_n^2
$$
$$
= tr\{\|\mathbf{x}_k - \hat{\mathbf{y}}\|_n^2 - \langle \mathbf{x}_k - \hat{\mathbf{y}}, \mathbf{y}_0 \rangle - \langle \mathbf{y}_0, \mathbf{x}_k - \hat{\mathbf{y}} \rangle + \|\mathbf{y}_0\|_n^2 \}
$$
$$
= tr\left\{ \|\mathbf{x}_k - \hat{\mathbf{y}}\|_n^2 - \sum_{j=0}^r \langle \mathbf{x}_k - \hat{\mathbf{y}}, \mathbf{w}_j \rangle P_{0j}^\top - \sum_{j=0}^r P_{0j} \langle \mathbf{x}_k - \hat{\mathbf{y}}, \mathbf{w}_j \rangle^\top + \|\mathbf{y}_0\|_n^2 \right\}
$$
$$
= tr\|\mathbf{x}_k - \hat{\mathbf{y}}\|_n^2 + tr\|\mathbf{y}_0\|_n^2
$$
$$
\geq tr\|\mathbf{x}_k - \hat{\mathbf{y}}\|_n^2 \,,
$$

so that $tr\|\mathbf{x}_k - \hat{\mathbf{y}}\|_n^2 = F_k$. Furthermore, equality is attained if and only if $tr\|\mathbf{y}_0\|_n^2 = 0$ or $\mathbf{y}_0 = 0$ so that $\mathbf{y} = \hat{\mathbf{y}}$ (cf. Exercise 3.1). This completes the proof of the lemma.

3.2 Innovations Sequences

To use the data information, we require an "orthogonalization" process.

Definition 3.1. Given a random q-vector data sequence $\{\mathbf{v}_j\}$, $j = 0, \cdots, k$. The *innovations sequence* $\{\mathbf{z}_j\}$, $j = 0, \cdots, k$, of $\{\mathbf{v}_j\}$ (i.e., a sequence obtained by changing the original data sequence $\{\mathbf{v}_j\}$) is defined by

$$
\mathbf{z}_j = \mathbf{v}_j - C_j \hat{\mathbf{y}}_{j-1}, \quad j = 0, 1, \cdots, k, \tag{3.4}
$$

with $\hat{\mathbf{y}}_{-1} = 0$ and

$$
\hat{\mathbf{y}}_{j-1} = \sum_{i=0}^{j-1} \hat{P}_{j-1,i} \mathbf{v}_i \in Y(\mathbf{v}_0, \cdots, \mathbf{v}_{j-1}), \quad j = 1, \cdots, k,
$$

where the $q \times n$ matrices C_j are the observation matrices in (3.1) and the $n \times q$ matrices $\hat{P}_{j-1,i}$ are chosen so that $\hat{\mathbf{y}}_{j-1}$ solves the minimization problem (3.2) with $Y(\mathbf{w}_0, \cdots, \mathbf{w}_r)$ replaced by $Y(\mathbf{v}_0, \cdots, \mathbf{v}_{j-1})$.

We first give the correlation property of the innovations sequence.

Lemma 3.2. *The innovations sequence $\{z_j\}$ of $\{v_j\}$ satisfies the following property:*

$$\langle z_j, z_\ell \rangle = (R_\ell + C_\ell \| x_\ell - \hat{y}_{\ell-1} \|_n^2 C_\ell^\top) \delta_{j\ell} ,$$

where $R_\ell = Var(\underline{\eta}_\ell) > 0$.

For convenience, we set

$$\hat{e}_j = C_j(x_j - \hat{y}_{j-1}). \tag{3.5}$$

To prove the lemma, we first observe that

$$z_j = \hat{e}_j + \underline{\eta}_j , \tag{3.6}$$

where $\{\underline{\eta}_k\}$ is the observation noise sequence, and

$$\langle \underline{\eta}_\ell, \hat{e}_j \rangle = O_{q \times q} \quad \text{for all} \quad \ell \geq j. \tag{3.7}$$

Clearly, (3.6) follows from (3.4), (3.5), and the observation equation in (3.1). The proof of (3.7) is left to the reader as an exercise (cf. Exercise 3.2). Now, for $j = \ell$, we have, by (3.6), (3.7), and (3.5) consecutively,

$$
\begin{aligned}
\langle z_\ell, z_\ell \rangle &= \langle \hat{e}_\ell + \underline{\eta}_\ell, \hat{e}_\ell + \underline{\eta}_\ell \rangle \\
&= \langle \hat{e}_\ell, \hat{e}_\ell \rangle + \langle \underline{\eta}_\ell, \underline{\eta}_\ell \rangle \\
&= C_\ell \| x_\ell - \hat{y}_{\ell-1} \|_n^2 C_\ell^\top + R_\ell .
\end{aligned}
$$

For $j \neq \ell$, since $\langle \hat{e}_\ell, \hat{e}_j \rangle^\top = \langle \hat{e}_j, \hat{e}_\ell \rangle$, we can assume without loss of generality that $j > \ell$. Hence, by (3.6), (3.7), and Lemma 3.1 we have

$$
\begin{aligned}
\langle z_j, z_\ell \rangle &= \langle \hat{e}_j, \hat{e}_\ell \rangle + \langle \hat{e}_j, \underline{\eta}_\ell \rangle + \langle \underline{\eta}_j, \hat{e}_\ell \rangle + \langle \underline{\eta}_j, \underline{\eta}_\ell \rangle \\
&= \langle \hat{e}_j, \hat{e}_\ell + \underline{\eta}_\ell \rangle \\
&= \langle \hat{e}_j, z_\ell \rangle \\
&= \langle \hat{e}_j, v_\ell - C_\ell \hat{y}_{\ell-1} \rangle \\
&= \left\langle C_j(x_j - \hat{y}_{j-1}), v_\ell - C_\ell \sum_{i=0}^{\ell-1} \hat{P}_{\ell-1,i} v_i \right\rangle \\
&= C_j \langle x_j - \hat{y}_{j-1}, v_\ell \rangle - C_j \sum_{i=0}^{\ell-1} \langle x_j - \hat{y}_{j-1}, v_i \rangle \hat{P}_{\ell-1,i}^\top C_\ell^\top \\
&= O_{q \times q} .
\end{aligned}
$$

This completes the proof of the lemma.

Since $R_j > 0$, Lemma 3.2 says that $\{\mathbf{z}_j\}$ is an "orthogonal" sequence of nonzero vectors which we can normalize by setting

$$\mathbf{e}_j = \|\mathbf{z}_j\|_q^{-1}\mathbf{z}_j. \tag{3.8}$$

Then $\{\mathbf{e}_j\}$ is an "orthonormal" sequence in the sense that $\langle \mathbf{e}_i, \mathbf{e}_j \rangle = \delta_{ij}I_q$ for all i and j. Furthermore, it should be clear that

$$Y(\mathbf{e}_0, \cdots, \mathbf{e}_k) = Y(\mathbf{v}_0, \cdots, \mathbf{v}_k) \tag{3.9}$$

(cf. Exercise 3.3).

3.3 Minimum Variance Estimates

We are now ready to give the minimum variance estimate $\check{\mathbf{x}}_k$ of the state vector $\mathbf{x}_k$ by introducing the "Fourier expansion"

$$\check{\mathbf{x}}_k = \sum_{i=0}^{k}\langle \mathbf{x}_k, \mathbf{e}_i \rangle \mathbf{e}_i \tag{3.10}$$

of $\mathbf{x}_k$ with respect to the "orthonormal" sequence $\{\mathbf{e}_j\}$. Since

$$\langle \check{\mathbf{x}}_k, \mathbf{e}_j \rangle = \sum_{i=0}^{k}\langle \mathbf{x}_k, \mathbf{e}_i \rangle \langle \mathbf{e}_i, \mathbf{e}_j \rangle = \langle \mathbf{x}_k, \mathbf{e}_j \rangle,$$

we have

$$\langle \mathbf{x}_k - \check{\mathbf{x}}_k, \mathbf{e}_j \rangle = O_{n \times q}, \quad j = 0, 1, \cdots, k. \tag{3.11}$$

It follows from Exercise 3.3 that

$$\langle \mathbf{x}_k - \check{\mathbf{x}}_k, \mathbf{v}_j \rangle = O_{n \times q}, \quad j = 0, 1, \cdots, k, \tag{3.12}$$

so that by Lemma 3.1,

$$tr\|\mathbf{x}_k - \check{\mathbf{x}}_k\|_n^2 = min\{tr\|\mathbf{x}_k - \mathbf{y}\|_n^2 : \mathbf{y} \in Y(\mathbf{v}_0, \cdots, \mathbf{v}_k)\}.$$

That is, $\check{\mathbf{x}}_k$ is a minimum variance estimate of $\mathbf{x}_k$.

3.4 Kalman Filtering Equations

This section is devoted to the derivation of the Kalman filtering equations. From Assumption 2.1, we first observe that

$$\langle \underline{\xi}_{k-1}, \mathbf{e}_j \rangle = O_{n \times q}, \quad j = 0, 1, \cdots, k-1,$$

(cf. Exercise 3.4), so that

$$
\begin{aligned}
\check{\mathbf{x}}_k &= \sum_{j=0}^{k} \langle \mathbf{x}_k, \mathbf{e}_j \rangle \mathbf{e}_j \\
&= \sum_{j=0}^{k-1} \langle \mathbf{x}_k, \mathbf{e}_j \rangle \mathbf{e}_j + \langle \mathbf{x}_k, \mathbf{e}_k \rangle \mathbf{e}_k \\
&= \sum_{j=0}^{k-1} \{ \langle A_{k-1} \mathbf{x}_{k-1}, \mathbf{e}_j \rangle \mathbf{e}_j + \langle \Gamma_{k-1} \underline{\xi}_{k-1}, \mathbf{e}_j \rangle \mathbf{e}_j \} + \langle \mathbf{x}_k, \mathbf{e}_k \rangle \mathbf{e}_k \\
&= A_{k-1} \sum_{j=0}^{k-1} \langle \mathbf{x}_{k-1}, \mathbf{e}_j \rangle \mathbf{e}_j + \langle \mathbf{x}_k, \mathbf{e}_k \rangle \mathbf{e}_k \\
&= A_{k-1} \check{\mathbf{x}}_{k-1} + \langle \mathbf{x}_k, \mathbf{e}_k \rangle \mathbf{e}_k .
\end{aligned}
$$

Hence, by defining

$$\check{\mathbf{x}}_{k|k-1} = A_{k-1} \check{\mathbf{x}}_{k-1}, \tag{3.13}$$

where $\check{\mathbf{x}}_{k-1} := \check{\mathbf{x}}_{k-1|k-1}$, we have

$$\check{\mathbf{x}}_k = \check{\mathbf{x}}_{k|k} = \check{\mathbf{x}}_{k|k-1} + \langle \mathbf{x}_k, \mathbf{e}_k \rangle \mathbf{e}_k . \tag{3.14}$$

Obviously, if we can show that there exists a constant $n \times q$ matrix $\check{G}_k$ such that

$$\langle \mathbf{x}_k, \mathbf{e}_k \rangle \mathbf{e}_k = \check{G}_k (\mathbf{v}_k - C_k \check{\mathbf{x}}_{k|k-1}),$$

then the "prediction-correction" formulation of the Kalman filter is obtained. To accomplish this, we consider the random vector $(\mathbf{v}_k - C_k \check{\mathbf{x}}_{k|k-1})$ and obtain the following:

Lemma 3.3. For $j = 0, 1, \cdots, k$,

$$\langle \mathbf{v}_k - C_k \check{\mathbf{x}}_{k|k-1}, \mathbf{e}_j \rangle = \|\mathbf{z}_k\|_q \delta_{kj} .$$

To prove the lemma, we first observe that

$$\langle \hat{\mathbf{y}}_j, \mathbf{z}_k \rangle = O_{n \times q}, \quad j = 0, 1, \cdots, k-1, \tag{3.15}$$

(cf. Exercise 3.4). Hence, using (3.14), (3.11), and (3.15), we have

$$\langle \mathbf{v}_k - C_k \check{\mathbf{x}}_{k|k-1}, \mathbf{e}_k \rangle$$
$$= \langle \mathbf{v}_k - C_k(\check{\mathbf{x}}_{k|k} - \langle \mathbf{x}_k, \mathbf{e}_k \rangle \mathbf{e}_k), \mathbf{e}_k \rangle$$
$$= \langle \mathbf{v}_k, \mathbf{e}_k \rangle - C_k\{\langle \check{\mathbf{x}}_{k|k}, \mathbf{e}_k \rangle - \langle \mathbf{x}_k, \mathbf{e}_k \rangle\}$$
$$= \langle \mathbf{v}_k, \mathbf{e}_k \rangle - C_k \langle \check{\mathbf{x}}_{k|k} - \mathbf{x}_k, \mathbf{e}_k \rangle$$
$$= \langle \mathbf{v}_k, \mathbf{e}_k \rangle$$
$$= \langle \mathbf{z}_k + C_k \hat{\mathbf{y}}_{k-1}, \|\mathbf{z}_k\|_q^{-1} \mathbf{z}_k \rangle$$
$$= \langle \mathbf{z}_k, \mathbf{z}_k \rangle \|\mathbf{z}_k\|_q^{-1} + C_k \langle \hat{\mathbf{y}}_{k-1}, \mathbf{z}_k \rangle \|\mathbf{z}_k\|_q^{-1}$$
$$= \|\mathbf{z}_k\|_q .$$

On the other hand, using (3.14), (3.11), and (3.7), we have

$$\langle \mathbf{v}_k - C_k \check{\mathbf{x}}_{k|k-1}, \mathbf{e}_j \rangle$$
$$= \langle C_k \mathbf{x}_k + \underline{\eta}_k - C_k(\check{\mathbf{x}}_{k|k} - \langle \mathbf{x}_k, \mathbf{e}_k \rangle \mathbf{e}_k), \mathbf{e}_j \rangle$$
$$= C_k \langle \mathbf{x}_k - \check{\mathbf{x}}_{k|k}, \mathbf{e}_j \rangle + \langle \underline{\eta}_k, \mathbf{e}_j \rangle + C_k \langle \mathbf{x}_k, \mathbf{e}_k \rangle \langle \mathbf{e}_k, \mathbf{e}_j \rangle$$
$$= O_{q \times q}$$

for $j = 0, 1, \cdots, k - 1$. This completes the proof of the Lemma.

It is clear, by using Exercise 3.3 and the definition of $\check{\mathbf{x}}_{k-1} = \check{\mathbf{x}}_{k-1|k-1}$, that the random q-vector $(\mathbf{v}_k - C_k \check{\mathbf{x}}_{k|k-1})$ can be expressed as $\sum_{i=0}^k M_i \mathbf{e}_i$ for some constant $q \times q$ matrices M_i. It follows now from Lemma 3.3 that for $j = 0, 1, \cdots, k$,

$$\left\langle \sum_{i=0}^k M_i \mathbf{e}_i, \mathbf{e}_j \right\rangle = \|\mathbf{z}_k\|_q \delta_{kj} ,$$

so that $M_0 = M_1 = \cdots = M_{k-1} = 0$ and $M_k = \|\mathbf{z}_k\|_q$. Hence,

$$\mathbf{v}_k - C_k \check{\mathbf{x}}_{k|k-1} = M_k \mathbf{e}_k = \|\mathbf{z}_k\|_q \mathbf{e}_k .$$

Define

$$\check{G}_k = \langle \mathbf{x}_k, \mathbf{e}_k \rangle \|\mathbf{z}_k\|_q^{-1} .$$

Then we obtain

$$\langle \mathbf{x}_k, \mathbf{e}_k \rangle \mathbf{e}_k = \check{G}_k(\mathbf{v}_k - C_k \check{\mathbf{x}}_{k|k-1}) .$$

This, together with (3.14), gives the "prediction-correction" equation:

$$\check{\mathbf{x}}_{k|k} = \check{\mathbf{x}}_{k|k-1} + \check{G}_k(\mathbf{v}_k - C_k \check{\mathbf{x}}_{k|k-1}) . \tag{3.16}$$

We remark that $\check{\mathbf{x}}_{k|k}$ is an unbiased estimate of $\mathbf{x}_k$ by choosing an appropriate initial estimate. In fact,

$$\mathbf{x}_k - \check{\mathbf{x}}_{k|k}$$
$$= A_{k-1}\mathbf{x}_{k-1} + \Gamma_{k-1}\underline{\xi}_{k-1} - A_{k-1}\check{\mathbf{x}}_{k-1|k-1} - \check{G}_k(\mathbf{v}_k - C_k A_{k-1}\check{\mathbf{x}}_{k-1|k-1}).$$

By using $\mathbf{v}_k = C_k\mathbf{x}_k + \underline{\eta}_k = C_k A_{k-1}\mathbf{x}_{k-1} + C_k\Gamma_{k-1}\underline{\xi}_{k-1} + \underline{\eta}_k$, we have

$$\mathbf{x}_k - \check{\mathbf{x}}_{k|k}$$
$$= (I - \check{G}_k C_k)A_{k-1}(\mathbf{x}_{k-1} - \check{\mathbf{x}}_{k-1|k-1})$$
$$+ (I - \check{G}_k C_k)\Gamma_{k-1}\underline{\xi}_{k-1} - \check{G}_k\underline{\eta}_k. \tag{3.17}$$

Since the noise sequences are of zero-mean, we have

$$E(\mathbf{x}_k - \check{\mathbf{x}}_{k|k}) = (I - \check{G}_k C_k)A_{k-1}E(\mathbf{x}_{k-1} - \check{\mathbf{x}}_{k-1|k-1}),$$

so that

$$E(\mathbf{x}_k - \check{\mathbf{x}}_{k|k}) = (I - \check{G}_k C_k)A_{k-1}\cdots(I - \check{G}_1 C_1)A_0 E(\mathbf{x}_0 - \check{\mathbf{x}}_{0|0}).$$

Hence, if we set
$$\check{\mathbf{x}}_{0|0} = E(\mathbf{x}_0), \tag{3.18}$$

then $E(\mathbf{x}_k - \check{\mathbf{x}}_{k|k}) = 0$ or $E(\check{\mathbf{x}}_{k|k}) = E(\mathbf{x}_k)$ for all k, i.e., $\check{\mathbf{x}}_{k|k}$ is indeed an unbiased estimate of $\mathbf{x}_k$.

Now what is left is to derive a recursive formula for $\check{G}_k$. Using (3.12) and (3.17), we first have

$$0 = \langle \mathbf{x}_k - \check{\mathbf{x}}_{k|k}, \mathbf{v}_k \rangle$$
$$= \langle (I - \check{G}_k C_k)A_{k-1}(\mathbf{x}_{k-1} - \check{\mathbf{x}}_{k-1|k-1}) + (I - \check{G}_k C_k)\Gamma_{k-1}\underline{\xi}_{k-1} - \check{G}_k\underline{\eta}_k,$$
$$C_k A_{k-1}((\mathbf{x}_{k-1} - \check{\mathbf{x}}_{k-1|k-1}) + \check{\mathbf{x}}_{k-1|k-1}) + C_k\Gamma_{k-1}\underline{\xi}_{k-1} + \underline{\eta}_k \rangle$$
$$= (I - \check{G}_k C_k)A_{k-1}\|\mathbf{x}_{k-1} - \check{\mathbf{x}}_{k-1|k-1}\|_n^2 A_{k-1}^{\top}C_k^{\top}$$
$$+ (I - \check{G}_k C_k)\Gamma_{k-1}Q_{k-1}\Gamma_{k-1}^{\top}C_k^{\top} - \check{G}_k R_k, \tag{3.19}$$

where we have used the facts that $\langle \mathbf{x}_{k-1} - \check{\mathbf{x}}_{k-1|k-1}, \check{\mathbf{x}}_{k-1|k-1} \rangle = O_{n\times n}$, a consequence of Lemma 3.1, and

$$\langle \mathbf{x}_k, \underline{\xi}_k \rangle = O_{n\times n}, \quad \langle \check{\mathbf{x}}_{k|k}, \underline{\xi}_j \rangle = O_{n\times n},$$
$$\langle \mathbf{x}_k, \underline{\eta}_j \rangle = O_{n\times q}, \quad \langle \check{\mathbf{x}}_{k-1|k-1}, \underline{\eta}_k \rangle = O_{n\times q}, \tag{3.20}$$

$j = 0, \cdots, k$ (cf. Exercise 3.5). Define

$$P_{k,k} = \|\mathbf{x}_k - \check{\mathbf{x}}_{k|k}\|_n^2$$

and
$$P_{k,k-1} = \|\mathbf{x}_k - \check{\mathbf{x}}_{k|k-1}\|_n^2 \,.$$

Then again by Exercise 3.5 we have

$$\begin{aligned} P_{k,k-1} &= \|A_{k-1}\mathbf{x}_{k-1} + \Gamma_{k-1}\underline{\xi}_{k-1} - A_{k-1}\check{\mathbf{x}}_{k-1|k-1}\|_n^2 \\ &= A_{k-1}\|\mathbf{x}_{k-1} - \check{\mathbf{x}}_{k-1|k-1}\|_n^2 A_{k-1}^\top + \Gamma_{k-1}Q_{k-1}\Gamma_{k-1}^\top \end{aligned}$$

or
$$P_{k,k-1} = A_{k-1}P_{k-1,k-1}A_{k-1}^\top + \Gamma_{k-1}Q_{k-1}\Gamma_{k-1}^\top \,. \tag{3.21}$$

On the other hand, from (3.19), we also obtain

$$\begin{aligned} (I - \check{G}_k C_k)A_{k-1}P_{k-1,k-1}A_{k-1}^\top C_k^\top \\ + (I - \check{G}_k C_k)\Gamma_{k-1}Q_{k-1}\Gamma_{k-1}^\top C_k^\top - \check{G}_k R_k = 0 \,. \end{aligned}$$

In solving for $\check{G}_k$ from this expression, we write

$$\begin{aligned} \check{G}_k[R_k + C_k(A_{k-1}P_{k-1,k-1}A_{k-1}^\top + \Gamma_{k-1}Q_{k-1}\Gamma_{k-1}^\top)C_k^\top] \\ = [A_{k-1}P_{k-1,k-1}A_{k-1}^\top + \Gamma_{k-1}Q_{k-1}\Gamma_{k-1}^\top]C_k^\top \\ = P_{k,k-1}C_k^\top \,. \end{aligned}$$

and obtain
$$\check{G}_k = P_{k,k-1}C_k^\top(R_k + C_k P_{k,k-1}C_k^\top)^{-1} \,, \tag{3.22}$$

where R_k is positive definite and $C_k P_{k,k-1}C_k^\top$ is non-negative definite so that their sum is positive definite (cf. Exercise 2.2).

Next, we wish to write $P_{k,k}$ in terms of $P_{k,k-1}$, so that together with (3.21), we will have a recursive scheme. This can be done as follows:

$$\begin{aligned} P_{k,k} &= \|\mathbf{x}_k - \check{\mathbf{x}}_{k|k}\|_n^2 \\ &= \|\mathbf{x}_k - (\check{\mathbf{x}}_{k|k-1} + \check{G}_k(\mathbf{v}_k - C_k\check{\mathbf{x}}_{k|k-1}))\|_n^2 \\ &= \|\mathbf{x}_k - \check{\mathbf{x}}_{k|k-1} - \check{G}_k(C_k\mathbf{x}_k + \underline{\eta}_k) + \check{G}_k C_k\check{\mathbf{x}}_{k|k-1}\|_n^2 \\ &= \|(I - \check{G}_k C_k)(\mathbf{x}_k - \check{\mathbf{x}}_{k|k-1}) - \check{G}_k\underline{\eta}_k\|_n^2 \\ &= (I - \check{G}_k C_k)\|\mathbf{x}_k - \check{\mathbf{x}}_{k|k-1}\|_n^2(I - \check{G}_k C_k)^\top + \check{G}_k R_k\check{G}_k^\top \\ &= (I - \check{G}_k C_k)P_{k,k-1}(I - \check{G}_k C_k)^\top + \check{G}_k R_k\check{G}_k^\top \,, \end{aligned}$$

where we have applied Exercise 3.5 to conclude that $\langle \mathbf{x}_k - \check{\mathbf{x}}_{k|k-1}, \underline{\eta}_k \rangle = O_{n \times q}$. This relation can be further simplified by using (3.22). Indeed, since

$$\begin{aligned} (I - \check{G}_k C_k)P_{k,k-1}(\check{G}_k C_k)^\top \\ = P_{k,k-1}C_k^\top\check{G}_k^\top - \check{G}_k C_k P_{k,k-1}C_k^\top\check{G}_k^\top \\ = \check{G}_k C_k P_{k,k-1}C_k^\top\check{G}_k^\top + \check{G}_k R_k\check{G}_k^\top - \check{G}_k C_k P_{k,k-1}C_k^\top\check{G}_k^\top \\ = \check{G}_k R_k\check{G}_k^\top \,, \end{aligned}$$

we have

$$
\begin{aligned}
P_{k,k} &= (I - \check{G}_k C_k) P_{k,k-1} (I - \check{G}_k C_k)^\top + (I - \check{G}_k C_k) P_{k,k-1} (\check{G}_k C_k)^\top \\
&= (I - \check{G}_k C_k) P_{k,k-1}.
\end{aligned} \tag{3.23}
$$

Therefore, combining (3.13), (3.16), (3.18), (3.21), (3.22) and (3.23), together with

$$
P_{0,0} = \|\mathbf{x}_0 - \check{\mathbf{x}}_{0|0}\|_n^2 = Var(\mathbf{x}_0), \tag{3.24}
$$

we obtain the Kalman filtering equations which agree with the ones we derived in Chapter 2. That is, we have $\check{\mathbf{x}}_{k|k} = \hat{\mathbf{x}}_{k|k}$, $\check{\mathbf{x}}_{k|k-1} = \hat{\mathbf{x}}_{k|k-1}$ and $\check{G}_k = G_k$ as follows:

$$
\begin{cases}
P_{0,0} = Var(\mathbf{x}_0) \\
P_{k,k-1} = A_{k-1} P_{k-1,k-1} A_{k-1}^\top + \Gamma_{k-1} Q_{k-1} \Gamma_{k-1}^\top \\
G_k = P_{k,k-1} C_k^\top (C_k P_{k,k-1} C_k^\top + R_k)^{-1} \\
P_{k,k} = (I - G_k C_k) P_{k,k-1} \\
\hat{\mathbf{x}}_{0|0} = E(\mathbf{x}_0) \\
\hat{\mathbf{x}}_{k|k-1} = A_{k-1} \hat{\mathbf{x}}_{k-1|k-1} \\
\hat{\mathbf{x}}_{k|k} = \hat{\mathbf{x}}_{k|k-1} + G_k(\mathbf{v}_k - C_k \hat{\mathbf{x}}_{k|k-1}) \\
k = 1, 2, \cdots.
\end{cases} \tag{3.25}
$$

Of course, the Kalman filtering equations (2.18) derived in Section 2.4 for the general linear deterministic/stochastic system

$$
\begin{cases}
\mathbf{x}_{k+1} = A_k \mathbf{x}_k + B_k \mathbf{u}_k + \Gamma_k \underline{\xi}_k \\
\mathbf{v}_k = C_k \mathbf{x}_k + D_k \mathbf{u}_k + \underline{\eta}_k
\end{cases}
$$

can also be obtained without the assumption on the invertibility of the matrices $A_k, Var(\underline{\xi}_{k,j})$, etc. (cf. Exercise 3.6).

3.5 Real-Time Tracking

To illustrate the application of the Kalman filtering algorithm described by (3.25), let us consider an example of real-time tracking. Let $\mathbf{x}(t)$, $0 \le t < \infty$, denote the trajectory in three-dimensional space of a flying object, where t denotes the time variable (cf. Fig.3.1). This vector-valued function is discretized by sampling and quantizing with sampling time $h > 0$ to yield

$$
\mathbf{x}_k \doteq \mathbf{x}(kh), \qquad k = 0, 1, \cdots.
$$

Fig. 3.1.

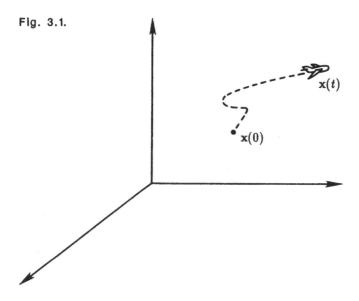

For practical purposes, $\mathbf{x}(t)$ can be assumed to have continuous first and second order derivatives, denoted by $\dot{\mathbf{x}}(t)$ and $\ddot{\mathbf{x}}(t)$, respectively, so that for small values of h, the position and velocity vectors $\mathbf{x}_k$ and $\dot{\mathbf{x}}_k \doteq \dot{\mathbf{x}}(kh)$ are governed by the equations

$$\begin{cases} \mathbf{x}_{k+1} = \mathbf{x}_k + h\dot{\mathbf{x}}_k + \frac{1}{2}h^2\ddot{\mathbf{x}}_k \\ \dot{\mathbf{x}}_{k+1} = \dot{\mathbf{x}}_k + h\ddot{\mathbf{x}}_k \,, \end{cases}$$

where $\ddot{\mathbf{x}}_k \doteq \ddot{\mathbf{x}}(kh)$ and $k = 0, 1, \cdots$. In addition, in many applications only the position (vector) of the flying object is observed at each time instant, so that $\mathbf{v}_k = C\mathbf{x}_k$ with $C = [\, I \quad 0 \quad 0 \,]$ is measured. In view of Exercise 3.8, to facilitate our discussion we only consider the tracking model

$$\begin{cases} \begin{bmatrix} x_{k+1}[1] \\ x_{k+1}[2] \\ x_{k+1}[3] \end{bmatrix} = \begin{bmatrix} 1 & h & h^2/2 \\ 0 & 1 & h \\ 0 & 0 & 1 \end{bmatrix} \begin{bmatrix} x_k[1] \\ x_k[2] \\ x_k[3] \end{bmatrix} + \begin{bmatrix} \xi_k[1] \\ \xi_k[2] \\ \xi_k[3] \end{bmatrix} \\ v_k = [\, 1 \quad 0 \quad 0 \,] \begin{bmatrix} x_k[1] \\ x_k[2] \\ x_k[3] \end{bmatrix} + \eta_k \,. \end{cases} \qquad (3.26)$$

Here, $\{\underline{\xi}_k\}$, with $\underline{\xi}_k := [\, \xi_k[1] \quad \xi_k[2] \quad \xi_k[3] \,]^\top$, and $\{\eta_k\}$ are assumed to be zero-mean Gaussian white noise sequences satisfying:

$$E(\underline{\xi}_k) = 0, \qquad E(\eta_k) = 0,$$

$$E(\underline{\xi}_k \underline{\xi}_\ell^\top) = Q_k \delta_{k\ell}, \qquad E(\eta_k \eta_\ell) = r_k \delta_{k\ell}, \qquad E(\underline{\xi}_k \eta_\ell) = 0,$$

$$E(\mathbf{x}_0 \underline{\xi}_k^\top) = 0, \qquad E(\mathbf{x}_0 \eta_k) = 0,$$

where Q_k is a non-negative definite symmetric matrix and $r_k > 0$ for all k. It is further assumed that initial conditions $E(\mathbf{x}_0)$ and $Var(\mathbf{x}_0)$ are given. For this tracking model, the Kalman filtering algorithm can be specified as follows: Let $P_k := P_{k,k}$ and let $P[i, j]$ denote the (i, j)th entry of P. Then we have

$$P_{k,k-1}[1, 1] = P_{k-1}[1, 1] + 2hP_{k-1}[1, 2] + h^2 P_{k-1}[1, 3] + h^2 P_{k-1}[2, 2]$$
$$+ h^3 P_{k-1}[2, 3] + \frac{h^4}{4} P_{k-1}[3, 3] + Q_{k-1}[1, 1],$$

$$P_{k,k-1}[1, 2] = P_{k,k-1}[2, 1]$$
$$= P_{k-1}[1, 2] + hP_{k-1}[1, 3] + hP_{k-1}[2, 2] + \frac{3h^2}{2} P_{k-1}[2, 3]$$
$$+ \frac{h^3}{2} P_{k-1}[3, 3] + Q_{k-1}[1, 2],$$

$$P_{k,k-1}[2, 2] = P_{k-1}[2, 2] + 2hP_{k-1}[2, 3] + h^2 P_{k-1}[3, 3] + Q_{k-1}[2, 2],$$

$$P_{k,k-1}[1, 3] = P_{k,k-1}[3, 1]$$
$$= P_{k-1}[1, 3] + hP_{k-1}[2, 3] + \frac{h^2}{2} P_{k-1}[3, 3] + Q_{k-1}[1, 3],$$

$$P_{k,k-1}[2, 3] = P_{k,k-1}[3, 2]$$
$$= P_{k-1}[2, 3] + hP_{k-1}[3, 3] + Q_{k-1}[2, 3],$$

$$P_{k,k-1}[3, 3] = P_{k-1}[3, 3] + Q_{k-1}[3, 3],$$

with $P_{0,0} = Var(\mathbf{x}_0)$,

$$G_k = \frac{1}{P_{k,k-1}[1, 1] + r_k} \begin{bmatrix} P_{k,k-1}[1, 1] \\ P_{k,k-1}[1, 2] \\ P_{k,k-1}[1, 3] \end{bmatrix},$$

$$P_k = P_{k,k-1} - \frac{1}{P_{k,k-1}[1, 1] + r_k} \times$$
$$\begin{bmatrix} P_{k,k-1}^2[1, 1] & P_{k,k-1}[1, 1]P_{k,k-1}[1, 2] & P_{k,k-1}[1, 1]P_{k,k-1}[1, 3] \\ P_{k,k-1}[1, 1]P_{k,k-1}[1, 2] & P_{k,k-1}^2[1, 2] & P_{k,k-1}[1, 2]P_{k,k-1}[1, 3] \\ P_{k,k-1}[1, 1]P_{k,k-1}[1, 3] & P_{k,k-1}[1, 2]P_{k,k-1}[1, 3] & P_{k,k-1}^2[1, 3] \end{bmatrix}$$

and the Kalman filtering algorithm is given by

$$\begin{bmatrix} \hat{x}_{k|k}[1] \\ \hat{x}_{k|k}[2] \\ \hat{x}_{k|k}[3] \end{bmatrix} = \begin{bmatrix} 1 - G_k[1] & (1 - G_k[1])h & (1 - G_k[1])h^2/2 \\ -G_k[2] & 1 - hG_k[2] & h - h^2 G_k[2]/2 \\ -G_k[3] & -hG_k[3] & 1 - h^2 G_k[3]/2 \end{bmatrix}$$

$$\cdot \begin{bmatrix} \hat{x}_{k-1|k-1}[1] \\ \hat{x}_{k-1|k-1}[2] \\ \hat{x}_{k-1|k-1}[3] \end{bmatrix} + \begin{bmatrix} G_k[1] \\ G_k[2] \\ G_k[3] \end{bmatrix} v_k \qquad (3.27)$$

with $\hat{x}_{0|0} = E(\mathbf{x}_0)$.

Exercises

3.1. Let $A \neq 0$ be a non-negative definite and symmetric constant matrix. Show that $\mathrm{tr} A > 0$. (Hint: Decompose A as $A = BB^\mathsf{T}$ with $B \neq 0$.)

3.2. Let

$$\hat{\mathbf{e}}_j = C_j(\mathbf{x}_j - \hat{\mathbf{y}}_{j-1}) = C_j\left(\mathbf{x}_j - \sum_{i=0}^{j-1} \hat{P}_{j-1,i}\mathbf{v}_i\right),$$

where $\hat{P}_{j-1,i}$ are some constant matrices. Use Assumption 2.1 to show that

$$\langle \underline{\eta}_\ell, \hat{\mathbf{e}}_j \rangle = O_{q \times q}$$

for all $\ell \geq j$.

3.3. For random vectors $\mathbf{w}_0, \cdots, \mathbf{w}_r$, define

$$Y(\mathbf{w}_0, \cdots, \mathbf{w}_r)$$
$$= \left\{ \mathbf{y} : \ \mathbf{y} = \sum_{i=0}^{r} P_i \mathbf{w}_i, \quad P_0, \cdots, P_r, \ constant \ matrices \right\}.$$

Let

$$\mathbf{z}_j = \mathbf{v}_j - C_j \sum_{i=0}^{j-1} \hat{P}_{j-1,i}\mathbf{v}_i$$

be defined as in (3.4) and $\mathbf{e}_j = \|\mathbf{z}_j\|^{-1}\mathbf{z}_j$. Show that

$$Y(\mathbf{e}_0, \cdots, \mathbf{e}_k) = Y(\mathbf{v}_0, \cdots, \mathbf{v}_k).$$

3.4. Let

$$\hat{\mathbf{y}}_{j-1} = \sum_{i=0}^{j-1} \hat{P}_{j-1,i}\mathbf{v}_i$$

and

$$\mathbf{z}_j = \mathbf{v}_j - C_j \sum_{i=0}^{j-1} \hat{P}_{j-1,i}\mathbf{v}_i.$$

Show that

$$\langle \hat{\mathbf{y}}_j, \mathbf{z}_k \rangle = O_{n \times q}, \qquad j = 0, 1, \cdots, k-1.$$

3.5. Let $\mathbf{e}_j$ be defined as in Exercise 3.3. Also define

$$\check{\mathbf{x}}_k = \sum_{i=0}^{k} \langle \mathbf{x}_k, \ \mathbf{e}_i \rangle \mathbf{e}_i$$

as in (3.10). Show that

$$\langle \mathbf{x}_k, \ \underline{\xi}_k \rangle = O_{n \times n}, \qquad \langle \check{\mathbf{x}}_{k|k}, \ \underline{\xi}_j \rangle = O_{n \times n},$$

$$\langle \mathbf{x}_k, \ \underline{\eta}_j \rangle = O_{n \times q}, \qquad \langle \check{\mathbf{x}}_{k-1|k-1}, \ \underline{\eta}_k \rangle = O_{n \times q},$$

for $j = 0, 1, \cdots, k$.

3.6. Consider the linear deterministic/stochastic system

$$\begin{cases} \mathbf{x}_{k+1} = A_k \mathbf{x}_k + B_k \mathbf{u}_k + \Gamma_k \underline{\xi}_k \\ \mathbf{v}_k = C_k \mathbf{x}_k + D_k \mathbf{u}_k + \underline{\eta}_k, \end{cases}$$

where $\{\mathbf{u}_k\}$ is a given sequence of deterministic control input m-vectors, $1 \leq m \leq n$. Suppose that Assumption 2.1 is satisfied. Derive the Kalman filtering algorithm for this model.

3.7. Consider a simplified radar tracking model where a large-amplitude and narrow-width impulse signal is transmitted by an antenna. The impulse signal propagates at the speed of light c, and is reflected by a flying object being tracked. The radar antenna receives the reflected signal so that a time-difference Δt is obtained. The range (or distance) d from the radar to the object is then given by $d = c\Delta t/2$. The impulse signal is transmitted periodically with period h. Assume that the object is traveling at a constant velocity w with random disturbance $\xi \sim N(0, q)$, so that the range d satisfies the difference equation

$$d_{k+1} = d_k + h(w_k + \xi_k).$$

Suppose also that the measured range using the formula $d = c\Delta t/2$ has an inherent error Δd and is contaminated with noise η where $\eta \sim N(0, r)$, so that

$$v_k = d_k + \Delta d_k + \eta_k.$$

Assume that the initial target range is d_0 which is independent of ξ_k and η_k, and that $\{\xi_k\}$ and $\{\eta_k\}$ are also independent (cf. Fig.3.2). Derive a Kalman filtering algorithm as a range-estimator for this radar tracking system.

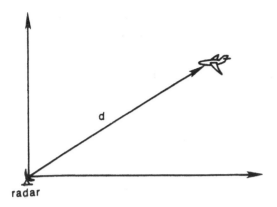

Fig. 3.2.

3.8. A linear stochastic system for radar tracking can be described as follows. Let Σ, ΔA, ΔE be the range, the azimuthal angular error, and the elevational angular error, respectively, of the target, with the radar being located at the origin (cf. Fig.3.3). Consider $\Sigma, \Delta A$, and ΔE as functions of time with first and second derivatives denoted by $\dot{\Sigma}$, $\Delta\dot{A}$, $\Delta\dot{E}$, $\ddot{\Sigma}$, $\Delta\ddot{A}$, $\Delta\ddot{E}$, respectively. Let $h > 0$ be the sampling time unit and set $\Sigma_k = \Sigma(kh)$, $\dot{\Sigma}_k = \dot{\Sigma}(kh)$, $\ddot{\Sigma}_k = \ddot{\Sigma}(kh)$, etc. Then, using the second degree Taylor polynomial approximation, the radar tracking model takes on the following linear stochastic state-space description:

$$\begin{cases} \mathbf{x}_{k+1} = \tilde{A}\mathbf{x}_k + \Gamma_k\underline{\xi}_k \\ \mathbf{v}_k = \tilde{C}\mathbf{x}_k + \underline{\eta}_k\,, \end{cases}$$

where

$$\mathbf{x}_k = \begin{bmatrix} \Sigma_k & \dot{\Sigma}_k & \ddot{\Sigma}_k & \Delta A_k & \Delta\dot{A}_k & \Delta\ddot{A}_k & \Delta E_k & \Delta\dot{E}_k & \Delta\ddot{E}_k \end{bmatrix}^\top,$$

$$\tilde{A} = \begin{bmatrix} 1 & h & h^2/2 & & & & & & \\ 0 & 1 & h & & & & & & \\ 0 & 0 & 1 & & & & & & \\ & & & 1 & h & h^2/2 & & & \\ & & & 0 & 1 & h & & & \\ & & & 0 & 0 & 1 & & & \\ & & & & & & 1 & h & h^2/2 \\ & & & & & & 0 & 1 & h \\ & & & & & & 0 & 0 & 1 \end{bmatrix},$$

$$\tilde{C} = \begin{bmatrix} 1 & 0 & 0 & 0 & 0 & 0 & 0 & 0 & 0 \\ 0 & 0 & 0 & 1 & 0 & 0 & 0 & 0 & 0 \\ 0 & 0 & 0 & 0 & 0 & 0 & 1 & 0 & 0 \end{bmatrix},$$

and $\{\underline{\xi}_k\}$ and $\{\underline{\eta}_k\}$ are independent zero-mean Gaussian white noise sequences with $Var(\underline{\xi}_k) = Q_k$ and $Var(\underline{\eta}_k) = R_k$. Assume that

$$\Gamma_k = \begin{bmatrix} \Gamma_k^1 & & \\ & \Gamma_k^2 & \\ & & \Gamma_k^3 \end{bmatrix},$$

$$Q_k = \begin{bmatrix} Q_k^1 & & \\ & Q_k^2 & \\ & & Q_k^3 \end{bmatrix}, \qquad R_k = \begin{bmatrix} R_k^1 & & \\ & R_k^2 & \\ & & R_k^3 \end{bmatrix},$$

where Γ_k^i are 3×3 submatrices, Q_k^i, 3×3 non-negative definite symmetric submatrices, and R_k^i, 3×3 positive definite symmetric submatrices, for $i = 1, 2, 3$. Show that this system can be decoupled into three subsystems with analogous state-space descriptions.

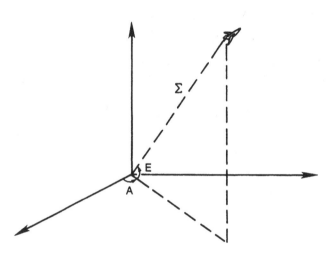

Fig. 3.3.

4. Correlated System and Measurement Noise Processes

In the previous two chapters, Kalman filtering for the model involving uncorrelated system and measurement noise processes was studied. That is, we have assumed all along that

$$E(\underline{\xi}_k \underline{\eta}_\ell^\top) = 0$$

for $k, \ell = 0, 1, \cdots$. However, in applications such as aircraft inertial navigation systems, where vibration of the aircraft induces a common source of noise for both the dynamic driving system and onboard radar measurement, the system and measurement noise sequences $\{\underline{\xi}_k\}$ and $\{\underline{\eta}_k\}$ are correlated in the statistical sense, with

$$E(\underline{\xi}_k \underline{\eta}_\ell^\top) = S_k \delta_{k\ell},$$

$k, \ell = 0, 1, \cdots$, where each S_k is a known non-negative definite matrix. This chapter is devoted to the study of Kalman filtering for the above model.

4.1 The Affine Model

Consider the linear stochastic state-space description

$$\begin{cases} \mathbf{x}_{k+1} = A_k \mathbf{x}_k + \Gamma_k \underline{\xi}_k \\ \mathbf{v}_k = C_k \mathbf{x}_k + \underline{\eta}_k \end{cases}$$

with initial state $\mathbf{x}_0$, where A_k, C_k and Γ_k are known constant matrices. We start with least-squares estimation as discussed in Section 1.3. Recall that least-squares estimates are linear functions of the data vectors; that is, if $\hat{\mathbf{x}}$ is the least-squares estimate of the state vector $\mathbf{x}$ using the data $\mathbf{v}$, then it follows that $\hat{\mathbf{x}} = H\mathbf{v}$ for some matrix H. To study Kalman filtering with correlated system and measurement noise processes, it is necessary to extend to a more general model in determining the estimator $\hat{\mathbf{x}}$. It turns out that the affine model

$$\hat{\mathbf{x}} = \mathbf{h} + H\mathbf{v} \tag{4.1}$$

which provides an extra parameter vector $\mathbf{h}$ is sufficient. Here, $\mathbf{h}$ is some constant n-vector and H some constant $n \times q$ matrix. Of course, the requirements on our optimal estimate $\hat{\mathbf{x}}$ of $\mathbf{x}$ are: $\hat{\mathbf{x}}$ *is an unbiased estimator of* $\mathbf{x}$ *in the sense that*

$$E(\hat{\mathbf{x}}) = E(\mathbf{x}) \tag{4.2}$$

and the estimate is of minimum (error) variance.

From (4.1) it follows that

$$\mathbf{h} = E(\mathbf{h}) = E(\hat{\mathbf{x}} - H\mathbf{v}) = E(\hat{\mathbf{x}}) - H(E(\mathbf{v})).$$

Hence, to satisfy the requirement (4.2), we must have

$$\mathbf{h} = E(\mathbf{x}) - H E(\mathbf{v}). \tag{4.3}$$

or, equivalently,

$$\hat{\mathbf{x}} = E(\mathbf{x}) - H(E(\mathbf{v}) - \mathbf{v}). \tag{4.4}$$

On the other hand, to satisfy the minimum variance requirement, we use the notation

$$F(H) = Var(\mathbf{x} - \hat{\mathbf{x}}) = \|\mathbf{x} - \hat{\mathbf{x}}\|_n^2 ,$$

so that by (4.4) and the fact that $\|\mathbf{v}\|_q^2 = Var(\mathbf{v})$ is positive definite, we obtain

$$\begin{aligned}
F(H) &= \langle \mathbf{x} - \hat{\mathbf{x}}, \mathbf{x} - \hat{\mathbf{x}} \rangle \\
&= \langle (\mathbf{x} - E(\mathbf{x})) - H(\mathbf{v} - E(\mathbf{v})), (\mathbf{x} - E(\mathbf{x})) - H(\mathbf{v} - E(\mathbf{v})) \rangle \\
&= \|\mathbf{x}\|_n^2 - H\langle \mathbf{v}, \mathbf{x} \rangle - \langle \mathbf{x}, \mathbf{v} \rangle H^{\top} + H\|\mathbf{v}\|_q^2 H^{\top} \\
&= \{\|\mathbf{x}\|_n^2 - \langle \mathbf{x}, \mathbf{v} \rangle [\|\mathbf{v}\|_q^2]^{-1} \langle \mathbf{v}, \mathbf{x} \rangle\} \\
&\quad + \{H\|\mathbf{v}\|_q^2 H^{\top} - H\langle \mathbf{v}, \mathbf{x} \rangle - \langle \mathbf{x}, \mathbf{v} \rangle H^{\top} + \langle \mathbf{x}, \mathbf{v} \rangle [\|\mathbf{v}\|_q^2]^{-1} \langle \mathbf{v}, \mathbf{x} \rangle\} \\
&= \{\|\mathbf{x}\|_n^2 - \langle \mathbf{x}, \mathbf{v} \rangle [\|\mathbf{v}\|_q^2]^{-1} \langle \mathbf{v}, \mathbf{x} \rangle\} \\
&\quad + [H - \langle \mathbf{x}, \mathbf{v} \rangle [\|\mathbf{v}\|_q^2]^{-1}] \|\mathbf{v}\|_q^2 [H - \langle \mathbf{x}, \mathbf{v} \rangle [\|\mathbf{v}\|_q^2]^{-1}]^{\top} ,
\end{aligned}$$

where the facts that $\langle \mathbf{x}, \mathbf{v} \rangle^{\top} = \langle \mathbf{v}, \mathbf{x} \rangle$ and that $Var(\mathbf{v})$ is nonsingular have been used.

Recall that minimum variance estimation means the existence of H^* such that $F(H) \geq F(H^*)$, or $F(H) - F(H^*)$ is non-negative definite, for all constant matrices H. This can be attained by simply setting

$$H^* = \langle \mathbf{x}, \mathbf{v} \rangle [\|\mathbf{v}\|_q^2]^{-1} , \tag{4.5}$$

so that

$$F(H) - F(H^*) = [H - \langle \mathbf{x}, \mathbf{v} \rangle [\|\mathbf{v}\|_q^2]^{-1}] \|\mathbf{v}\|_q^2 [H - \langle \mathbf{x}, \mathbf{v} \rangle [\|\mathbf{v}\|_q^2]^{-1}]^{\top},$$

which is non-negative definite for all constant matrices H. Furthermore, H^* is unique in the sense that $F(H) - F(H^*) = 0$ if and only if $H = H^*$. Hence, we can conclude that $\hat{\mathbf{x}}$ can be uniquely expressed as

$$\hat{\mathbf{x}} = \mathbf{h} + H^* \mathbf{v},$$

where H^* is given by (4.5). We will also use the notation $\hat{\mathbf{x}} = L(\mathbf{x}, \mathbf{v})$ for the optimal estimate of $\mathbf{x}$ with data $\mathbf{v}$, so that by using (4.4) and (4.5), it follows that this "optimal estimate operator" satisfies:

$$L(\mathbf{x}, \mathbf{v}) = E(\mathbf{x}) + \langle \mathbf{x}, \mathbf{v} \rangle [\|\mathbf{v}\|_q^2]^{-1}(\mathbf{v} - E(\mathbf{v})). \qquad (4.6)$$

4.2 Optimal Estimate Operators

First, we remark that for any fixed data vector $\mathbf{v}$, $L(\cdot, \mathbf{v})$ is a linear operator in the sense that

$$L(A\mathbf{x} + B\mathbf{y}, \mathbf{v}) = AL(\mathbf{x}, \mathbf{v}) + BL(\mathbf{y}, \mathbf{v}) \qquad (4.7)$$

for all constant matrices A and B and state vectors $\mathbf{x}$ and $\mathbf{y}$ (cf. Exercise 4.1). In addition, if the state vector is a constant vector $\mathbf{a}$, then

$$L(\mathbf{a}, \mathbf{v}) = \mathbf{a} \qquad (4.8)$$

(cf. Exercise 4.2). This means that if $\mathbf{x}$ is a constant vector, so that $E(\mathbf{x}) = \mathbf{x}$, then $\hat{\mathbf{x}} = \mathbf{x}$, or the estimate is exact.

We need some additional properties of $L(\mathbf{x}, \mathbf{v})$. For this purpose we first establish the following.

Lemma 4.1. *Let $\mathbf{v}$ be a given data vector and $\mathbf{y} = \mathbf{h} + H\mathbf{v}$, where $\mathbf{h}$ is determined by the condition $E(\mathbf{y}) = E(\mathbf{x})$, so that $\mathbf{y}$ is uniquely determined by the constant matrix H. If $\mathbf{x}^*$ is one of the $\mathbf{y}$'s such that*

$$tr\|\mathbf{x} - \mathbf{x}^*\|_n^2 = \min_H tr\|\mathbf{x} - \mathbf{y}\|_n^2,$$

then it follows that $\mathbf{x}^* = \hat{\mathbf{x}}$, where $\hat{\mathbf{x}} = L(\mathbf{x}, \mathbf{v})$ is given by (4.6).

This lemma says that the minimum variance estimate $\hat{x}$ and the "minimum trace variance" estimate x^* of x from the same data v are identical over all affine models.

To prove the lemma, let us consider

$$tr\|x - y\|_n^2$$
$$= trE((x - y)(x - y)^\top)$$
$$= E((x - y)^\top(x - y))$$
$$= E((x - E(x)) - H(v - E(v))^\top((x - E(x)) - H(v - E(v)),$$

where (4.3) has been used. Taking

$$\frac{\partial}{\partial H}(tr\|x - y\|_n^2) = 0,$$

we arrive at

$$x^* = E(x) - \langle x, v \rangle [\|v\|_q^2]^{-1}(E(v) - v) \tag{4.9}$$

which is the same as the $\hat{x}$ given in (4.6) (cf. Exercise 4.3). This completes the proof of the Lemma.

4.3 Effect on Optimal Estimation with Additional Data

Now, let us recall from Lemma 3.1 in the previous chapter that $\hat{y} \in Y = Y(w_0, \cdots, w_r)$ satisfies

$$tr\|x - \hat{y}\|_n^2 = \min_{y \in Y} tr\|x - y\|_n^2$$

if and only if

$$\langle x - \hat{y}, w_j \rangle = O_{n \times q}, \quad j = 0, 1, \cdots, r.$$

Set $Y = Y(v - E(v))$ and $\hat{x} = L(x, v) = E(x) + H^*(v - E(v))$, where $H^* = \langle x, v \rangle [\|v\|_q^2]^{-1}$. If we use the notation

$$\tilde{x} = x - E(x) \quad \text{and} \quad \tilde{v} = v - E(v),$$

then we obtain

$$\|x - \hat{x}\|_n^2 = \|(x - E(x)) - H^*(v - E(v))\|_n^2 = \|\tilde{x} - H^*\tilde{v}\|_n^2.$$

But H^* was chosen such that $F(H^*) \leq F(H)$ for all H, and this implies that $trF(H^*) \leq trF(H)$ for all H. Hence, it follows that

$$tr\|\tilde{\mathbf{x}} - H^*\tilde{\mathbf{v}}\|_n^2 \leq tr\|\tilde{\mathbf{x}} - \mathbf{y}\|_n^2$$

for all $\mathbf{y} \in Y(\mathbf{v} - E(\mathbf{v})) = Y(\tilde{\mathbf{v}})$. By Lemma 3.1, we have

$$\langle \tilde{\mathbf{x}} - H^*\tilde{\mathbf{v}}, \tilde{\mathbf{v}} \rangle = O_{n \times q}.$$

Since $E(\mathbf{v})$ is a constant, $\langle \tilde{\mathbf{x}} - H^*\tilde{\mathbf{v}}, E(\mathbf{v}) \rangle = O_{n \times q}$, so that

$$\langle \tilde{\mathbf{x}} - H^*\tilde{\mathbf{v}}, \mathbf{v} \rangle = O_{n \times q},$$

or

$$\langle \mathbf{x} - \hat{\mathbf{x}}, \mathbf{v} \rangle = O_{n \times q}. \tag{4.10}$$

Consider two random data vectors $\mathbf{v}^1$ and $\mathbf{v}^2$ and set

$$\begin{cases} \mathbf{x}^{\#} = \mathbf{x} - L(\mathbf{x}, \mathbf{v}^1) \\ \mathbf{v}^{2\#} = \mathbf{v}^2 - L(\mathbf{v}^2, \mathbf{v}^1). \end{cases} \tag{4.11}$$

Then from (4.10) and the definition of the optimal estimate operator L, we have

$$\langle \mathbf{x}^{\#}, \mathbf{v}^1 \rangle = 0, \tag{4.12}$$

and similarly,

$$\langle \mathbf{v}^{2\#}, \mathbf{v}^1 \rangle = 0. \tag{4.13}$$

The following lemma is essential for further investigation.

Lemma 4.2. *Let* $\mathbf{x}$ *be a state vector and* $\mathbf{v}^1$, $\mathbf{v}^2$ *be random observation data vectors with nonzero finite variances. Set*

$$\mathbf{v} = \begin{bmatrix} \mathbf{v}^1 \\ \mathbf{v}^2 \end{bmatrix}.$$

Then the minimum variance estimate $\hat{\mathbf{x}}$ *of* $\mathbf{x}$ *using the data* $\mathbf{v}$ *can be approximated by the minimum variance estimate* $L(\mathbf{x}, \mathbf{v}^1)$ *of* $\mathbf{x}$ *using the data* $\mathbf{v}^1$ *in the sense that*

$$\hat{\mathbf{x}} = L(\mathbf{x}, \mathbf{v}) = L(\mathbf{x}, \mathbf{v}^1) + e(\mathbf{x}, \mathbf{v}^2) \tag{4.14}$$

with the error

$$\begin{aligned} e(\mathbf{x}, \mathbf{v}^2) &:= L(\mathbf{x}^{\#}, \mathbf{v}^{2\#}) \\ &= \langle \mathbf{x}^{\#}, \mathbf{v}^{2\#} \rangle [\|\mathbf{v}^{2\#}\|^2]^{-1} \mathbf{v}^{2\#}. \end{aligned} \tag{4.15}$$

We first verify (4.15). Since $L(\mathbf{x}, \mathbf{v}^1)$ is an unbiased estimate of $\mathbf{x}$ (cf. (4.6)),

$$E(\mathbf{x}^\#) = E(\mathbf{x} - L(\mathbf{x}, \mathbf{v}^1)) = 0.$$

Similarly, $E(\mathbf{v}^{2\#}) = 0$. Hence, by (4.6), we have

$$
\begin{aligned}
L(\mathbf{x}^\#, \mathbf{v}^{2\#}) &= E(\mathbf{x}^\#) + \langle \mathbf{x}^\#, \mathbf{v}^{2\#} \rangle [\|\mathbf{v}^{2\#}\|^2]^{-1}(\mathbf{v}^{2\#} - E(\mathbf{v}^{2\#})) \\
&= \langle \mathbf{x}^\#, \mathbf{v}^{2\#} \rangle [\|\mathbf{v}^{2\#}\|^2]^{-1} \mathbf{v}^{2\#},
\end{aligned}
$$

yielding (4.15). To prove (4.14), it is equivalent to showing that

$$\mathbf{x}^0 := L(\mathbf{x}, \mathbf{v}^1) + L(\mathbf{x}^\#, \mathbf{v}^{2\#})$$

is an affine unbiased minimum variance estimate of $\mathbf{x}$ from the data $\mathbf{v}$, so that by the uniqueness of $\hat{\mathbf{x}}$, $\mathbf{x}^0 = \hat{\mathbf{x}} = L(\mathbf{x}, \mathbf{v})$. First, note that

$$
\begin{aligned}
\mathbf{x}^0 &= L(\mathbf{x}, \mathbf{v}^1) + L(\mathbf{x}^\#, \mathbf{v}^{2\#}) \\
&= (\mathbf{h}_1 + H_1 \mathbf{v}^1) + (\mathbf{h}_2 + H_2(\mathbf{v}^2 - L(\mathbf{v}^2, \mathbf{v}^1))) \\
&= (\mathbf{h}_1 + H_1 \mathbf{v}^1) + \mathbf{h}_2 + H_2(\mathbf{v}^2 - (\mathbf{h}_3 + H_3 \mathbf{v}^1)) \\
&= (\mathbf{h}_1 + \mathbf{h}_2 - H_2 \mathbf{h}_3) + H \begin{bmatrix} \mathbf{v}^1 \\ \mathbf{v}^2 \end{bmatrix} \\
&:= \mathbf{h} + H\mathbf{v},
\end{aligned}
$$

where $H = [H_1 - H_2 H_3 \ \ H_2]$. Hence, $\mathbf{x}^0$ is an affine transformation of $\mathbf{v}$. Next, since $E(L(\mathbf{x}, \mathbf{v}^1)) = E(\mathbf{x})$ and $E(L(\mathbf{x}^\#, \mathbf{v}^{2\#})) = E(\mathbf{x}^\#) = 0$, we have

$$E(\mathbf{x}^0) = E(L(\mathbf{x}, \mathbf{v}^1)) + E(L(\mathbf{x}^\#, \mathbf{v}^{2\#})) = E(\mathbf{x}).$$

Hence, $\mathbf{x}^0$ is an unbiased estimate of $\mathbf{x}$. Finally, to prove that $\mathbf{x}^0$ is a minimum variance estimate of $\mathbf{x}$, we note that by using Lemmas 4.1 and 3.1, it is sufficient to establish the orthogonality property

$$\langle \mathbf{x} - \mathbf{x}^0, \mathbf{v} \rangle = O_{n \times q}.$$

This can be done as follows. By (4.15), (4.11), (4.12), and (4.13), we have

$$
\begin{aligned}
\langle \mathbf{x} - \mathbf{x}^0, \mathbf{v} \rangle &= \langle \mathbf{x}^\# - \langle \mathbf{x}^\#, \mathbf{v}^{2\#} \rangle [\|\mathbf{v}^{2\#}\|^2]^{-1} \mathbf{v}^{2\#}, \mathbf{v} \rangle \\
&= \left\langle \mathbf{x}^\#, \begin{bmatrix} \mathbf{v}^1 \\ \mathbf{v}^2 \end{bmatrix} \right\rangle - \langle \mathbf{x}^\#, \mathbf{v}^{2\#} \rangle [\|\mathbf{v}^{2\#}\|^2]^{-1} \left\langle \mathbf{v}^{2\#}, \begin{bmatrix} \mathbf{v}^1 \\ \mathbf{v}^2 \end{bmatrix} \right\rangle \\
&= \langle \mathbf{x}^\#, \mathbf{v}^2 \rangle - \langle \mathbf{x}^\#, \mathbf{v}^{2\#} \rangle [\|\mathbf{v}^{2\#}\|^2]^{-1} \langle \mathbf{v}^{2\#}, \mathbf{v}^2 \rangle.
\end{aligned}
$$

But since $\mathbf{v}^2 = \mathbf{v}^{2\#} + L(\mathbf{v}^2, \mathbf{v}^1)$, it follows that

$$\langle \mathbf{v}^{2\#}, \mathbf{v}^2 \rangle = \langle \mathbf{v}^{2\#}, \mathbf{v}^{2\#} \rangle + \langle \mathbf{v}^{2\#}, L(\mathbf{v}^2, \mathbf{v}^1) \rangle$$

from which, by using (4.6), (4.13), and the fact that $\langle \mathbf{v}^{2\#}, E(\mathbf{v}^1) \rangle = \langle \mathbf{v}^{2\#}, E(\mathbf{v}^2) \rangle = 0$, we arrive at

$$\begin{aligned}
&\langle \mathbf{v}^{2\#}, L(\mathbf{v}^2, \mathbf{v}^1) \rangle \\
&= \langle \mathbf{v}^{2\#}, E(\mathbf{v}^2) + \langle \mathbf{v}^2, \mathbf{v}^1 \rangle [\|\mathbf{v}^1\|^2]^{-1} (\mathbf{v}^1 - E(\mathbf{v}^1)) \rangle \\
&= ((\langle \mathbf{v}^{2\#}, \mathbf{v}^1 \rangle - \langle \mathbf{v}^{2\#}, E(\mathbf{v}^1) \rangle) [\|\mathbf{v}^1\|^2]^{-1} \langle \mathbf{v}^1, \mathbf{v}^2 \rangle \\
&= 0,
\end{aligned}$$

so that

$$\langle \mathbf{v}^{2\#}, \mathbf{v}^2 \rangle = \langle \mathbf{v}^{2\#}, \mathbf{v}^{2\#} \rangle .$$

Similarly, we also have

$$\langle \mathbf{x}^{\#}, \mathbf{v}^2 \rangle = \langle \mathbf{x}^{\#}, \mathbf{v}^{2\#} \rangle .$$

Hence, indeed, we have the orthogonality property:

$$\begin{aligned}
&\langle \mathbf{x} - \mathbf{x}^0, \mathbf{v} \rangle \\
&= \langle \mathbf{x}^{\#}, \mathbf{v}^{2\#} \rangle - \langle \mathbf{x}^{\#}, \mathbf{v}^{2\#} \rangle [\|\mathbf{v}^{2\#}\|^2]^{-1} \langle \mathbf{v}^{2\#}, \mathbf{v}^{2\#} \rangle \\
&= O_{n \times q} .
\end{aligned}$$

This completes the proof of the Lemma.

4.4 Derivation of Kalman Filtering Equations

We are now ready to study Kalman filtering with correlated system and measurement noises. Let us again consider the linear stochastic system described by

$$\begin{cases} \mathbf{x}_{k+1} = A_k \mathbf{x}_k + \Gamma_k \underline{\xi}_k \\ \mathbf{v}_k = C_k \mathbf{x}_k + \underline{\eta}_k \end{cases} \tag{4.16}$$

with initial state $\mathbf{x}_0$, where A_k, C_k and Γ_k are known constant matrices. We will adopt Assumption 2.1 here with the exception that the two noise sequences $\{\underline{\xi}_k\}$ and $\{\underline{\eta}_k\}$ may be correlated, namely: we assume that $\{\underline{\xi}_k\}$ and $\{\underline{\eta}_k\}$ are zero-mean Gaussian white noise sequences satisfying

$$\begin{aligned}
E(\underline{\xi}_k \mathbf{x}_0^\top) &= O_{p \times n}, \quad E(\underline{\eta}_k \mathbf{x}_0^\top) = O_{q \times n}, \\
E(\underline{\xi}_k \underline{\xi}_\ell^\top) &= Q_k \delta_{k\ell}, \quad E(\underline{\eta}_k \underline{\eta}_\ell^\top) = R_k \delta_{k\ell}, \\
E(\underline{\xi}_k \underline{\eta}_\ell^\top) &= S_k \delta_{k\ell},
\end{aligned}$$

where Q_k, R_k are, respectively, known non-negative definite and positive definite matrices and S_k is a known, but not necessarily zero, non-negative definite matrix.

The problem is to determine the optimal estimate $\hat{\mathbf{x}}_k = \hat{\mathbf{x}}_{k|k}$ of the state vector $\mathbf{x}_k$ from the data vectors $\mathbf{v}_0, \mathbf{v}_1, \cdots, \mathbf{v}_k$, using the initial information $E(\mathbf{x}_0)$ and $Var(\mathbf{x}_0)$. We have the following result.

Theorem 4.1. *The optimal estimate $\hat{\mathbf{x}}_k = \hat{\mathbf{x}}_{k|k}$ of $\mathbf{x}_k$ from the data $\mathbf{v}_0, \mathbf{v}_1, \cdots, \mathbf{v}_k$ can be computed recursively as follows: Define*

$$P_{0,0} = Var(\mathbf{x}_0).$$

Then, for $k = 1, 2, \cdots$, compute

$$P_{k,k-1} = (A_{k-1} - K_{k-1}C_{k-1})P_{k-1,k-1}(A_{k-1} - K_{k-1}C_{k-1})^\top \\ + \Gamma_{k-1}Q_{k-1}\Gamma_{k-1}^\top - K_{k-1}R_{k-1}K_{k-1}^\top, \tag{a}$$

where

$$K_{k-1} = \Gamma_{k-1}S_{k-1}R_{k-1}^{-1}, \tag{b}$$

and the Kalman gain matrix

$$G_k = P_{k,k-1}C_k^\top (C_k P_{k,k-1}C_k^\top + R_k)^{-1} \tag{c}$$

with

$$P_{k,k} = (I - G_k C_k)P_{k,k-1}. \tag{d}$$

Then, with the initial condition

$$\hat{\mathbf{x}}_{0|0} = E(\mathbf{x}_0),$$

compute, for $k = 1, 2, \cdots$, the prediction estimates

$$\hat{\mathbf{x}}_{k|k-1} = A_{k-1}\hat{\mathbf{x}}_{k-1|k-1} + K_{k-1}(\mathbf{v}_{k-1} - C_{k-1}\hat{\mathbf{x}}_{k-1|k-1}) \tag{e}$$

and the correction estimates

$$\hat{\mathbf{x}}_{k|k} = \hat{\mathbf{x}}_{k|k-1} + G_k(\mathbf{v}_k - C_k \hat{\mathbf{x}}_{k|k-1}), \tag{f}$$

(cf. Fig.4.1).

These are the Kalman filtering equations for correlated system and measurement noise processes. We remark that if the system noise and measurement noise are uncorrelated; that is, $S_{k-1} = O_{p \times q}$, so that $K_{k-1} = O_{n \times q}$ for all $k = 1, 2, \cdots$, then the above Kalman filtering equations reduce to the ones discussed in Chapters 2 and 3.

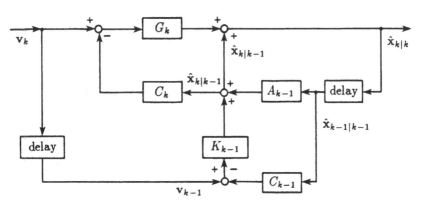

Fig. 4.1.

We will first derive the prediction-correction formulas (e) and (f). In this process, the matrices $P_{k,k-1}$, $P_{k,k}$, and G_k will be defined, and their computational schemes (a), (b), (c), and (d) will be determined. Let

$$\mathbf{v}^{k-1} = \begin{bmatrix} \mathbf{v}_0 \\ \vdots \\ \mathbf{v}_{k-1} \end{bmatrix} \quad \text{and} \quad \mathbf{v}^k = \begin{bmatrix} \mathbf{v}^{k-1} \\ \mathbf{v}_k \end{bmatrix}.$$

Then, $\mathbf{v}^k$, $\mathbf{v}^{k-1}$, and $\mathbf{v}_k$ can be considered as the data vectors $\mathbf{v}$, $\mathbf{v}^1$, and $\mathbf{v}^2$ in Lemma 4.2, respectively. Also, set

$$\hat{\mathbf{x}}_{k|k-1} = L(\mathbf{x}_k, \mathbf{v}^{k-1}),$$
$$\hat{\mathbf{x}}_{k|k} = L(\mathbf{x}_k, \mathbf{v}^k),$$

and

$$\mathbf{x}_k^{\#} = \mathbf{x}_k - \hat{\mathbf{x}}_{k|k-1} = \mathbf{x}_k - L(\mathbf{x}_k, \mathbf{v}^{k-1}).$$

Then we have the following properties

$$\begin{cases} \langle \underline{\xi}_{k-1}, \mathbf{v}^{k-2} \rangle = 0, & \langle \underline{\eta}_{k-1}, \mathbf{v}^{k-2} \rangle = 0, \\ \langle \underline{\xi}_{k-1}, \mathbf{x}_{k-1} \rangle = 0, & \langle \underline{\eta}_{k-1}, \mathbf{x}_{k-1} \rangle = 0, \\ \langle \mathbf{x}_{k-1}^{\#}, \underline{\xi}_{k-1} \rangle = 0, & \langle \mathbf{x}_{k-1}^{\#}, \underline{\eta}_{k-1} \rangle = 0, \\ \langle \hat{\mathbf{x}}_{k-1|k-2}, \underline{\xi}_{k-1} \rangle = 0, & \langle \hat{\mathbf{x}}_{k-1|k-2}, \underline{\eta}_{k-1} \rangle = 0, \end{cases} \tag{4.17}$$

(cf. Exercise 4.4). To derive the prediction formula, the idea is to add the "zero term" $K_{k-1}(\mathbf{v}_{k-1} - C_{k-1}\mathbf{x}_{k-1} - \underline{\eta}_{k-1})$ to the estimate

$$\hat{\mathbf{x}}_{k|k-1} = L(A_{k-1}\mathbf{x}_{k-1} + \Gamma_{k-1}\underline{\xi}_{k-1}, \mathbf{v}^{k-1}).$$

For an appropriate matrix K_{k-1}, we could eliminate the noise correlation in the estimate by absorbing it in K_{k-1}. More precisely, since $L(\cdot, \mathbf{v}^{k-1})$ is linear, it follows that

$$\hat{\mathbf{x}}_{k|k-1}$$
$$= L(A_{k-1}\mathbf{x}_{k-1} + \Gamma_{k-1}\underline{\xi}_{k-1} + K_{k-1}(\mathbf{v}_{k-1} - C_{k-1}\mathbf{x}_{k-1} - \underline{\eta}_{k-1}), \mathbf{v}^{k-1})$$
$$= L((A_{k-1} - K_{k-1}C_{k-1})\mathbf{x}_{k-1} + K_{k-1}\mathbf{v}_{k-1}$$
$$\quad + (\Gamma_{k-1}\underline{\xi}_{k-1} - K_{k-1}\underline{\eta}_{k-1}), \mathbf{v}^{k-1})$$
$$= (A_{k-1} - K_{k-1}C_{k-1})L(\mathbf{x}_{k-1}, \mathbf{v}^{k-1}) + K_{k-1}L(\mathbf{v}_{k-1}, \mathbf{v}^{k-1})$$
$$\quad + L(\Gamma_{k-1}\underline{\xi}_{k-1} - K_{k-1}\underline{\eta}_{k-1}, \mathbf{v}^{k-1})$$
$$:= I_1 + I_2 + I_3.$$

We will force the noise term I_3 to be zero by choosing K_{k-1} appropriately. To accomplish this, observe that from (4.6) and (4.17), we first have

$$I_3 = L(\Gamma_{k-1}\underline{\xi}_{k-1} - K_{k-1}\underline{\eta}_{k-1}, \mathbf{v}^{k-1})$$
$$= E(\Gamma_{k-1}\underline{\xi}_{k-1} - K_{k-1}\underline{\eta}_{k-1})$$
$$\quad + \langle \Gamma_{k-1}\underline{\xi}_{k-1} - K_{k-1}\underline{\eta}_{k-1}, \mathbf{v}^{k-1} \rangle [\|\mathbf{v}^{k-1}\|^2]^{-1}(\mathbf{v}^{k-1} - E(\mathbf{v}^{k-1}))$$
$$= \left\langle \Gamma_{k-1}\underline{\xi}_{k-1} - K_{k-1}\underline{\eta}_{k-1}, \begin{bmatrix} \mathbf{v}^{k-2} \\ \mathbf{v}_{k-1} \end{bmatrix} \right\rangle [\|\mathbf{v}^{k-1}\|^2]^{-1}(\mathbf{v}^{k-1} - E(\mathbf{v}^{k-1}))$$
$$= \langle \Gamma_{k-1}\underline{\xi}_{k-1} - K_{k-1}\underline{\eta}_{k-1}, \mathbf{v}_{k-1} \rangle [\|\mathbf{v}^{k-1}\|^2]^{-1}(\mathbf{v}^{k-1} - E(\mathbf{v}^{k-1}))$$
$$= \langle \Gamma_{k-1}\underline{\xi}_{k-1} - K_{k-1}\underline{\eta}_{k-1}, C_{k-1}\mathbf{x}_{k-1} + \underline{\eta}_{k-1} \rangle [\|\mathbf{v}^{k-1}\|^2]^{-1}(\mathbf{v}^{k-1} - E(\mathbf{v}^{k-1}))$$
$$= (\Gamma_{k-1}S_{k-1} - K_{k-1}R_{k-1})[\|\mathbf{v}^{k-1}\|^2]^{-1}(\mathbf{v}^{k-1} - E(\mathbf{v}^{k-1})).$$

Hence, by choosing

$$K_{k-1} = \Gamma_{k-1}S_{k-1}R_{k-1}^{-1},$$

so that (b) is satisfied, we have $I_3 = 0$. Next, I_1 and I_2 can be determined as follows:

$$I_1 = (A_{k-1} - K_{k-1}C_{k-1})L(\mathbf{x}_{k-1}, \mathbf{v}^{k-1})$$
$$= (A_{k-1} - K_{k-1}C_{k-1})\hat{\mathbf{x}}_{k-1|k-1}$$

and, by Lemma 4.2 with $\mathbf{v}_{k-1}^{\#} = \mathbf{v}_{k-1} - L(\mathbf{v}_{k-1}, \mathbf{v}^{k-2})$, we have

$$I_2 = K_{k-1}L(\mathbf{v}_{k-1}, \mathbf{v}^{k-1})$$
$$= K_{k-1}L\left(\mathbf{v}_{k-1}, \begin{bmatrix} \mathbf{v}^{k-2} \\ \mathbf{v}_{k-1} \end{bmatrix}\right)$$
$$= K_{k-1}(L(\mathbf{v}_{k-1}, \mathbf{v}^{k-2}) + \langle \mathbf{v}_{k-1}^{\#}, \mathbf{v}_{k-1}^{\#} \rangle [\|\mathbf{v}_{k-1}^{\#}\|^2]^{-1}\mathbf{v}_{k-1}^{\#})$$
$$= K_{k-1}(L(\mathbf{v}_{k-1}, \mathbf{v}^{k-2}) + \mathbf{v}_{k-1}^{\#})$$
$$= K_{k-1}\mathbf{v}_{k-1}.$$

Hence, it follows that

$$\hat{\mathbf{x}}_{k|k-1} = I_1 + I_2$$
$$= (A_{k-1} - K_{k-1}C_{k-1})\hat{\mathbf{x}}_{k-1|k-1} + K_{k-1}\mathbf{v}_{k-1}$$
$$= A_{k-1}\hat{\mathbf{x}}_{k-1|k-1} + K_{k-1}(\mathbf{v}_{k-1} - C_{k-1}\hat{\mathbf{x}}_{k-1|k-1})$$

which is the prediction formula (e).

To derive the correction formula, we use Lemma 4.2 again to conclude that

$$\hat{\mathbf{x}}_{k|k} = L(\mathbf{x}_k, \mathbf{v}^{k-1}) + \langle \mathbf{x}_k^{\#}, \mathbf{v}_k^{\#} \rangle [\|\mathbf{v}_k^{\#}\|^2]^{-1}\mathbf{v}_k^{\#}$$
$$= \hat{\mathbf{x}}_{k|k-1} + \langle \mathbf{x}_k^{\#}, \mathbf{v}_k^{\#} \rangle [\|\mathbf{v}_k^{\#}\|^2]^{-1}\mathbf{v}_k^{\#}, \tag{4.18}$$

where

$$\mathbf{x}_k^{\#} = \mathbf{x}_k - \hat{\mathbf{x}}_{k|k-1},$$

and, using (4.6) and (4.17), we arrive at

$$\mathbf{v}_k^{\#} = \mathbf{v}_k - L(\mathbf{v}_k, \mathbf{v}^{k-1})$$
$$= C_k\mathbf{x}_k + \underline{\eta}_k - L(C_k\mathbf{x}_k + \underline{\eta}_k, \mathbf{v}^{k-1})$$
$$= C_k\mathbf{x}_k + \underline{\eta}_k - C_kL(\mathbf{x}_k, \mathbf{v}^{k-1}) - L(\underline{\eta}_k, \mathbf{v}^{k-1})$$
$$= C_k(\mathbf{x}_k - L(\mathbf{x}_k, \mathbf{v}^{k-1})) + \underline{\eta}_k - E(\underline{\eta}_k)$$
$$\quad - \langle \underline{\eta}_k, \mathbf{v}^{k-1} \rangle [\|\mathbf{v}^{k-1}\|^2]^{-1}(\mathbf{v}^{k-1} - E(\mathbf{v}^{k-1}))$$
$$= C_k(\mathbf{x}_k - L(\mathbf{x}_k, \mathbf{v}^{k-1})) + \underline{\eta}_k$$
$$\quad - C_k(\mathbf{x}_k - \hat{\mathbf{x}}_{k|k-1}) + \underline{\eta}_k$$

Hence, by applying (4.17) again, it follows from (4.18) that

$$
\begin{aligned}
\hat{\mathbf{x}}_{k|k} &= \hat{\mathbf{x}}_{k|k-1} + \langle \mathbf{x}_k - \hat{\mathbf{x}}_{k|k-1}, C_k(\mathbf{x}_k - \hat{\mathbf{x}}_{k|k-1}) + \underline{\eta}_k \rangle \\
&\quad \cdot [\|C_k(\mathbf{x}_k - \hat{\mathbf{x}}_{k|k-1}) + \underline{\eta}_k\|_q^2]^{-1}(C_k(\mathbf{x}_k - \hat{\mathbf{x}}_{k|k-1}) + \underline{\eta}_k) \\
&= \hat{\mathbf{x}}_{k|k-1} + \|\mathbf{x}_k - \hat{\mathbf{x}}_{k|k-1}\|_n^2 C_k^\top \\
&\quad \cdot [C_k\|\mathbf{x}_k - \hat{\mathbf{x}}_{k|k-1}\|_n^2 C_k^\top + R_k]^{-1}(\mathbf{v}_k - C_k\hat{\mathbf{x}}_{k|k-1}) \\
&= \hat{\mathbf{x}}_{k|k-1} + G_k(\mathbf{v}_k - C_k\hat{\mathbf{x}}_{k|k-1}),
\end{aligned}
$$

which is the correction formula (f), if we set

$$
P_{k,j} = \|\mathbf{x}_k - \hat{\mathbf{x}}_{k|j}\|_n^2
$$

and

$$
G_k = P_{k,k-1}C_k^\top (C_k P_{k,k-1} C_k^\top + R_k)^{-1}. \tag{4.19}
$$

What is left is to verify the recursive relations (a) and (d) for $P_{k,k-1}$ and $P_{k,k}$. To do so, we need the following two formulas, the justification of which is left to the reader:

$$
(I - G_k C_k)P_{k,k-1}C_k = G_k R_k \tag{4.20}
$$

and

$$
\langle \mathbf{x}_{k-1} - \hat{\mathbf{x}}_{k-1|k-1}, \Gamma_{k-1}\underline{\xi}_{k-1} - K_{k-1}\underline{\eta}_{k-1} \rangle = O_{n \times n}, \tag{4.21}
$$

(cf. Exercise 4.5).

Now, using (e), (b), and (4.21) consecutively, we have

$$
\begin{aligned}
&P_{k,k-1} \\
&= \|\mathbf{x}_k - \hat{\mathbf{x}}_{k|k-1}\|_n^2 \\
&= \|A_{k-1}\mathbf{x}_{k-1} + \Gamma_{k-1}\underline{\xi}_{k-1} - A_{k-1}\hat{\mathbf{x}}_{k-1|k-1} \\
&\quad - K_{k-1}(\mathbf{v}_{k-1} - C_{k-1}\hat{\mathbf{x}}_{k-1|k-1})\|_n^2 \\
&= \|A_{k-1}\mathbf{x}_{k-1} + \Gamma_{k-1}\underline{\xi}_{k-1} - A_{k-1}\hat{\mathbf{x}}_{k-1|k-1} \\
&\quad - K_{k-1}(C_{k-1}\mathbf{x}_{k-1} + \underline{\eta}_{k-1} - C_{k-1}\hat{\mathbf{x}}_{k-1|k-1})\|_n^2 \\
&= \|(A_{k-1} - K_{k-1}C_{k-1})(\mathbf{x}_{k-1} - \hat{\mathbf{x}}_{k-1|k-1}) \\
&\quad + (\Gamma_{k-1}\underline{\xi}_{k-1} - K_{k-1}\underline{\eta}_{k-1})\|_n^2 \\
&= (A_{k-1} - K_{k-1}C_{k-1})P_{k-1|k-1}(A_{k-1} - K_{k-1}C_{k-1})^\top + \Gamma_{k-1}Q_{k-1}\Gamma_{k-1}^\top \\
&\quad + K_{k-1}R_{k-1}K_{k-1}^\top - \Gamma_{k-1}S_{k-1}K_{k-1}^\top - K_{k-1}S_{k-1}^\top\Gamma_{k-1}^\top \\
&= (A_{k-1} - K_{k-1}C_{k-1})P_{k-1,k-1}(A_{k-1} - K_{k-1}C_{k-1})^\top \\
&\quad + \Gamma_{k-1}Q_{k-1}\Gamma_{k-1}^\top - K_{k-1}R_{k-1}K_{k-1}^\top,
\end{aligned}
$$

which is (a).

Finally, using (f), (4.17), and (4.20) consecutively, we also have

$$
\begin{aligned}
P_{k,k} & \\
&= \|\mathbf{x}_k - \hat{\mathbf{x}}_{k|k}\|_n^2 \\
&= \|\mathbf{x}_k - \hat{\mathbf{x}}_{k|k-1} - G_k(\mathbf{v}_k - C_k\hat{\mathbf{x}}_{k|k-1})\|_n^2 \\
&= \|(\mathbf{x}_k - \hat{\mathbf{x}}_{k|k-1}) - G_k(C_k\mathbf{x}_k + \underline{\eta}_k - C_k\hat{\mathbf{x}}_{k|k-1})\|_n^2 \\
&= \|(I - G_kC_k)(\mathbf{x}_k - \hat{\mathbf{x}}_{k|k-1}) - G_k\underline{\eta}_k\|_n^2 \\
&= (I - G_kC_k)P_{k,k-1}(I - G_kC_k)^\top + G_kR_kG_k^\top \\
&= (I - G_kC_k)P_{k,k-1} - (I - G_kC_k)P_{k,k-1}C_k^\top G_k^\top + G_kR_kG_k^\top \\
&= (I - G_kC_k)P_{k,k-1}\,,
\end{aligned}
$$

which is (d). This completes the proof of the theorem.

4.5 Real-Time Applications

An aircraft radar guidance system provides a typical application. This system can be described by using the model (3.26) considered in the last chapter, with only one modification, namely: the tracking radar is now onboard to provide the position data information. Hence, both the system and measurement noise processes come from the same source such as vibration, and will be correlated. For instance, let us consider the following state-space description

$$
\left\{
\begin{aligned}
\begin{bmatrix} x_{k+1}[1] \\ x_{k+1}[2] \\ x_{k+1}[3] \end{bmatrix} &= \begin{bmatrix} 1 & h & h^2/2 \\ 0 & 1 & h \\ 0 & 0 & 1 \end{bmatrix} \begin{bmatrix} x_k[1] \\ x_k[2] \\ x_k[3] \end{bmatrix} + \begin{bmatrix} \xi_k[1] \\ \xi_k[2] \\ \xi_k[3] \end{bmatrix} \\
v_k &= [\, 1 \quad 0 \quad 0 \,] \begin{bmatrix} x_k[1] \\ x_k[2] \\ x_k[3] \end{bmatrix} + \eta_k\,,
\end{aligned}
\right.
\tag{4.22}
$$

where $\{\underline{\xi}_k\}$, with $\underline{\xi}_k := [\, \xi_k[1] \quad \xi_k[2] \quad \xi_k[3] \,]^\top$, and $\{\eta_k\}$ are assumed to be correlated zero-mean Gaussian white noise sequences satisfying

$$
\begin{aligned}
E(\underline{\xi}_k) &= 0\,, & E(\eta_k) &= 0\,, \\
E(\underline{\xi}_k\underline{\xi}_\ell^\top) &= Q_k\delta_{k\ell}\,, & E(\eta_k\eta_\ell) &= r_k\delta_{k\ell}\,, & E(\underline{\xi}_k\eta_\ell) &= \mathbf{s}_k\delta_{k\ell}\,, \\
E(\mathbf{x}_0\underline{\xi}_k^\top) &= 0\,, & E(\mathbf{x}_0\eta_k) &= 0\,,
\end{aligned}
$$

with $Q_k \geq 0$, $r_k > 0$, $\mathbf{s}_k := [\, s_k[1] \quad s_k[2] \quad s_k[3] \,]^\top \geq 0$ for all k, and $E(\mathbf{x}_0)$ and $Var(\mathbf{x}_0)$ are both assumed to be given.

An application of Theorem 4.1 to this system yields the following:

$$P_{k,k-1}[1,1]$$
$$= P_{k-1}[1,1] + 2hP_{k-1}[1,2] + h^2 P_{k-1}[1,3] + h^2 P_{k-1}[2,2]$$
$$+ h^3 P_{k-1}[2,3] + \frac{h^4}{4} P_{k-1}[3,3] + Q_{k-1}[1,1]$$
$$+ \frac{s_{k-1}[1]}{r_{k-1}} \left\{ \frac{s_{k-1}[1]}{r_{k-1}} P_{k-1}[1,1] - 2\left(P_{k-1}[1,1] \right. \right.$$
$$\left. \left. + hP_{k-1}[1,2] + \frac{h^2}{2} P_{k-1}[1,3] \right) - s_{k-1}[1] \right\},$$

$$P_{k,k-1}[1,2] = P_{k,k-1}[2,1]$$
$$= P_{k-1}[1,2] + hP_{k-1}[1,3] + hP_{k-1}[2,2] + \frac{3h^2}{2} P_{k-1}[2,3]$$
$$+ \frac{h^3}{2} P_{k-1}[3,3] + Q_{k-1}[1,2] + \left\{ \frac{s_{k-1}[1]s_{k-1}[2]}{r_{k-1}^2} P_{k-1}[1,1] \right.$$
$$- \frac{s_{k-1}[1]}{r_{k-1}} \left(P_{k-1}[1,2] + hP_{k-1}[1,3] \right) - \frac{s_{k-1}[2]}{r_{k-1}} \left(P_{k-1}[1,1] \right.$$
$$\left. + hP_{k-1}[1,2] + \frac{h^2}{2} P_{k-1}[1,3] \right) - \frac{s_{k-1}[1]s_{k-1}[2]}{r_{k-1}} \right\},$$

$$P_{k,k-1}[2,2]$$
$$= P_{k-1}[2,2] + 2hP_{k-1}[2,3] + h^2 P_{k-1}[3,3] + Q_{k-1}[2,2]$$
$$+ \frac{s_{k-1}[2]}{r_{k-1}} \left\{ \frac{s_{k-1}[2]}{r_{k-1}} P_{k-1}[2,2] - 2\left(P_{k-1}[1,2] + hP_{k-1}[1,3] \right) - s_{k-1}[2] \right\},$$

$$P_{k,k-1}[1,3] = P_{k,k-1}[3,1]$$
$$= P_{k-1}[1,3] + hP_{k-1}[2,3] + \frac{h^2}{2} P_{k-1}[3,3] + Q_{k-1}[1,3]$$
$$+ \left\{ \frac{s_{k-1}[1]s_{k-1}[3]}{r_{k-1}^2} P_{k-1}[1,1] - \frac{s_{k-1}[1]}{r_{k-1}} P_{k-1}[1,3] - \frac{s_{k-1}[3]}{r_{k-1}} \left(P_{k-1}[1,1] \right. \right.$$
$$\left. \left. + hP_{k-1}[1,2] + \frac{h^2}{2} P_{k-1}[1,3] \right) - \frac{s_{k-1}[1]s_{k-1}[3]}{r_{k-1}} \right\},$$

$$P_{k,k-1}[2,3] = P_{k,k-1}[3,2]$$
$$= P_{k-1}[2,3] + hP_{k-1}[3,3] + Q_{k-1}[2,3] + \left\{ \frac{s_{k-1}[2]s_{k-1}[3]}{r_{k-1}^2} P_{k-1}[1,1] \right.$$
$$\left. - \frac{s_{k-1}[2]s_{k-1}[3]}{r_{k-1}} - \frac{s_{k-1}[3]}{r_{k-1}} \left(P_{k-1}[1,2] + hP_{k-1}[1,3] \right) - \frac{s_{k-1}[2]s_{k-1}[3]}{r_{k-1}} \right\},$$

$$P_{k,k-1}[3,3] = P_{k-1}[3,3] + Q_{k-1}[3,3]$$

$$+ \frac{s_{k-1}[3]}{r_{k-1}} \left\{ \frac{s_{k-1}[3]}{r_{k-1}} P_{k-1}[1,1] - 2P_{k-1}[1,3] - s_{k-1}[3] \right\},$$

where $P_{k-1} = P_{k,k-1}$ and $P[i,j]$ denotes the $(i,j)th$ entry of P. In addition,

$$G_k = \frac{1}{P_{k,k-1}[1,1] + r_k} \begin{bmatrix} P_{k,k-1}[1,1] \\ P_{k,k-1}[1,2] \\ P_{k,k-1}[1,3] \end{bmatrix},$$

$$P_k = P_{k,k-1} - \frac{1}{P_{k,k-1}[1,1] + r_k} \cdot$$

$$\begin{bmatrix} P_{k,k-1}{}^2[1,1] & P_{k,k-1}[1,1]P_{k,k-1}[1,2] & P_{k,k-1}[1,1]P_{k,k-1}[1,3] \\ P_{k,k-1}[1,1]P_{k,k-1}[1,2] & P_{k,k-1}{}^2[1,2] & P_{k,k-1}[1,2]P_{k,k-1}[1,3] \\ P_{k,k-1}[1,1]P_{k,k-1}[1,3] & P_{k,k-1}[1,2]P_{k,k-1}[1,3] & P_{k,k-1}{}^2[1,3] \end{bmatrix}$$

with $P_0 = Var(\mathbf{x}_0)$, and

$$\begin{bmatrix} \hat{x}_{k|k-1}[1] \\ \hat{x}_{k|k-1}[2] \\ \hat{x}_{k|k-1}[3] \end{bmatrix} = \begin{bmatrix} 1 & h & h^2/2 \\ 0 & 1 & h \\ 0 & 0 & 1 \end{bmatrix} \begin{bmatrix} \hat{x}_{k-1|k-1}[1] \\ \hat{x}_{k-1|k-1}[2] \\ \hat{x}_{k-1|k-1}[3] \end{bmatrix}$$

$$+ \frac{v_k - \hat{x}_{k-1|k-1}[1]}{r_{k-1}} \begin{bmatrix} s_{k-1}[1] \\ s_{k-1}[2] \\ s_{k-1}[3] \end{bmatrix},$$

$$\begin{bmatrix} \hat{x}_{k|k}[1] \\ \hat{x}_{k|k}[2] \\ \hat{x}_{k|k}[3] \end{bmatrix} = \begin{bmatrix} 1 - G_k[1] & (1-G_k[1])h & (1-G_k[1])h^2/2 \\ -G_k[2] & 1 - hG_k[2] & h - h^2G_k[2]/2 \\ -G_k[3] & -hG_k[3] & 1 - h^2G_k[3]/2 \end{bmatrix} \begin{bmatrix} \hat{x}_{k-1|k-1}[1] \\ \hat{x}_{k-1|k-1}[2] \\ \hat{x}_{k-1|k-1}[3] \end{bmatrix}$$

$$+ \begin{bmatrix} G_k[1] \\ G_k[2] \\ G_k[3] \end{bmatrix} v_k,$$

with $\hat{\mathbf{x}}_{0|0} = E(\mathbf{x}_0)$.

4.6 Linear Deterministic/Stochastic Systems

Finally, let us discuss the general linear stochastic system with deterministic control input $\mathbf{u}_k$ being incorporated. More precisely, we have the following state-space description:

$$\begin{cases} \mathbf{x}_{k+1} = A_k \mathbf{x}_k + B_k \mathbf{u}_k + \Gamma_k \underline{\xi}_k \\ \mathbf{v}_k = C_k \mathbf{x}_k + D_k \mathbf{u}_k + \underline{\eta}_k, \end{cases} \tag{4.23}$$

where $\mathbf{u}_k$ is a deterministic control input m-vector with $1 \le m \le n$. It can be proved (cf. Exercise 4.6) that the Kalman filtering algorithm for this system is given by

$$\begin{cases}
P_{0,0} = Var(\mathbf{x}_0), \quad K_{k-1} = \Gamma_{k-1}S_{k-1}R_{k-1}^{-1}, \\
P_{k,k-1} = (A_{k-1} - K_{k-1}C_{k-1})P_{k-1,k-1}(A_{k-1} - K_{k-1}C_{k-1})^\top \\
\quad\quad + \Gamma_{k-1}Q_{k-1}\Gamma_{k-1}^\top - K_{k-1}R_{k-1}K_{k-1}^\top \\
G_k = P_{k,k-1}C_k^\top(C_kP_{k,k-1}C_k^\top + R_k)^{-1} \\
P_{k,k} = (I - G_kC_k)P_{k,k-1} \\
\hat{\mathbf{x}}_{0|0} = E(\mathbf{x}_0) \\
\hat{\mathbf{x}}_{k|k-1} = A_{k-1}\hat{\mathbf{x}}_{k-1|k-1} + B_{k-1}\mathbf{u}_{k-1} \\
\quad\quad + K_{k-1}(\mathbf{v}_{k-1} - D_{k-1}\mathbf{u}_{k-1} - C_{k-1}\hat{\mathbf{x}}_{k-1|k-1}) \\
\hat{\mathbf{x}}_{k|k} = \hat{\mathbf{x}}_{k|k-1} + G_k(\mathbf{v}_k - D_k\mathbf{u}_k - C_k\hat{\mathbf{x}}_{k|k-1}) \\
k = 1, 2, \cdots,
\end{cases} \tag{4.24}$$

(cf. Fig.4.2).

We remark that if the system and measurement noise processes are uncorrelated so that $S_k = 0$ for all $k = 0, 1, 2, \cdots$, then (4.24) reduces to (2.18) or (3.25), as expected.

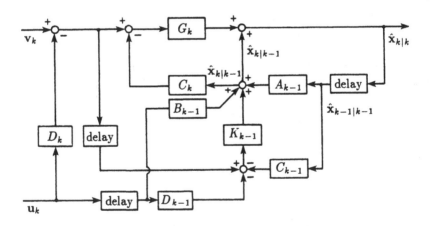

Fig. 4.2.

Exercises

4.1. Let $\mathbf{v}$ be a random vector and define

$$L(\mathbf{x},\ \mathbf{v}) = E(\mathbf{x}) + \langle \mathbf{x},\ \mathbf{v}\rangle \big[\|\mathbf{v}\|^2\big]^{-1}(\mathbf{v} - E(\mathbf{v}))\,.$$

Show that $L(\cdot,\ \mathbf{v})$ is a linear operator in the sense that

$$L(A\mathbf{x} + B\mathbf{y},\ \mathbf{v}) = AL(\mathbf{x},\ \mathbf{v}) + BL(\mathbf{y},\ \mathbf{v})$$

for all constant matrices A and B and random vectors $\mathbf{x}$ and $\mathbf{y}$.

4.2. Let $\mathbf{v}$ be a random vector and $L(\cdot,\ \mathbf{v})$ be defined as above. Show that if $\mathbf{a}$ is a constant vector then $L(\mathbf{a},\ \mathbf{v}) = \mathbf{a}$.

4.3. For a real-valued function f and a matrix $A = [a_{ij}]$, define

$$\frac{df}{dA} = \left[\frac{\partial f}{\partial a_{ij}}\right]^{\mathsf{T}}.$$

By taking

$$\frac{\partial}{\partial H}(tr\|\mathbf{x} - \mathbf{y}\|^2) = 0\,,$$

show that the solution $\mathbf{x}^*$ of the minimization problem

$$tr\|\mathbf{x}^* - \mathbf{y}\|^2 = \min_{H} tr\|\mathbf{x} - \mathbf{y}\|^2\,,$$

where $\mathbf{y} = E(\mathbf{x}) + H(\mathbf{v} - E(\mathbf{v}))$, can be obtained by setting

$$\mathbf{x}^* = E(\mathbf{x}) - \langle \mathbf{x},\ \mathbf{v}\rangle \big[\|\mathbf{v}\|^2\big]^{-1}(E(\mathbf{v}) - \mathbf{v})\,,$$

where

$$H^* = \langle \mathbf{x},\ \mathbf{v}\rangle \big[\|\mathbf{v}\|^2\big]^{-1}.$$

4.4. Consider the linear stochastic system (4.16). Let

$$\mathbf{v}^{k-1} = \begin{bmatrix} \mathbf{v}_0 \\ \vdots \\ \mathbf{v}_{k-1} \end{bmatrix} \quad and \quad \mathbf{v}^k = \begin{bmatrix} \mathbf{v}^{k-1} \\ \mathbf{v}_k \end{bmatrix}.$$

Define $L(\mathbf{x},\ \mathbf{v})$ as in Exercise 4.1 and let $\mathbf{x}_k^{\#} = \mathbf{x}_k - \hat{\mathbf{x}}_{k|k-1}$ with $\hat{\mathbf{x}}_{k|k-1} := L(\mathbf{x}_k, \mathbf{v}^{k-1})$. Prove that

$$\langle \underline{\xi}_{k-1},\ \mathbf{v}^{k-2}\rangle = 0\,, \qquad \langle \underline{\eta}_{k-1},\ \mathbf{v}^{k-2}\rangle = 0\,,$$

$$\langle \underline{\xi}_{k-1},\ \mathbf{x}_{k-1}\rangle = 0\,, \qquad \langle \underline{\eta}_{k-1},\ \mathbf{x}_{k-1}\rangle = 0\,,$$

$$\langle \mathbf{x}_{k-1}^{\#},\ \underline{\xi}_{k-1}\rangle = 0\,, \qquad \langle \mathbf{x}_{k-1}^{\#},\ \underline{\eta}_{k-1}\rangle = 0\,,$$

$$\langle \hat{\mathbf{x}}_{k-1|k-2},\ \underline{\xi}_{k-1}\rangle = 0\,, \qquad \langle \hat{\mathbf{x}}_{k-1|k-2},\ \underline{\eta}_{k-1}\rangle = 0\,.$$

4.5. Verify that
$$(I - G_k C_k) P_{k,k-1} C_k = G_k R_k$$
and
$$\langle \mathbf{x}_{k-1} - \hat{\mathbf{x}}_{k-1|k-1}, \ \Gamma_{k-1}\underline{\xi}_{k-1} - K_{k-1}\underline{\eta}_{k-1} \rangle = O_{n \times n} .$$

4.6. Consider the linear deterministic/stochastic system
$$\begin{cases} \mathbf{x}_{k+1} = A_k \mathbf{x}_k + B_k \mathbf{u}_k + \Gamma_k \underline{\xi}_k \\ \mathbf{v}_k = C_k \mathbf{x}_k + D_k \mathbf{u}_k + \underline{\eta}_k, \end{cases}$$
where $\{\mathbf{u}_k\}$ is a given sequence of deterministic control inputs. Suppose that the same assumption for (4.16) is satisfied. Derive the Kalman filtering algorithm for this model.

4.7. Consider the following so-called *ARMAX* (*auto-regressive moving-average model with exogeneous inputs*) model in signal processing:
$$v_k = -a_1 v_{k-1} - a_2 v_{k-2} - a_3 v_{k-3} + b_0 u_k + b_1 u_{k-1} + b_2 u_{k-2} + c_0 e_k + c_1 e_{k-1},$$
where $\{v_j\}$ and $\{u_j\}$ are output and input signals, respectively, $\{e_j\}$ is a zero-mean Gaussian white noise sequence with $Var(e_j) = s_j > 0$, and a_j, b_j, c_j are constants.
(a) Derive a state-space description for this *ARMAX* model.
(b) Specify the Kalman filtering algorithm for this state-space description.

4.8. More generally, consider the following *ARMAX* model in signal processing:
$$v_k = -\sum_{j=1}^{n} a_j v_{k-j} + \sum_{j=0}^{m} b_j u_{k-j} + \sum_{j=0}^{\ell} c_j e_{k-j},$$
where $0 \le m, \ell \le n$, $\{v_j\}$ and $\{u_j\}$ are output and input signals, respectively, $\{e_j\}$ is a zero-mean Gaussian white noise sequence with $Var(e_j) = s_j > 0$, and a_j, b_j, c_j are constants.
(a) Derive a state-space description for this *ARMAX* model.
(b) Specify the Kalman filtering algorithm for this state-space description.

5. Colored Noise

Consider the linear stochastic system with the following state-space description:

$$\begin{cases} \mathbf{x}_{k+1} = A_k\mathbf{x}_k + \Gamma_k\underline{\xi}_k \\ \mathbf{v}_k = C_k\mathbf{x}_k + \underline{\eta}_k, \end{cases} \tag{5.1}$$

where A_k, Γ_k, and C_k are known $n \times n, n \times p$, and $q \times n$ constant matrices, respectively, with $1 \leq p,q \leq n$. The problem is to give a linear unbiased minimum variance estimate of $\mathbf{x}_k$ with initial quantities $E(\mathbf{x}_0)$ and $Var(\mathbf{x}_0)$ under the assumption that

(i) $\underline{\xi}_k = M_{k-1}\underline{\xi}_{k-1} + \underline{\beta}_k$

(ii) $\underline{\eta}_k = N_{k-1}\underline{\eta}_{k-1} + \underline{\gamma}_k$,

where $\underline{\xi}_{-1} = \underline{\eta}_{-1} = 0$, $\{\underline{\beta}_k\}$ and $\{\underline{\gamma}_k\}$ are uncorrelated zero-mean Gaussian white noise sequences satisfying

$$E(\underline{\beta}_k\underline{\gamma}_\ell^\top) = 0, \quad E(\underline{\beta}_k\underline{\beta}_\ell^\top) = Q_k\delta_{k\ell}, \quad E(\underline{\gamma}_k\underline{\gamma}_\ell^\top) = R_k\delta_{k\ell},$$

and M_{k-1} and N_{k-1} are known $p \times p$ and $q \times q$ constant matrices. The noise sequences $\{\underline{\xi}_k\}$ and $\{\underline{\eta}_k\}$ satisfying (i) and (ii) will be called *colored noise processes*. This chapter is devoted to the study of Kalman filtering with this assumption on the noise sequences.

5.1 Outline of Procedure

The idea in dealing with the colored model (5.1) is first to make the system equation in (5.1) white. To accomplish this, we simply set

$$\mathbf{z}_k = \begin{bmatrix} \mathbf{x}_k \\ \underline{\xi}_k \end{bmatrix}, \qquad \tilde{A}_k = \begin{bmatrix} A_k & \Gamma_k \\ 0 & M_k \end{bmatrix}, \qquad \underline{\tilde{\beta}}_k = \begin{bmatrix} 0 \\ \underline{\beta}_k \end{bmatrix},$$

and arrive at

$$\mathbf{z}_{k+1} = \tilde{A}_k\mathbf{z}_k + \underline{\tilde{\beta}}_{k+1}. \tag{5.2}$$

Note that the observation equation in (5.1) becomes

$$\mathbf{v}_k = \tilde{C}_k \mathbf{z}_k + \underline{\eta}_k ,\tag{5.3}$$

where $\tilde{C}_k = [C_k \quad 0]$.

We will use the same model as was used in Chapter 4, by considering

$$\hat{\mathbf{z}}_{k|j} = L(\mathbf{z}_k, \mathbf{v}^j) ,$$

where

$$\mathbf{v}^j = \begin{bmatrix} \mathbf{v}_0 \\ \vdots \\ \mathbf{v}_j \end{bmatrix} \quad \text{and} \quad \hat{\mathbf{z}}_k := \hat{\mathbf{z}}_{k|k} .$$

The second step is to derive a recursive relation (instead of the prediction-correction relation) for $\hat{\mathbf{z}}_k$. To do so, we have, using the linearity of L,

$$\hat{\mathbf{z}}_k = \tilde{A}_{k-1} L(\mathbf{z}_{k-1}, \mathbf{v}^k) + L(\underline{\tilde{\beta}}_k, \mathbf{v}^k) .$$

From the noise assumption, it can be shown that

$$L(\underline{\tilde{\beta}}_k, \mathbf{v}^k) = 0 ,\tag{5.4}$$

so that

$$\hat{\mathbf{z}}_k = \tilde{A}_{k-1} \hat{\mathbf{z}}_{k-1|k} ,\tag{5.5}$$

(cf. Exercise 5.1) and in order to obtain a recursive relation for $\hat{\mathbf{z}}_k$, we have to express $\hat{\mathbf{z}}_{k-1|k}$ in terms of $\hat{\mathbf{z}}_{k-1}$. This can be done by using Lemma 4.2, namely:

$$\begin{aligned} \hat{\mathbf{z}}_{k-1|k} &= L\left(\mathbf{z}_{k-1}, \begin{bmatrix} \mathbf{v}^{k-1} \\ \mathbf{v}_k \end{bmatrix}\right) \\ &= \hat{\mathbf{z}}_{k-1} + \langle \mathbf{z}_{k-1} - \hat{\mathbf{z}}_{k-1}, \mathbf{v}_k^{\#} \rangle [\|\mathbf{v}_k^{\#}\|^2]^{-1} \mathbf{v}_k^{\#} , \end{aligned}\tag{5.6}$$

where $\hat{\mathbf{z}}_{k-1} = L(\mathbf{z}_{k-1}, \mathbf{v}^{k-1})$ and $\mathbf{v}_k^{\#} = \mathbf{v}_k - L(\mathbf{v}_k, \mathbf{v}^{k-1})$.

5.2 Error Estimates

It is now important to understand the error term in (5.6). First, we will derive a formula for $\mathbf{v}_k^{\#}$. By the definition of $\mathbf{v}_k^{\#}$ and the observation equation (5.3) and noting that

$$\tilde{C}_k \underline{\tilde{\beta}}_k = [C_k \quad 0] \begin{bmatrix} 0 \\ \underline{\beta}_k \end{bmatrix} = 0 ,$$

we have

$$\begin{aligned}
\mathbf{v}_k &= \tilde{C}_k \mathbf{z}_k + \underline{\eta}_k \\
&= \tilde{C} \mathbf{z}_k + N_{k-1} \underline{\eta}_{k-1} + \underline{\gamma}_k \\
&= \tilde{C}_k (\tilde{A}_{k-1} \mathbf{z}_{k-1} + \underline{\tilde{\beta}}_k) + N_{k-1} (\mathbf{v}_{k-1} - \tilde{C}_{k-1} \mathbf{z}_{k-1}) + \underline{\gamma}_k \\
&= H_{k-1} \mathbf{z}_{k-1} + N_{k-1} \mathbf{v}_{k-1} + \underline{\gamma}_k \,,
\end{aligned} \tag{5.7}$$

with

$$\begin{aligned}
H_{k-1} &= \tilde{C}_k \tilde{A}_{k-1} - N_{k-1} \tilde{C}_{k-1} \\
&= [C_k A_{k-1} - N_{k-1} C_{k-1} \quad C_k \Gamma_{k-1}] \,.
\end{aligned}$$

Hence, by the linearity of L, it follows that

$$\begin{aligned}
&L(\mathbf{v}_k, \mathbf{v}^{k-1}) \\
&= H_{k-1} L(\mathbf{z}_{k-1}, \mathbf{v}^{k-1}) + N_{k-1} L(\mathbf{v}_{k-1}, \mathbf{v}^{k-1}) + L(\underline{\gamma}_k, \mathbf{v}^{k-1}) \,.
\end{aligned}$$

Since $L(\mathbf{z}_{k-1}, \mathbf{v}^{k-1}) = \hat{\mathbf{z}}_{k-1}$,

$$L(\mathbf{v}_{k-1}, \mathbf{v}^{k-1}) = \mathbf{v}_{k-1} \tag{5.8}$$

(cf. Exercise 5.2), and

$$L(\underline{\gamma}_k, \mathbf{v}^{k-1}) = 0 \,, \tag{5.9}$$

we obtain

$$\begin{aligned}
\mathbf{v}_k^\# &= \mathbf{v}_k - L(\mathbf{v}_k, \mathbf{v}^{k-1}) \\
&= \mathbf{v}_k - (H_{k-1} \hat{\mathbf{z}}_{k-1} + N_{k-1} \mathbf{v}_{k-1}) \\
&= \mathbf{v}_k - N_{k-1} \mathbf{v}_{k-1} - H_{k-1} \hat{\mathbf{z}}_{k-1} \,.
\end{aligned} \tag{5.10}$$

In addition, by using (5.7) and (5.10), we also have

$$\mathbf{v}_k^\# = H_{k-1} (\mathbf{z}_{k-1} - \hat{\mathbf{z}}_{k-1}) + \underline{\gamma}_k \,. \tag{5.11}$$

Now, let us return to (5.6). Using (5.11), (5.10), and $\langle \mathbf{z}_{k-1} - \hat{\mathbf{z}}_{k-1}, \underline{\gamma}_k \rangle = 0$ (cf. Exercise 5.3), we arrive at

$$\begin{aligned}
&\hat{\mathbf{z}}_{k-1|k} \\
&= \hat{\mathbf{z}}_{k-1} + \langle \mathbf{z}_{k-1} - \hat{\mathbf{z}}_{k-1}, \mathbf{v}_k^\# \rangle [\| \mathbf{v}_k^\# \|^2]^{-1} \mathbf{v}_k^\# \\
&= \hat{\mathbf{z}}_{k-1} + \| \mathbf{z}_{k-1} - \hat{\mathbf{z}}_{k-1} \|^2 H_{k-1}^\mathsf{T} (H_{k-1} \| \mathbf{z}_{k-1} - \hat{\mathbf{z}}_{k-1} \|^2 H_{k-1}^\mathsf{T} + R_k)^{-1} \\
&\quad \cdot (\mathbf{v}_k - N_{k-1} \mathbf{v}_{k-1} - H_{k-1} \hat{\mathbf{z}}_{k-1}) \,.
\end{aligned}$$

Putting this into (5.5) gives

$$\hat{\mathbf{z}}_k = \tilde{A}_{k-1} \hat{\mathbf{z}}_{k-1} + G_k (\mathbf{v}_k - N_{k-1} \mathbf{v}_{k-1} - H_{k-1} \hat{\mathbf{z}}_{k-1}) \,,$$

or

$$\begin{bmatrix} \hat{\mathbf{x}}_k \\ \hat{\underline{\xi}}_k \end{bmatrix} = \begin{bmatrix} A_{k-1} & \Gamma_{k-1} \\ 0 & M_{k-1} \end{bmatrix} \begin{bmatrix} \hat{\mathbf{x}}_{k-1} \\ \hat{\underline{\xi}}_{k-1} \end{bmatrix}$$
$$+ G_k \left(\mathbf{v}_k - N_{k-1}\mathbf{v}_{k-1} - H_{k-1} \begin{bmatrix} \hat{\mathbf{x}}_{k-1} \\ \hat{\underline{\xi}}_{k-1} \end{bmatrix} \right), \tag{5.12}$$

where

$$G_k = \tilde{A}_{k-1} P_{k-1} H_{k-1}^\top (H_{k-1} P_{k-1} H_{k-1}^\top + R_k)^{-1}$$
$$= \begin{bmatrix} A_{k-1} & \Gamma_{k-1} \\ 0 & M_{k-1} \end{bmatrix} P_{k-1} H_{k-1}^\top (H_{k-1} P_{k-1} H_{k-1}^\top + R_k)^{-1} \tag{5.13}$$

with

$$P_k = \|\mathbf{z}_k - \hat{\mathbf{z}}_k\|^2 = Var(\mathbf{z}_k - \hat{\mathbf{z}}_k). \tag{5.14}$$

5.3 Kalman Filtering Process

What is left is to derive an algorithm to compute P_k and the initial condition $\hat{\mathbf{z}}_0$. Using (5.7), we have

$$\mathbf{z}_k - \hat{\mathbf{z}}_k$$
$$= (\tilde{A}_{k-1}\mathbf{z}_{k-1} + \underline{\tilde{\beta}}_k) - (\tilde{A}_{k-1}\hat{\mathbf{z}}_{k-1} + G_k(\mathbf{v}_k - N_{k-1}\mathbf{v}_{k-1} - H_{k-1}\hat{\mathbf{z}}_{k-1}))$$
$$= \tilde{A}_{k-1}(\mathbf{z}_{k-1} - \hat{\mathbf{z}}_{k-1}) + \underline{\tilde{\beta}}_k - G_k(H_{k-1}\mathbf{z}_{k-1} + \underline{\gamma}_k - H_{k-1}\hat{\mathbf{z}}_{k-1})$$
$$= (\tilde{A}_{k-1} - G_k H_{k-1})(\mathbf{z}_{k-1} - \hat{\mathbf{z}}_{k-1}) + (\underline{\tilde{\beta}}_k - G_k\underline{\gamma}_k).$$

In addition, it follows from Exercise 5.3 and $\langle \mathbf{z}_{k-1} - \hat{\mathbf{z}}_{k-1}, \underline{\tilde{\beta}}_k \rangle = 0$ (cf. Exercise 5.4) that

$$P_k = (\tilde{A}_{k-1} - G_k H_{k-1}) P_{k-1} (\tilde{A}_{k-1} - G_k H_{k-1})^\top$$
$$+ \begin{bmatrix} 0 & 0 \\ 0 & Q_k \end{bmatrix} + G_k R_k G_k^\top$$
$$= (\tilde{A}_{k-1} - G_k H_{k-1}) P_{k-1} \tilde{A}_{k-1} + \begin{bmatrix} 0 & 0 \\ 0 & Q_k \end{bmatrix}, \tag{5.15}$$

where the identity

$$- (\tilde{A}_{k-1} - G_k H_{k-1}) P_{k-1} (G_k H_{k-1})^\top + G_k R_k G_k^\top$$
$$= - \tilde{A}_{k-1} P_{k-1} H_{k-1}^\top G_k^\top + G_k H_{k-1} P_{k-1} H_{k-1}^\top G_k^\top + G_k R_k G_k^\top$$
$$= - \tilde{A}_{k-1} P_{k-1} H_{k-1}^\top G_k^\top + G_k(H_{k-1} P_{k-1} H_{k-1}^\top + R_k) G_k^\top$$
$$= - \tilde{A}_{k-1} P_{k-1} H_{k-1}^\top G_k^\top + \tilde{A}_{k-1} P_{k-1} H_{k-1}^\top G_k^\top$$
$$= 0,$$

which is a consequence of (5.13), has been used.

The initial estimate is

$$\hat{z}_0 = L(z_0, v_0) = E(z_0) - \langle z_0, v_0 \rangle [\|v_0\|^2]^{-1}(E(v_0) - v_0)$$
$$= \begin{bmatrix} E(x_0) \\ 0 \end{bmatrix} - \begin{bmatrix} Var(x_0)C_0^\top \\ 0 \end{bmatrix} [C_0 Var(x_0)C_0^\top + R_0]^{-1}$$
$$\cdot (C_0 E(x_0) - v_0). \tag{5.16}$$

We remark that

$$E(\hat{z}_0) = \begin{bmatrix} E(x_0) \\ 0 \end{bmatrix} = E(z_0),$$

and since

$$z_k - \hat{z}_k$$
$$= (\tilde{A}_{k-1} - G_k H_{k-1})(z_{k-1} - \hat{z}_{k-1}) + (\underline{\tilde{\beta}}_k - G_k \underline{\gamma}_k)$$
$$= (\tilde{A}_{k-1} - G_k H_{k-1}) \cdots (\tilde{A}_0 - G_1 H_0)(z_0 - \hat{z}_0) + \text{noise},$$

we have $E(z_k - \hat{z}_k) = 0$. That is, $\hat{z}_k$ is an unbiased linear minimum (error) variance estimate of z_k.

In addition, from (5.16) we also have the initial condition

$$P_0 = Var(z_0 - \hat{z}_0)$$
$$= Var\left(\begin{bmatrix} x_0 - E(x_0) \\ \underline{\xi}_0 \end{bmatrix} \right.$$
$$\left. + \begin{bmatrix} Var(x_0)C_0^\top \\ 0 \end{bmatrix} [C_0 Var(x_0)C_0^\top + R_0]^{-1}[C_0 E(x_0) - v_0] \right)$$
$$= \begin{bmatrix} Var(x_0 - E(x_0)) & 0 \\ +Var(x_0)C_0^\top [C_0 Var(x_0)C_0^\top + R_0]^{-1}[C_0 E(x_0) - v_0]) & \\ 0 & Q_0 \end{bmatrix}$$
$$= \begin{bmatrix} Var(x_0) & 0 \\ -[Var(x_0)]C_0^\top [C_0 Var(x_0)C_0^\top + R_0]^{-1}C_0[Var(x_0)] & \\ 0 & Q_0 \end{bmatrix} \tag{5.17a}$$

(cf. Exercise 5.5). If $Var(x_0)$ is nonsingular, then using the matrix inversion lemma (cf. Lemma 1.2), we have

$$P_0 = \begin{bmatrix} ([Var(x_0)]^{-1} + C_0^\top R_0^{-1} C_0)^{-1} & 0 \\ 0 & Q_0 \end{bmatrix}. \tag{5.17b}$$

Finally, returning to the definition $\mathbf{z}_k = \begin{bmatrix} \mathbf{x}_k \\ \underline{\xi}_k \end{bmatrix}$, we have

$$
\begin{aligned}
\hat{\mathbf{z}}_k &= L(\mathbf{z}_k, \mathbf{v}^k) \\
&= E(\mathbf{z}_k) + \langle \mathbf{z}_k, \mathbf{v}^k \rangle [\|\mathbf{v}^k\|^2]^{-1}(\mathbf{v}^k - E(\mathbf{v}^k)) \\
&= \begin{bmatrix} E(\mathbf{x}_k) \\ E(\underline{\xi}_k) \end{bmatrix} + \begin{bmatrix} \langle \mathbf{x}_k, \mathbf{v}^k \rangle \\ \langle \underline{\xi}_k, \mathbf{v}^k \rangle \end{bmatrix} [\|\mathbf{v}^k\|^2]^{-1}(\mathbf{v}^k - E(\mathbf{v}^k)) \\
&= \begin{bmatrix} E(\mathbf{x}_k) + \langle \mathbf{x}_k, \mathbf{v}^k \rangle [\|\mathbf{v}^k\|^2]^{-1}(\mathbf{v}^k - E(\mathbf{v}^k)) \\ E(\underline{\xi}_k) + \langle \underline{\xi}_k, \mathbf{v}^k \rangle [\|\mathbf{v}^k\|^2]^{-1}(\mathbf{v}^k - E(\mathbf{v}^k)) \end{bmatrix} \\
&= \begin{bmatrix} \hat{\mathbf{x}}_k \\ \hat{\underline{\xi}}_k \end{bmatrix}.
\end{aligned}
$$

In summary, the Kalman filtering process for the colored noise model (5.1) is given by

$$
\begin{bmatrix} \hat{\mathbf{x}}_k \\ \hat{\underline{\xi}}_k \end{bmatrix} = \begin{bmatrix} A_{k-1} & \Gamma_{k-1} \\ 0 & M_{k-1} \end{bmatrix} \begin{bmatrix} \hat{\mathbf{x}}_{k-1} \\ \hat{\underline{\xi}}_{k-1} \end{bmatrix}
$$
$$
+ G_k \left(\mathbf{v}_k - N_{k-1}\mathbf{v}_{k-1} - H_{k-1} \begin{bmatrix} \hat{\mathbf{x}}_{k-1} \\ \hat{\underline{\xi}}_{k-1} \end{bmatrix} \right), \tag{5.18}
$$

where

$$
H_{k-1} = [C_k A_{k-1} - N_{k-1}C_{k-1} \quad C_k\Gamma_{k-1}]
$$

and

$$
G_k = \begin{bmatrix} A_{k-1} & \Gamma_{k-1} \\ 0 & M_{k-1} \end{bmatrix} P_{k-1}H_{k-1}^\top (H_{k-1}P_{k-1}H_{k-1}^\top + R_k)^{-1} \tag{5.19}
$$

with P_{k-1} given by the formula

$$
P_k = \left(\begin{bmatrix} A_{k-1} & \Gamma_{k-1} \\ 0 & M_{k-1} \end{bmatrix} - G_k H_{k-1} \right) P_{k-1} \begin{bmatrix} A_{k-1}^\top & 0 \\ \Gamma_{k-1}^\top & M_{k-1}^\top \end{bmatrix}
$$
$$
+ \begin{bmatrix} 0 & 0 \\ 0 & Q_k \end{bmatrix}, \tag{5.20}
$$

$k = 1, 2, \cdots$. The initial conditions are given by (5.16) and (5.17a or 5.17b).

We remark that if the colored noise processes become white (that is, both $M_k = 0$ and $N_k = 0$ for all k), then this Kalman filtering algorithm reduces to the one derived in Chapters 2 and 3, by simply setting

$$
\hat{\mathbf{x}}_{k|k-1} = A_{k-1}\hat{\mathbf{x}}_{k-1|k-1},
$$

so that the recursive relation for $\hat{\mathbf{x}}_k$ is decoupled into two equations: the prediction and correction equations. In addition, by defining $P_k = P_{k|k}$ and

$$
P_{k,k-1} = A_{k-1}P_{k-1}A_{k-1}^\top + \Gamma_k Q_k \Gamma_k^\top,
$$

it can be shown that (5.20) reduces to the algorithm for computing the matrices $P_{k,k-1}$ and $P_{k,k}$. We leave the verification as an exercise to the reader (cf. Exercise 5.6).

5.4 White System Noise

If only the measurement noise is colored but the system noise is white, i.e. $M_k = 0$ and $N_k \neq 0$, then it is not necessary to obtain the extra estimate $\hat{\xi}_k$ in order to derive the Kalman filtering equations. In this case, the filtering algorithm is given as follows (cf. Exercise 5.7):

$$
\left\{
\begin{aligned}
&P_0 = \left[[Var(\mathbf{x}_0)]^{-1} + C_0^\top R_0^{-1} C_0 \right]^{-1} \\
&H_{k-1} = [C_k A_{k-1} - N_{k-1} C_{k-1}] \\
&G_k = (A_{k-1} P_{k-1} H_{k-1}^\top + \Gamma_{k-1} Q_{k-1} \Gamma_{k-1}^\top C_k^\top)(H_{k-1} P_{k-1} H_{k-1}^\top \\
&\qquad + C_k \Gamma_{k-1} Q_{k-1} \Gamma_{k-1}^\top C_k^\top + R_{k-1})^{-1} \\
&P_k = (A_{k-1} - G_k H_{k-1}) P_{k-1} A_{k-1}^\top + (I - G_k C_k) \Gamma_{k-1} Q_{k-1} \Gamma_{k-1}^\top \\
&\hat{\mathbf{x}}_0 = E(\mathbf{x}_0) - [Var(\mathbf{x}_0)] C_0^\top [C_0 Var(\mathbf{x}_0) C_0^\top + R_0]^{-1} [C_0 E(\mathbf{x}_0) - \mathbf{v}_0] \\
&\hat{\mathbf{x}}_k = A_{k-1} \hat{\mathbf{x}}_{k-1} + G_k (\mathbf{v}_k - N_{k-1} \mathbf{v}_{k-1} - H_{k-1} \hat{\mathbf{x}}_{k-1}) \\
&k = 1, 2, \cdots,
\end{aligned}
\right.
$$

$$(5.21)$$

(cf. Fig.5.1).

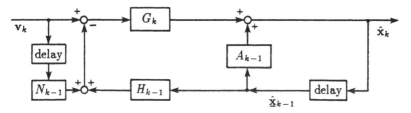

Fig. 5.1.

5.5 Real-Time Applications

Now, let us consider a tracking system (cf. Exercise 3.8 and (3.26)) with colored input, namely: the state-space description

$$
\left\{
\begin{aligned}
\mathbf{x}_{k+1} &= A\mathbf{x}_k + \underline{\xi}_k \\
v_k &= C\mathbf{x}_k + \eta_k,
\end{aligned}
\right.
$$

$$(5.22)$$

where

$$
A = \begin{bmatrix} 1 & h & h^2/2 \\ 0 & 1 & h \\ 0 & 0 & 1 \end{bmatrix}, \qquad C = [\,1 \ \ 0 \ \ 0\,],
$$

with sampling time $h > 0$, and

$$\begin{cases} \underline{\xi}_k = F\underline{\xi}_{k-1} + \underline{\beta}_k \\ \eta_k = g\eta_{k-1} + \gamma_k , \end{cases} \tag{5.23}$$

where $\{\underline{\beta}_k\}$ and $\{\gamma_k\}$ are both zero-mean Gaussian white noise sequences satisfying the following assumptions:

$$E(\underline{\beta}_k\underline{\beta}_\ell^\top) = Q_k\delta_{k\ell}, \quad E(\eta_k\eta_\ell) = r_k\delta_{k\ell}, \quad E(\underline{\beta}_k\gamma_\ell) = 0,$$

$$E(\mathbf{x}_0\underline{\xi}_0^\top) = 0, \quad E(\mathbf{x}_0\eta_0) = 0, \quad E(\mathbf{x}_0\underline{\beta}_k^\top) = 0,$$

$$E(\mathbf{x}_0\gamma_k) = 0, \quad E(\underline{\xi}_0\underline{\beta}_k^\top) = 0, \quad E(\underline{\xi}_0\gamma_k) = 0,$$

$$E(\eta_0\underline{\beta}_k) = 0, \quad \underline{\xi}_{-1} = 0, \quad \eta_{-1} = 0.$$

Set

$$\mathbf{z}_k = \begin{bmatrix} \mathbf{x}_k \\ \underline{\xi}_k \end{bmatrix}, \quad \tilde{A}_k = \begin{bmatrix} A_k & I \\ 0 & F \end{bmatrix}, \quad \tilde{\underline{\beta}}_k = \begin{bmatrix} 0 \\ \underline{\beta}_k \end{bmatrix}, \quad \tilde{C}_k = [\, C_k \quad 0\,].$$

Then we have

$$\begin{cases} \mathbf{z}_{k+1} = \tilde{A}_k\mathbf{z}_k + \tilde{\underline{\beta}}_{k+1} \\ v_k = \tilde{C}_k\mathbf{z}_k + \eta_k \end{cases}$$

as described in (5.2) and (5.3). The associated Kalman filtering algorithm is then given by formulas (5.18-20) as follows:

$$\begin{cases} \begin{bmatrix} \hat{\mathbf{x}}_k \\ \hat{\underline{\xi}}_k \end{bmatrix} = \begin{bmatrix} A & I \\ 0 & F \end{bmatrix}\begin{bmatrix} \hat{\mathbf{x}}_{k-1} \\ \hat{\underline{\xi}}_{k-1} \end{bmatrix} + G_k\left(v_k - gv_{k-1} - \mathbf{h}_{k-1}^\top\begin{bmatrix} \hat{\mathbf{x}}_{k-1} \\ \hat{\underline{\xi}}_{k-1} \end{bmatrix}\right) \\ \begin{bmatrix} \hat{\mathbf{x}}_0 \\ \hat{\underline{\xi}}_0 \end{bmatrix} = \begin{bmatrix} E(\mathbf{x}_0) \\ 0 \end{bmatrix}, \end{cases}$$

where

$$\mathbf{h}_{k-1} = [\, CA - gC \quad C\,]^\top = [\, (1-g) \quad h \quad h^2/2 \quad 1 \quad 0 \quad 0\,]^\top,$$

and

$$G_k = \begin{bmatrix} A & I \\ 0 & F \end{bmatrix}P_{k-1}\mathbf{h}_{k-1}(\mathbf{h}_{k-1}^\top P_{k-1}\mathbf{h}_{k-1} + r_k)^{-1}$$

with P_{k-1} given by the formula

$$\begin{cases} P_k = \left(\begin{bmatrix} A & I \\ 0 & F \end{bmatrix} - G_k\mathbf{h}_{k-1}^\top\right)P_{k-1}\begin{bmatrix} A^\top & 0 \\ I & F^2 \end{bmatrix} + \begin{bmatrix} 0 & 0 \\ 0 & Q_k \end{bmatrix} \\ P_0 = \begin{bmatrix} Var(\mathbf{x}_0) & 0 \\ 0 & Var(\underline{\xi}_0) \end{bmatrix}. \end{cases}$$

Exercises

5.1. Let $\{\underline{\beta}_k\}$ be a sequence of zero-mean Gaussian white noise and $\{\mathbf{v}_k\}$ a sequence of observation data as in the system (5.1). Set

$$\tilde{\underline{\beta}}_k = \begin{bmatrix} 0 \\ \underline{\beta}_k \end{bmatrix}, \qquad \mathbf{v}^k = \begin{bmatrix} \mathbf{v}_0 \\ \vdots \\ \mathbf{v}_k \end{bmatrix},$$

and define $L(\mathbf{x}, \mathbf{v})$ as in (4.6). Show that

$$L(\tilde{\underline{\beta}}_k, \mathbf{v}^k) = 0.$$

5.2. Let $\{\underline{\gamma}_k\}$ be a sequence of zero-mean Gaussian white noise and $\mathbf{v}^k$ and $L(\mathbf{x}, \mathbf{v})$ be defined as above. Show that

$$L(\mathbf{v}_{k-1}, \mathbf{v}^{k-1}) = \mathbf{v}_{k-1}$$

and

$$L(\underline{\gamma}_k, \mathbf{v}^{k-1}) = 0.$$

5.3. Let $\{\underline{\gamma}_k\}$ be a sequence of zero-mean Gaussian white noise and $\mathbf{v}^k$ and $L(\mathbf{x}, \mathbf{v})$ be defined as in Exercise 5.1. Furthermore, set

$$\hat{\mathbf{z}}_{k-1} = L(\mathbf{z}_{k-1}, \mathbf{v}^{k-1}) \qquad and \qquad \mathbf{z}_{k-1} = \begin{bmatrix} \mathbf{x}_{k-1} \\ \underline{\xi}_{k-1} \end{bmatrix}.$$

Show that

$$\langle \mathbf{z}_{k-1} - \hat{\mathbf{z}}_{k-1}, \underline{\gamma}_k \rangle = 0.$$

5.4. Let $\{\underline{\beta}_k\}$ be a sequence of zero-mean Gaussian white noise and set

$$\tilde{\underline{\beta}}_k = \begin{bmatrix} 0 \\ \underline{\beta}_k \end{bmatrix}.$$

Furthermore, define $\hat{\mathbf{z}}_{k-1}$ as in Exercise 5.3. Show that

$$\langle \mathbf{z}_{k-1} - \hat{\mathbf{z}}_{k-1}, \tilde{\underline{\beta}}_k \rangle = 0.$$

5.5. Let $L(\mathbf{x}, \mathbf{v})$ be defined as in (4.6) and set $\hat{\mathbf{z}}_0 = L(\mathbf{z}_0, \mathbf{v}_0)$ with

$$\mathbf{z}_0 = \begin{bmatrix} \mathbf{x}_0 \\ \underline{\xi}_0 \end{bmatrix}.$$

Show that

$$Var(\mathbf{z}_0 - \hat{\mathbf{z}}_0)$$
$$= \begin{bmatrix} Var(\mathbf{x}_0) & 0 \\ -[Var(\mathbf{x}_0)]C_0^\top[C_0Var(\mathbf{x}_0)C_0^\top + R_0]^{-1}C_0[Var(\mathbf{x}_0)] & \\ 0 & Q_0 \end{bmatrix}.$$

5.6. Verify that if the matrices M_k and N_k defined in (5.1) are identically zero for all k, then the Kalman filtering algorithm given by (5.18-20) reduces to the one derived in Chapters 2 and 3 for the linear stochastic system with uncorrelated system and measurement white noise processes.

5.7. Simplify the Kalman filtering algorithm for the system (5.1) where $M_k = 0$ but $N_k \neq 0$.

5.8. Consider the tracking system (5.22) with colored input (5.23) .

(a) Reformulate this system with colored input as a new augmented system with Gaussian white input by setting

$$\underline{X}_k = \begin{bmatrix} \mathbf{x}_k \\ \underline{\xi}_k \\ \eta_k \end{bmatrix}, \quad \underline{\zeta}_k = \begin{bmatrix} 0 \\ \underline{\beta}_{k+1} \\ \gamma_{k+1} \end{bmatrix},$$

$$A_c = \begin{bmatrix} A & I & 0 \\ 0 & F & 0 \\ 0 & 0 & g \end{bmatrix} \quad and \quad C_c = [\, C \quad 0 \quad 0 \quad 1 \,].$$

(b) By formally applying formulas (3.25) to this augmented system, give the Kalman filtering algorithm to the tracking system (5.22) with colored input (5.23).

(c) What are the major disadvantages of this approach ?

6. Limiting Kalman Filter

In this chapter, we consider the special case where all known constant matrices are independent of time. That is, we are going to study the time-invariant linear stochastic system with the state-space description:

$$\begin{cases} \mathbf{x}_{k+1} = A\mathbf{x}_k + \Gamma\underline{\xi}_k \\ \mathbf{v}_k = C\mathbf{x}_k + \underline{\eta}_k \,. \end{cases} \tag{6.1}$$

Here, A, Γ, and C are known $n \times n, n \times p$, and $q \times n$ constant matrices, respectively, with $1 \le p, q \le n$, $\{\underline{\xi}_k\}$ and $\{\underline{\eta}_k\}$ are zero-mean Gaussian white noise sequences with

$$E(\underline{\xi}_k \underline{\xi}_\ell^\top) = Q\delta_{k\ell}, \quad E(\underline{\eta}_k \underline{\eta}_\ell^\top) = R\delta_{k\ell}, \quad \text{and} \quad E(\underline{\xi}_k \underline{\eta}_\ell^\top) = 0,$$

where Q and R are known $p \times p$ and $q \times q$ non-negative and positive definite symmetric matrices, respectively, independent of k.

The Kalman filtering algorithm for this special case can be described as follows (cf. Fig. 6.1):

$$\begin{cases} \hat{\mathbf{x}}_{k|k} = \hat{\mathbf{x}}_{k|k-1} + G_k(\mathbf{v}_k - C\hat{\mathbf{x}}_{k|k-1}) \\ \hat{\mathbf{x}}_{k|k-1} = A\hat{\mathbf{x}}_{k-1|k-1} \\ \hat{\mathbf{x}}_{0|0} = E(\mathbf{x}_0) \end{cases} \tag{6.2}$$

with

$$\begin{cases} P_{0,0} = Var(\mathbf{x}_0) \\ P_{k,k-1} = AP_{k-1,k-1}A^\top + \Gamma Q\Gamma^\top \\ G_k = P_{k,k-1}C^\top(CP_{k,k-1}C^\top + R)^{-1} \\ P_{k,k} = (I - G_kC)P_{k,k-1} \,. \end{cases} \tag{6.3}$$

Note that even for this simple model, it is necessary to invert a matrix at every instant to obtain the Kalman gain matrix G_k in (6.3) before the prediction-correction filtering (6.2) can be carried out. In real-time applications, it is sometimes necessary to replace G_k in (6.2) by a constant gain matrix in order to save computation time.

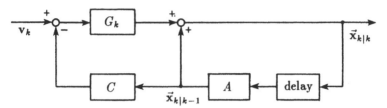

Fig. 6.1.

The *limiting (or steady-state) Kalman filter* will be defined by replacing G_k with its "limit" G as $k \to \infty$, where G is called the *limiting Kalman gain matrix*, so that the prediction-correction equations in (6.2) become

$$\begin{cases} \vec{\mathbf{x}}_{k|k} = \vec{\mathbf{x}}_{k|k-1} + G(\mathbf{v}_k - C\vec{\mathbf{x}}_{k|k-1}) \\ \vec{\mathbf{x}}_{k|k-1} = A\vec{\mathbf{x}}_{k-1|k-1} \\ \vec{\mathbf{x}}_{0|0} = E(\mathbf{x}_0). \end{cases} \qquad (6.4)$$

Under very mild conditions on the linear system (6.1), we will see that the sequence $\{G_k\}$ does converge and, in fact, $tr\|\vec{\mathbf{x}}_{k|k} - \hat{\mathbf{x}}_{k|k}\|_n^2$ tends to zero exponentially fast. Hence, replacing G_k by G does not change the actual optimal estimates by too much.

6.1 Outline of Procedure

In view of the definition of G_k in (6.3), in order to study the convergence of G_k, it is sufficient to study the convergence of

$$P_k := P_{k,k-1}.$$

We will first establish a recursive relation for P_k. Since

$$\begin{aligned} P_k = P_{k,k-1} &= AP_{k-1,k-1}A^\top + \Gamma Q\Gamma^\top \\ &= A(I - G_{k-1}C)P_{k-1,k-2}A^\top + \Gamma Q\Gamma^\top \\ &= A(I - P_{k-1,k-2}C^\top(CP_{k-1,k-2}C^\top + R)^{-1}C)P_{k-1,k-2}A^\top + \Gamma Q\Gamma^\top \\ &= A(P_{k-1} - P_{k-1}C^\top(CP_{k-1}C^\top + R)^{-1}CP_{k-1})A^\top + \Gamma Q\Gamma^\top, \end{aligned}$$

it follows that by setting

$$\Psi(T) = A(T - TC^\top(CTC^\top + R)^{-1}CT)A^\top + \Gamma Q\Gamma^\top,$$

P_k indeed satisfies the recurrence relation

$$P_k = \Psi(P_{k-1}) \,. \tag{6.5}$$

This relation is called a *matrix Riccati equation*. If $P_k \to P$ as $k \to \infty$, then P would satisfy the matrix Riccati equation

$$P = \Psi(P) \,. \tag{6.6}$$

Consequently, we can solve (6.6) for P and define

$$G = PC^\top (CPC^\top + R)^{-1} \,,$$

so that $G_k \to G$ as $k \to \infty$. Note that since P_k is symmetric, so is $\Psi(P_k)$.

Our procedure in showing that $\{P_k\}$ actually converges is as follows:

(i) $P_k \leq W$ for all k and some constant symmetric matrix W (that is, $W - P_k$ is non-negative definite symmetric for all k);

(ii) $P_k \leq P_{k+1}$,   for $k = 0, 1, \cdots$;

and

(iii) $lim_{k\to\infty} P_k = P$ for some constant symmetric matrix P.

6.2 Preliminary Results

To obtain a unique P, we must show that it is independent of the initial condition $P_0 := P_{0,-1}$, as long as $P_0 \geq 0$.

Lemma 6.1. *Suppose that the linear system (6.1) is observable (that is, the matrix*

$$N_{CA} = \begin{bmatrix} C \\ CA \\ \vdots \\ CA^{n-1} \end{bmatrix}$$

has full rank). Then there exists a non-negative definite symmetric constant matrix W independent of P_0 such that

$$P_k \leq W$$

for all $k \geq n + 1$.

Since $\langle \mathbf{x}_k - \hat{\mathbf{x}}_k, \underline{\xi}_k \rangle = 0$ (cf. (3.20)), we first observe that

$$
\begin{aligned}
P_k := P_{k,k-1} &= \|\mathbf{x}_k - \hat{\mathbf{x}}_{k|k-1}\|_n^2 \\
&= \|A\mathbf{x}_{k-1} + \Gamma\underline{\xi}_{k-1} - A\hat{\mathbf{x}}_{k-1|k-1}\|_n^2 \\
&= A\|\mathbf{x}_{k-1} - \hat{\mathbf{x}}_{k-1|k-1}\|_n^2 A^\top + \Gamma Q \Gamma^\top .
\end{aligned}
$$

Also, since $\hat{\mathbf{x}}_{k-1|k-1}$ is the minimum variance estimate of $\mathbf{x}_{k-1}$, we have

$$
\|\mathbf{x}_{k-1} - \hat{\mathbf{x}}_{k-1|k-1}\|_n^2 \leq \|\mathbf{x}_{k-1} - \tilde{\mathbf{x}}_{k-1}\|_n^2
$$

for any other linear unbiased estimate $\tilde{\mathbf{x}}_{k-1}$ of $\mathbf{x}_{k-1}$. From the assumption that N_{CA} is of full rank, it follows that $N_{CA}^\top N_{CA}$ is nonsingular. In addition,

$$
N_{CA}^\top N_{CA} = \sum_{i=0}^{n-1} (A^\top)^i C^\top C A^i ,
$$

and this motivates the following choice of $\tilde{\mathbf{x}}_{k-1}$:

$$
\tilde{\mathbf{x}}_{k-1} = A^n [N_{CA}^\top N_{CA}]^{-1} \sum_{i=0}^{n-1} (A^\top)^i C^\top \mathbf{v}_{k-n-1+i}, \quad k \geq n+1. \tag{6.7}
$$

Clearly, $\tilde{\mathbf{x}}_{k-1}$ is linear in the data, and it can be shown that it is also unbiased (cf. Exercise 6.1). Furthermore,

$$
\begin{aligned}
\tilde{\mathbf{x}}_{k-1} \\
= A^n [N_{CA}^\top N_{CA}]^{-1} \sum_{i=0}^{n-1} (A^\top)^i C^\top (C\mathbf{x}_{k-n-1+i} + \underline{\eta}_{k-n-1+i}) \\
= A^n [N_{CA}^\top N_{CA}]^{-1} \sum_{i=0}^{n-1} (A^\top)^i C^\top \Big(C A^i \mathbf{x}_{k-n-1} \\
+ \sum_{j=0}^{i-1} C A^j \Gamma \underline{\xi}_{i-1-j} + \underline{\eta}_{k-n-1+i} \Big) \\
= A^n \mathbf{x}_{k-n-1} + A^n [N_{CA}^\top N_{CA}]^{-1} \\
\cdot \sum_{i=0}^{n-1} (A^\top)^i C^\top \Big(\sum_{j=0}^{i-1} C A^j \Gamma \underline{\xi}_{i-1-j} + \underline{\eta}_{k-n-1+i} \Big) .
\end{aligned}
$$

Since

$$
\mathbf{x}_{k-1} = A^n \mathbf{x}_{k-n-1} + \sum_{i=0}^{n-1} A^i \Gamma \underline{\xi}_{n-1+i} ,
$$

we have

$$\mathbf{x}_{k-1} - \tilde{\mathbf{x}}_{k-1} = \sum_{i=0}^{n-1} A^i \Gamma \underline{\xi}_{n-1+i}$$

$$- A^n [N_{CA}^\top N_{CA}]^{-1} \sum_{i=0}^{n-1} (A^\top)^i C^\top \left(\sum_{j=0}^{i-1} C A^j \underline{\xi}_{i-1-j} + \underline{\eta}_{k-n-1+i} \right).$$

Observe that $E(\underline{\xi}_m \underline{\eta}_\ell^\top) = 0$ and $E(\underline{\eta}_\ell \underline{\eta}_\ell^\top) = R$ for all m and ℓ, so that $\|\mathbf{x}_{k-1} - \tilde{\mathbf{x}}_{k-1}\|_n^2$ is independent of k for all $k \geq n+1$. Hence,

$$P_k = A\|\mathbf{x}_{k-1} - \hat{\mathbf{x}}_{k-1|k-1}\|_n^2 A^\top + \Gamma Q \Gamma^\top$$
$$\leq A\|\mathbf{x}_{k-1} - \tilde{\mathbf{x}}_{k-1|k-1}\|_n^2 A^\top + \Gamma Q \Gamma^\top$$
$$= A\|\mathbf{x}_n - \tilde{\mathbf{x}}_{n|n}\|_n^2 A^\top + \Gamma Q \Gamma^\top$$

for all $k \geq n+1$. Pick

$$W = A\|\mathbf{x}_n - \tilde{\mathbf{x}}_{n|n}\|_n^2 A^\top + \Gamma Q \Gamma^\top.$$

Then $P_k \leq W$ for all $k \geq n+1$. Note that W is independent of the initial condition $P_0 = P_{0,-1} = \|\mathbf{x}_0 - \hat{\mathbf{x}}_{0,-1}\|_n^2$. This completes the proof of the Lemma.

Lemma 6.2. *If P and Q are both non-negative definite and symmetric with $P \geq Q$, then $\Psi(P) \geq \Psi(Q)$.*

To prove this lemma, we will use the formula

$$\frac{d}{ds} A^{-1}(s) = -A^{-1}(s) \left[\frac{d}{ds} A(s) \right] A^{-1}(s)$$

(cf. Exercise 6.2). Denoting $T(s) = Q + s(P - Q)$, we have

$$\Psi(P) - \Psi(Q)$$
$$= \int_0^1 \frac{d}{ds} \Psi(Q + s(P - Q)) ds$$
$$= A \left\{ \int_0^1 \frac{d}{ds} \left\{ (Q + s(P - Q)) - (Q + s(P - Q)) C^\top \right. \right.$$
$$\left. \left. \cdot [C(Q + s(P - Q)) C^\top + R]^{-1} C(Q + s(P - Q)) \right\} ds \right\} A^\top$$
$$= A \left\{ \int_0^1 [P - Q - (P - Q) C^\top (CT(s) C^\top + R)^{-1} CT(s) \right.$$

$$- T(s)C^\top(CT(s)C^\top + R)^{-1}C(P - Q) + T(s)C^\top(CT(s)C^\top + R)^{-1}$$
$$\cdot\, C(P - Q)C^\top(CT(s)C^\top + R)^{-1}CT(s)]ds\Big\}A^\top$$

$$= A\Big\{\int_0^1 [T(s)C^\top(CT(s)C^\top + R)^{-1}C](P - Q)$$
$$\cdot\, [T(s)C^\top(CT(s)C^\top + R)^{-1}C]^\top ds\Big\}A$$

$$\geq 0.$$

Hence, $\Psi(P) \geq \Psi(Q)$ as required.

We also have the following:

Lemma 6.3. *Suppose that the linear system (6.1) is observable. Then with the initial condition $P_0 = P_{0,-1} = 0$, the sequence $\{P_k\}$ converges componentwise to some symmetric matrix $P \geq 0$ as $k \to \infty$.*

Since

$$P_1 := \|\mathbf{x}_1 - \hat{\mathbf{x}}_{1|0}\|_n^2 \geq 0 = P_0$$

and both P_0 and P_1 are symmetric, Lemma 6.2 yields

$$P_2 = \Psi(P_1) \geq \Psi(P_0) = P_1$$
$$\cdots\cdots$$
$$P_{k+1} \geq P_k, \quad k = 0, 1, \cdots.$$

Hence, $\{P_k\}$ is monotonic nondecreasing and bounded above by W (cf. Lemma 6.1). For any n-vector $\mathbf{y}$, we have

$$0 \leq \mathbf{y}^\top P_k \mathbf{y} \leq \mathbf{y}^\top W \mathbf{y},$$

so that the sequence $\{\mathbf{y}^\top P_k \mathbf{y}\}$ is a bounded non-negative monotonic nondecreasing sequence of real numbers and must converge to some non-negative real number. If we choose

$$\mathbf{y} = [0 \cdots 0\ 1\ 0 \cdots 0]^\top$$

with 1 being placed at the ith component, then setting $P_k = [p_{ij}^{(k)}]$, we have

$$\mathbf{y}^\top P_k \mathbf{y} = p_{ii}^{(k)} \to p_{ii} \quad \text{as } k \to \infty$$

for some non-negative number p_{ii}. Next, if we choose

$$\mathbf{y} = [0 \cdots 0\ 1\ 0 \cdots 0\ 1\ 0 \cdots 0]^\top$$

with the two 1's being placed at the ith and jth components, then we have

$$\mathbf{y}^\top P_k \mathbf{y} = p_{ii}^{(k)} + p_{ij}^{(k)} + p_{ji}^{(k)} + p_{jj}^{(k)}$$
$$= p_{ii}^{(k)} + 2p_{ij}^{(k)} + p_{jj}^{(k)} \to q \quad \text{as} \ \ k \to \infty$$

for some non-negative number q. Since $p_{ii}^{(k)} \to p_{ii}$, we have

$$p_{ij}^{(k)} \to \frac{1}{2}(q - p_{ii} - p_{jj}) \quad \text{as} \ \ k \to \infty.$$

That is, $P_k \to P$. Since $P_k \geq 0$ and is symmetric, so is P. This completes the proof of the lemma.

We now define

$$G = \lim_{k \to \infty} G_k,$$

where $G_k = P_k C^\top (C P_k C^\top + R)^{-1}$. Then

$$G = PC^\top (CPC^\top + R)^{-1}. \tag{6.8}$$

Next, we will show that for any non-negative definite symmetric matrix P_0 as an initial choice, $\{P_k\}$ still converges to the same P. Hence, from now on, we will use an arbitrary non-negative definite symmetric matrix P_0, and recall that $P_k = \Psi(P_{k-1})$, $k = 1, 2, \cdots$, and $P = \Psi(P)$. We first need the following.

Lemma 6.4. *Let the linear system (6.1) be observable so that P can be defined using Lemma 6.3. Then the following relation*

$$P - P_k = A(I - GC)(P - P_{k-1})(I - G_{k-1}C)^\top A^\top \tag{6.9}$$

holds for all $k = 1, 2, \cdots$, and any non-negative definite symmetric initial condition P_0.

Since $G_{k-1} = P_{k-1}C^\top (C P_{k-1} C^\top + R)^{-1}$ and $P_{k-1}^\top = P_{k-1}$, the matrix $G_{k-1}C P_{k-1}$ is non-negative definite and symmetric, so that $G_{k-1}C P_{k-1} = P_{k-1}C^\top G_{k-1}^\top$. Hence, using (6.5) and (6.6), we have

$$P - P_k$$
$$= \Psi(P) - \Psi(P_{k-1})$$
$$= (APA^\top - AGCPA^\top) - (AP_{k-1}A^\top - AG_{k-1}CP_{k-1}A^\top)$$
$$= APA^\top - AGCPA^\top - AP_{k-1}A^\top + AP_{k-1}C^\top G_{k-1}^\top A^\top. \tag{6.10}$$

Now,

$$(I - GC)(P - P_{k-1})(I - G_{k-1}C)^\top$$
$$= P - P_{k-1} + P_{k-1}C^\top G_{k-1}^\top - GCP + R_e, \qquad (6.11)$$

where

$$R_e = GCP_{k-1} - PC^\top G_{k-1}^\top + GCPC^\top G_{k-1}^\top - GCP_{k-1}C^\top G_{k-1}^\top. \quad (6.12)$$

Hence, if we can show that $R_e = 0$, then (6.9) follows from (6.10) and (6.11). From the definition of G_{k-1}, we have $G_{k-1}(CP_{k-1}C^\top + R) = P_{k-1}C^\top$ or $(CP_{k-1}C^\top + R)G_{k-1} = CP_{k-1}$, so that

$$G_{k-1}CP_{k-1}C^\top = P_{k-1}C^\top - G_{k-1}R \qquad (6.13)$$

or

$$CP_{k-1}C^\top G_{k-1}^\top = CP_{k-1} - RG_{k-1}^\top. \qquad (6.14)$$

Taking $k \to \infty$ in (6.13) with initial condition $P_0 := P_{0,-1} = 0$, we have

$$GCPC^\top = PC^\top - GR, \qquad (6.15)$$

and putting (6.14) and (6.15) into (6.12), we can indeed conclude that $R_e = 0$. This completes the proof of the lemma.

Lemma 6.5.

$$P_k = [A(I - G_{k-1}C)]P_{k-1}[A(I - G_{k-1}C)]^\top$$
$$+ [AG_{k-1}]R[AG_{k-1}]^\top + \Gamma Q \Gamma^\top \qquad (6.16)$$

and consequently, for an observable system with $P_0 := P_{0,-1} = 0$,

$$P = [A(I - GC)]P[A(I - GC)]^\top + [AG]R[AG]^\top + \Gamma Q \Gamma^\top. \qquad (6.17)$$

Since $G_{k-1}(CP_{k-1}C^\top + R) = P_{k-1}C^\top$ from the definition, we have

$$G_{k-1}R = (I - G_{k-1}C)P_{k-1}C^\top$$

and hence,

$$AG_{k-1}RG_{k-1}^\top A^\top = A(I - G_{k-1}C)P_{k-1}C^\top G_{k-1}^\top A^\top.$$

Therefore, from the matrix Riccati equation $P_k = \Psi(P_{k-1})$, we may conclude that

$$P_k = A(I - G_{k-1}C)P_{k-1}A^\top + \Gamma Q \Gamma^\top$$
$$= A(I - G_{k-1}C)P_{k-1}(I - G_{k-1}C)^\top A^\top$$
$$+ A(I - G_{k-1}C)P_{k-1}C^\top G_{k-1}^\top A^\top + \Gamma Q \Gamma^\top$$
$$= A(I - G_{k-1}C)P_{k-1}(I - G_{k-1}C)^\top A^\top$$
$$+ AG_{k-1}RG_{k-1}^\top A^\top + \Gamma Q \Gamma^\top$$

which is (6.16).

Lemma 6.6. *Let the linear system (6.1) be (completely) controllable (that is, the matrix*

$$M_{A\Gamma} = [\Gamma \quad A\Gamma \quad \cdots \quad A^{n-1}\Gamma]$$

has full rank). Then for any non-negative definite symmetric initial matrix P_0, we have $P_k > 0$ for $k \geq n+1$. Consequently, $P > 0$.

Using (6.16) k times, we first have

$$
\begin{aligned}
P_k = & \Gamma Q \Gamma^\top + [A(I - G_{k-1}C)]\Gamma Q \Gamma^\top [A(I - G_{k-1}C)]^\top + \cdots + \{[A(I- \\
& G_{k-1}C)] \cdots [A(I - G_2 C)]\}\Gamma Q \Gamma^\top \{[A(I - G_{k-1}C)] \cdots [A(I - G_2 C)]\}^\top \\
& + [AG_{k-1}]R[AG_{k-1}]^\top + [A(I - G_{k-1}C)][AG_{k-2}]R[AG_{k-2}]^\top \\
& \cdot [A(I - G_{k-1}C)]^\top + \cdots + \{[A(I - G_{k-1}C)] \\
& \cdots [A(I - G_2 C)][AG_1]\}R\{[A(I - G_{k-1}C)] \\
& \cdots [A(I - G_2 C)][AG_1]\}^\top + \{[A(I - G_k C)] \\
& \cdots [A(I - G_1 C)]\}P_0 \{[A(I - G_k C)] \cdots [A(I - G_1 C)]\}^\top .
\end{aligned}
$$

To prove that $P_k > 0$ for $k \geq n+1$, it is sufficient to show that $\mathbf{y}^\top P_k \mathbf{y} = 0$ implies $\mathbf{y} = 0$. Let $\mathbf{y}$ be any n-vector such that $\mathbf{y}^\top P_k \mathbf{y} = 0$. Then, since Q, R and P_0 are non-negative definite, each term on the right hand side of the above identity must be zero. Hence, we have

$$\mathbf{y}^\top \Gamma Q \Gamma^\top \mathbf{y} = 0, \tag{6.18}$$

$$\mathbf{y}^\top [A(I - G_{k-1}C)]\Gamma Q \Gamma^\top [A(I - G_{k-1}C)]^\top \mathbf{y} = 0, \tag{6.19}$$

$$\cdots$$

$$
\begin{aligned}
& \mathbf{y}^\top \{[A(I - G_{k-1}C)] \cdots [A(I - G_2 C)]\}\Gamma Q \Gamma^\top \\
& \cdot \{[A(I - G_{k-1}C)] \cdots [A(I - G_2 C)]\}^\top \mathbf{y} = 0
\end{aligned}
\tag{6.20}
$$

and

$$\mathbf{y}^\top [AG_{k-1}]R[AG_{k-1}]^\top \mathbf{y} = 0, \tag{6.21}$$

$$\cdots$$

$$
\begin{aligned}
& \mathbf{y}^\top \{[A(I - G_{k-1}C)] \cdots [A(I - G_3 C)][AG_2]\}R \\
& \cdot \{[A(I - G_{k-1}C)] \cdots [A(I - G_3 C)][AG_2]\}^\top \mathbf{y} = 0.
\end{aligned}
\tag{6.22}
$$

Since $R > 0$, from (6.21) and (6.22), we have

$$\mathbf{y}^\top AG_{k-1} = 0, \tag{6.23}$$

$$\cdots$$

$$\mathbf{y}^\top [A(I - G_{k-1}C)] \cdots [A(I - G_2C)][AG_1] = 0. \qquad (6.24)$$

Now, it follows from $Q > 0$ and (6.18) that

$$\mathbf{y}^\top \Gamma = 0,$$

and then using (6.19) and (6.23) we obtain

$$\mathbf{y}^\top A\Gamma = 0,$$

and so on. Finally, we have

$$\mathbf{y}^\top A^j \Gamma = 0, \quad j = 0, 1, \cdots, n-1,$$

as long as $k \geq n + 1$. That is,

$$\mathbf{y}^\top M_{A\Gamma} \mathbf{y} = \mathbf{y}^\top [\Gamma \ A\Gamma \ \cdots \ A^{n-1}\Gamma] \mathbf{y} = 0.$$

Since the system is (completely) controllable, $M_{A\Gamma}$ is of full rank, and we must have $\mathbf{y} = 0$. Hence, $P_k > 0$ for all $k \geq n + 1$. This completes the proof of the lemma.

Now, using (6.9) repeatedly, we have

$$P - P_k = [A(I - GC)]^{k-n-1}(P - P_{n+1})B_k^\top, \qquad (6.25)$$

where

$$B_k = [A(I - G_{k-1}C)] \cdots [A(I - G_{n+1}C)], \quad k = n+2, n+3, \cdots,$$

with $B_{n+1} := I$. In order to show that $P_k \to P$ as $k \to \infty$, it is sufficient to show that $[A(I - GC)]^{k-n-1} \to 0$ as $k \to \infty$ and B_k is "bounded." In this respect, we have the following two lemmas.

Lemma 6.7. *Let the linear system (6.1) be observable. Then*

$$B_k B_k^\top \leq M, \quad k \geq n + 1,$$

for some constant matrix M. Consequently, if $B_k = [b_{ij}^{(k)}]$ then

$$\left| b_{ij}^{(k)} \right| \leq m,$$

for some constant m and for all i, j and k.

By Lemma 6.1, $P_k \leq W$ for $k \geq n + 1$. Hence, using Lemma 6.5 repeatedly, we have

$$
\begin{aligned}
W \geq P_k &\geq [A(I - G_{k-1}C)]P_{k-1}[A(I - G_{k-1}C)]^\top \\
&\geq [A(I - G_{k-1}C)][A(I - G_{k-2}C)]P_{k-2} \\
&\quad \cdot [A(I - G_{k-2}C)]^\top [A(I - G_{k-1}C)]^\top \\
&\quad \cdots \\
&\geq B_k P_{n+1} B_k^\top .
\end{aligned}
$$

Since P_{n+1} is real, symmetric, and positive definite, by Lemma 6.6, all its eigenvalues are real, and in fact, positive. Let λ_{min} be the smallest eigenvalue of P_{n+1} and note that $P_{n+1} \geq \lambda_{min}I$ (cf. Exercise 6.3). Then we have

$$
W \geq B_k \lambda_{min} I B_k^\top = \lambda_{min} B_k B_k^\top .
$$

Setting $M = \lambda_{min}^{-1} W$ completes the proof of the lemma.

Lemma 6.8. *Let λ be an arbitrary eigenvalue of $A(I - GC)$. If the system (6.1) is both (completely) controllable and observable, then $|\lambda| < 1$.*

Observe that λ is also an eigenvalue of $(I - GC)^\top A^\top$. Let $\mathbf{y}$ be a corresponding eigenvector. Then

$$
(I - GC)^\top A^\top \mathbf{y} = \lambda \mathbf{y}. \tag{6.26}
$$

Using (6.17), we have

$$
\bar{\mathbf{y}}^\top P \mathbf{y} = \bar{\lambda} \bar{\mathbf{y}}^\top P \lambda \mathbf{y} + \bar{\mathbf{y}}^\top [AG]R[AG]^\top \mathbf{y} + \bar{\mathbf{y}}^\top \Gamma Q \Gamma^\top \mathbf{y}.
$$

Hence,

$$
(1 - |\lambda|^2)\bar{\mathbf{y}}^\top P \mathbf{y} = \bar{\mathbf{y}}^\top [(AG)R(AG)^\top + \Gamma Q \Gamma^\top]\mathbf{y}.
$$

Since the right-hand side is non-negative and $\bar{\mathbf{y}}^\top P \mathbf{y} \geq 0$, we must have $1 - |\lambda|^2 \geq 0$ or $|\lambda| \leq 1$. Suppose that $|\lambda| = 1$. Then

$$
\bar{\mathbf{y}}^\top (AG)R(AG)^\top \mathbf{y} = 0 \quad \text{or} \quad \overline{[(AG)^\top \mathbf{y}]}^\top R[(AG)^\top \mathbf{y}] = 0
$$

and

$$
\bar{\mathbf{y}}^\top \Gamma Q \Gamma^\top \mathbf{y} = 0 \quad \text{or} \quad \overline{(\Gamma^\top \mathbf{y})}^\top Q(\Gamma^\top \mathbf{y}) = 0.
$$

Since $Q > 0$ and $R > 0$, we have

$$
G^\top A^\top \mathbf{y} = 0 \tag{6.27}
$$

and

$$\Gamma^\top \mathbf{y} = 0, \tag{6.28}$$

so that (6.26) implies $A^\top \mathbf{y} = \lambda \mathbf{y}$. Hence,

$$\Gamma^\top (A^j)^\top \mathbf{y} = \lambda^j \Gamma^\top \mathbf{y} = 0, \quad j = 0, 1, \cdots, n-1.$$

This gives

$$\mathbf{y}^\top M_{A\Gamma} = \mathbf{y}^\top [\Gamma \ \ A\Gamma \ \cdots \ A^{n-1}\Gamma] = 0.$$

Taking real and imaginary parts, we have

$$[Re(\mathbf{y})]^\top M_{A\Gamma} = 0 \quad \text{and} \quad [Im(\mathbf{y})]^\top M_{A\Gamma} = 0.$$

Since $\mathbf{y} \neq 0$, at least one of $Re(\mathbf{y})$ and $Im(\mathbf{y})$ is not zero. Hence, $M_{A\Gamma}$ is row dependent, contradicting the complete controllability hypothesis. Hence $|\lambda| < 1$. This completes the proof of the lemma.

6.3 Geometric Convergence

Combining the above results, we now have the following.

Theorem 6.1. *Let the linear stochastic system (6.1) be both (completely) controllable and observable. Then, for any initial state $\mathbf{x}_0$ such that $P_0 := P_{0,-1} = Var(\mathbf{x}_0)$ is non-negative definite and symmetric, $P_k := P_{k,k-1} \to P$ as $k \to \infty$, where $P > 0$ is symmetric and is independent of $\mathbf{x}_0$. Furthermore, the order of convergence is geometric; that is,*

$$tr(P_k - P)(P_k - P)^\top \leq Cr^k, \tag{6.29}$$

where $0 < r < 1$ and $C > 0$, independent of k. Consequently,

$$tr(G_k - G)(G_k - G)^\top \leq Cr^k. \tag{6.30}$$

To prove the theorem, let $F = A(I - GC)$. Using Lemma 6.7 and (6.25), we have

$$(P_k - P)(P_k - P)^\top$$
$$= F^{k-n-1}(P_{n+1} - P)B_k B_k^\top (P_{n+1} - P)(F^{k-n-1})^\top$$
$$\leq F^{k-n-1}\Omega(F^{k-n-1})^\top$$

for some non-negative definite symmetric constant matrix Ω. From Lemma 6.8, all eigenvalues of F are of absolute value less than 1. Hence, $F^k \to 0$ so that $P_k \to P$ as $k \to \infty$ (cf. Exercise 6.4). On the other hand, by Lemma 6.6, P is positive definite symmetric and is independent of P_0.

Using Lemmas 1.7 and 1.10, we have

$$tr(P_k - P)(P_k - P)^\top \le tr F^{k-n-1}(F^{k-n-1})^\top \cdot tr\,\Omega \le Cr^k,$$

where $0 < r < 1$ and C is independent of k. To prove (6.30), we first rewrite

$$
\begin{aligned}
G_k &- G \\
&= P_k C^\top (CP_k C^\top + R)^{-1} - PC^\top (CPC^\top + R)^{-1} \\
&= (P_k - P)C^\top (CP_k C^\top + R)^{-1} \\
&\quad + PC^\top [(CP_k C^\top + R)^{-1} - (CPC^\top + R)^{-1}] \\
&= (P_k - P)C^\top (CP_k C^\top + R)^{-1} + PC^\top (CP_k C^\top + R)^{-1} \\
&\quad \cdot [(CPC^\top + R) - (CP_k C^\top + R)](CPC^\top + R)^{-1} \\
&= (P_k - P)C^\top (CP_k C^\top + R)^{-1} \\
&\quad + PC^\top (CP_k C^\top + R)^{-1} C(P - P_k)C^\top (CPC^\top + R)^{-1}.
\end{aligned}
$$

Since for any $n \times n$ matrices A and B,

$$(A + B)(A + B)^\top \le 2(AA^\top + BB^\top)$$

(cf. Exercise 6.5), we have

$$
\begin{aligned}
(G_k &- G)(G_k - G)^\top \\
&\le 2(P_k - P)C^\top (CP_k C^\top + R)^{-1}(CP_k C^\top + R)^{-1}C(P_k - P) \\
&\quad + 2PC^\top (CP_k C^\top + R)^{-1}C(P - P_k)C^\top (CPC^\top + R)^{-1} \\
&\quad \cdot (CPC^\top + R)^{-1}C(P - P_k)C^\top (CP_k C^\top + R)^{-1}CP. \qquad (6.31)
\end{aligned}
$$

And since $P_0 \le P_k$, we have $CP_0 C^\top + R \le CP_k C^\top + R$, so that by Lemma 1.3,

$$(CP_k C^\top + R)^{-1} \le (CP_0 C^\top + R)^{-1},$$

and hence, by Lemma 1.9,

$$tr(CP_k C^\top + R)^{-1}(CP_k C^\top + R)^{-1} \le (tr(CP_0 C^\top + R)^{-1})^2.$$

Finally, by Lemma 1.7, it follows from (6.31) that

$$
\begin{aligned}
tr(G_k &- G)(G_k - G)^\top \\
&\le 2tr(P_k - P)(P_k - P)^\top \cdot tr\, C^\top C(tr(CP_0 C^\top + R)^{-1})^2 \\
&\quad + 2tr PP^\top \cdot tr\, C^\top C(tr(CP_0 C^\top + R)^{-1})^2 \cdot tr\, CC^\top \\
&\quad \cdot tr(P - P_k)(P - P_k)^\top \cdot tr\, C^\top C \cdot tr(CPC^\top + R)^{-1}(CPC^\top + R)^{-1} \\
&\le C_1 tr(P_k - P)(P_k - P)^\top \\
&\le Cr^k,
\end{aligned}
$$

where C_1 and C are constants, independent of k. This completes the proof of the theorem.

The following result shows that $\bar{\mathbf{x}}_k$ is an asymptotically optimal estimate of $\mathbf{x}_k$.

Theorem 6.2. *Let the linear system (6.1) be both (completely) controllable and observable. Then*

$$\lim_{k \to \infty} \|\mathbf{x}_k - \bar{\mathbf{x}}_k\|_n^2 = (P^{-1} + C^\top R^{-1} C)^{-1} = \lim_{k \to \infty} \|\mathbf{x}_k - \hat{\mathbf{x}}_k\|_n^2 \,.$$

The second equality can be easily verified. Indeed, using Lemma 1.2 (the matrix inversion lemma), we have

$$\lim_{k \to \infty} \|\mathbf{x}_k - \hat{\mathbf{x}}_k\|_n^2$$

$$= \lim_{k \to \infty} P_{k,k}$$

$$= \lim_{k \to \infty} (I - G_k C) P_{k,k-1}$$

$$= (I - GC)P$$

$$= P - PC^\top (CPC^\top + R)^{-1} CP$$

$$= (P^{-1} + C^\top R^{-1} C)^{-1} > 0 \,.$$

Hence, to verify the first equality, it is equivalent to showing that $\|\mathbf{x}_k - \bar{\mathbf{x}}_k\|_n^2 \to (I - GC)P$ as $k \to \infty$. We first rewrite

$$\mathbf{x}_k - \bar{\mathbf{x}}_k$$
$$= (A\mathbf{x}_{k-1} + \Gamma\underline{\xi}_{k-1}) - (A\bar{\mathbf{x}}_{k-1} + G\mathbf{v}_k - GCA\bar{\mathbf{x}}_{k-1})$$
$$= (A\mathbf{x}_{k-1} + \Gamma\underline{\xi}_{k-1}) - A\bar{\mathbf{x}}_{k-1}$$
$$\quad - G(CA\mathbf{x}_{k-1} + C\Gamma\underline{\xi}_{k-1} + \underline{\eta}_k) + GCA\bar{\mathbf{x}}_{k-1}$$
$$= (I - GC)A(\mathbf{x}_{k-1} - \bar{\mathbf{x}}_{k-1}) + (I - GC)\Gamma\underline{\xi}_{k-1} - G\underline{\eta}_k \,. \qquad (6.32)$$

Since

$$\langle \mathbf{x}_{k-1} - \bar{\mathbf{x}}_{k-1}, \underline{\xi}_{k-1} \rangle = 0 \qquad (6.33)$$

and

$$\langle \mathbf{x}_{k-1} - \bar{\mathbf{x}}_{k-1}, \underline{\eta}_k \rangle = 0 \qquad (6.34)$$

(cf. Exercise 6.6), we have

$$\|\mathbf{x}_k - \bar{\mathbf{x}}_k\|_n^2$$
$$= (I - GC)A\|\mathbf{x}_{k-1} - \bar{\mathbf{x}}_{k-1}\|_n^2 A^\top (I - GC)^\top$$
$$\quad + (I - GC)\Gamma Q\Gamma^\top (I - GC)^\top + GRG^\top \,. \qquad (6.35)$$

On the other hand, it can be proved that

$$P_{k,k} = (I - G_k C)AP_{k-1,k-1}A^\top (I - G_k C)^\top$$
$$\quad + (I - G_k C)\Gamma Q\Gamma^\top (I - G_k C)^\top + G_k RG_k^\top \qquad (6.36)$$

(cf. Exercise 6.7). Since $P_{k,k} = (I - G_kC)P_{k,k-1} \to (I - GC)P$ as $k \to \infty$, taking the limit gives

$$(I - GC)P =(I - GC)A[(I - GC)P]A^\top(I - GC)^\top \\ + (I - GC)\Gamma Q\Gamma^\top(I - GC)^\top + GRG^\top. \qquad (6.37)$$

Now, subtracting of (6.37) from (6.35) yields

$$\|\mathbf{x}_k - \vec{\mathbf{x}}_k\|_n^2 - (I - GC)P \\ =(I - GC)A[\|\mathbf{x}_{k-1} - \vec{\mathbf{x}}_{k-1}\|_n^2 - (I - GC)P]A^\top(I - GC)^\top.$$

By repeating this formula $k - 1$ times, we obtain

$$\|\mathbf{x}_k - \vec{\mathbf{x}}_k\|_n^2 - (I - GC)P \\ =[(I - GC)A]^k[\|\mathbf{x}_0 - \vec{\mathbf{x}}_0\|_n^2 - (I - GC)P][A^\top(I - GC)^\top]^k.$$

Finally, by imitating the proof of Lemma 6.8, it can be shown that all eigenvalues of $(I - GC)A$ are of absolute value less than 1 (cf. Exercise 6.8). Hence, using Exercise 6.4, we have $\|\mathbf{x}_k - \vec{\mathbf{x}}_k\|_n^2 - (I - GC)P \to 0$ as $k \to \infty$. This completes the proof of the theorem.

In the following, we will show that the error $\hat{\mathbf{x}}_k - \vec{\mathbf{x}}_k$ also tends to zero exponentially fast.

Theorem 6.3. *Let the linear system (6.1) be both (completely) controllable and observable. Then there exist a real number r, $0 < r < 1$, and a positive constant C, independent of k, such that*

$$tr\|\hat{\mathbf{x}}_k - \vec{\mathbf{x}}_k\|_n^2 \le Cr^k.$$

Denote $\underline{\epsilon}_k := \hat{\mathbf{x}}_k - \vec{\mathbf{x}}_k$ and $\underline{\delta}_k := \mathbf{x}_k - \hat{\mathbf{x}}_k$. Then, from the identities

$$\hat{\mathbf{x}}_k = A\hat{\mathbf{x}}_{k-1} + G_k(\mathbf{v}_k - CA\hat{\mathbf{x}}_{k-1}) \\ = A\hat{\mathbf{x}}_{k-1} + G(\mathbf{v}_k - CA\hat{\mathbf{x}}_{k-1}) + (G_k - G)(\mathbf{v}_k - CA\hat{\mathbf{x}}_{k-1})$$

and

$$\vec{\mathbf{x}}_k = A\vec{\mathbf{x}}_{k-1} + G(\mathbf{v}_k - CA\vec{\mathbf{x}}_{k-1}),$$

we have

$$\underline{\epsilon}_k = \hat{\mathbf{x}}_k - \vec{\mathbf{x}}_k \\ = A(\hat{\mathbf{x}}_{k-1} - \vec{\mathbf{x}}_{k-1}) - GCA(\hat{\mathbf{x}}_{k-1} - \vec{\mathbf{x}}_{k-1}) \\ + (G_k - G)(CA\mathbf{x}_{k-1} + C\Gamma\underline{\xi}_{k-1} + \underline{\eta}_k - CA\hat{\mathbf{x}}_{k-1}) \\ = (I - GC)A\underline{\epsilon}_{k-1} + (G_k - G)(CA\underline{\delta}_{k-1} + C\Gamma\underline{\xi}_{k-1} + \underline{\eta}_k).$$

Since

$$\begin{cases} \langle \underline{\varepsilon}_{k-1}, \underline{\xi}_{k-1} \rangle = 0, & \langle \underline{\varepsilon}_{k-1}, \underline{\eta}_k \rangle = 0, \\ \langle \underline{\delta}_{k-1}, \underline{\xi}_{k-1} \rangle = 0, & \langle \underline{\delta}_{k-1}, \underline{\eta}_k \rangle = 0, \end{cases} \tag{6.38}$$

and $\langle \underline{\xi}_{k-1}, \underline{\eta}_k \rangle = 0$ (cf. Exercise 6.9), we obtain

$$\begin{aligned} \|\underline{\varepsilon}_k\|_n^2 &= [(I - GC)A]\|\underline{\varepsilon}_{k-1}\|_n^2[(I - GC)A]^\top \\ &\quad + (G_k - G)CA\|\underline{\delta}_{k-1}\|_n^2 A^\top C^\top (G_k - G)^\top \\ &\quad + (G_k - G)C\Gamma Q \Gamma^\top C^\top (G_k - G)^\top + (G_k - G)R(G_k - G)^\top \\ &\quad + (I - GC)A\langle \underline{\varepsilon}_{k-1}, \underline{\delta}_{k-1} \rangle A^\top C^\top (G_k - G)^\top \\ &\quad + (G_k - G)CA\langle \underline{\delta}_{k-1}, \underline{\varepsilon}_{k-1} \rangle A^\top (I - GC)^\top \\ &= F\|\underline{\varepsilon}_{k-1}\|_n^2 F^\top + (G_k - G)\Omega_{k-1}(G_k - G)^\top \\ &\quad + FB_{k-1}(G_k - G)^\top + (G_k - G)B_{k-1}^\top F^\top, \end{aligned} \tag{6.39}$$

where

$$F = (I - GC)A,$$
$$B_{k-1} = \langle \underline{\varepsilon}_{k-1}, \underline{\delta}_{k-1} \rangle A^\top C^\top,$$

and

$$\Omega_{k-1} = CA\|\underline{\delta}_{k-1}\|_n^2 A^\top C^\top + C\Gamma Q \Gamma^\top C^\top + R.$$

Hence, using (6.39) repeatedly, we obtain

$$\|\underline{\varepsilon}_k\|_n^2 = F^k\|\underline{\varepsilon}_0\|_n^2(F^k)^\top + \sum_{i=0}^{k-1} F^i(G_{k-i} - G)\Omega_{k-1-i}(G_{k-i} - G)^\top (F^i)^\top$$

$$+ \sum_{i=0}^{k-1} F^i[FB_{k-1-i}(G_{k-i} - G)^\top + (G_{k-i} - G)B_{k-1-i}^\top F^\top](F^i)^\top. \tag{6.40}$$

On the other hand, since the B_j's are componentwise uniformly bounded (cf. Exercise 6.10), it can be proved, by using Lemmas 1.6, 1.7 and 1.10 and Theorem 6.1, that

$$tr[FB_{k-1-i}(G_{k-i} - G)^\top + (G_{k-i} - G)B_{k-1-i}^\top F^\top] \leq C_1 r_1^{k-i+1} \tag{6.41}$$

for some r_1, $0 < r_1 < 1$, and some positive constant C_1 independent of k and i (cf. Exercise 6.11). Hence, we obtain, again using Lemmas 1.7 and 1.10 and Theorem 6.1,

$$tr\|\underline{\varepsilon}_k\|_n^2 \leq tr\|\underline{\varepsilon}_0\|_n^2 \cdot tr\, F^k(F^k)^\top + \sum_{i=0}^{k-1} tr\, F^i(F^i)^\top$$

$$\cdot tr(G_{k-i} - G)(G_{k-i} - G)^\top \cdot tr\, \Omega_{k-1-i}$$

$$+ \sum_{i=0}^{k-1} tr \, F^i (F^i)^\top \cdot tr[FB_{k-1-i}(G_{k-i} - G)^\top$$

$$+ (G_{k-i} - G)B_{k-1-i}^\top F^\top]$$

$$\leq tr\|\underline{\varepsilon}_0\|_n^2 C_2 r_2^k + \sum_{i=0}^{k-1} C_3 r_3^i C_4 r_4^{k-i} + \sum_{i=0}^{k-1} C_5 r_5^i C_1 r_1^{k-i+1}$$

$$\leq p(k) r_6^k , \tag{6.42}$$

where $0 < r_2, r_3, r_4, r_5 < 1$, $r_6 = max(r_1, r_2, r_3, r_4, r_5) < 1$, C_2, C_3, C_4, C_5 are positive constants independent of i and k, and $p(k)$ is a polynomial of k. Hence, there exist a real number r, $r_6 < r < 1$, and a positive constant C, independent of k and satisfying $p(k)(r_6/r)^k \leq C$, such that

$$tr\|\underline{\varepsilon}_k\|_n^2 \leq Cr^k .$$

This completes the proof of the theorem.

6.4 Real-Time Applications

Now, let us re-examine the tracking model (3.26), namely: the state-space description

$$\begin{cases} \mathbf{x}_{k+1} = \begin{bmatrix} 1 & h & h^2/2 \\ 0 & 1 & h \\ 0 & 0 & 1 \end{bmatrix} \mathbf{x}_k + \underline{\xi}_k \\ v_k = [\, 1 \ \ 0 \ \ 0 \,] \mathbf{x}_k + \eta_k , \end{cases} \tag{6.43}$$

where $h > 0$ is the sampling time, $\{\underline{\xi}_k\}$ and $\{\eta_k\}$ are both zero-mean Gaussian white noise sequences satisfying the assumption that

$$E(\underline{\xi}_k \underline{\xi}_\ell^\top) = \begin{bmatrix} \sigma_p & 0 & 0 \\ 0 & \sigma_v & 0 \\ 0 & 0 & \sigma_a \end{bmatrix} \delta_{k\ell} , \qquad E(\eta_k \eta_\ell) = \sigma_m \delta_{k\ell} ,$$

$$E(\underline{\xi}_k \eta_\ell) = 0 , \qquad E(\underline{\xi}_k \mathbf{x}_0^\top) = 0 , \qquad E(\mathbf{x}_0 \eta_k) = 0 ,$$

and $\sigma_p, \sigma_v, \sigma_a \geq 0$, with $\sigma_p + \sigma_v + \sigma_a > 0$, and $\sigma_m > 0$. Since the matrices

$$M_{A\Gamma} = [\, \Gamma \ \ A\Gamma \ \ A^2\Gamma \,]$$

$$= \begin{bmatrix} 1 & 0 & 0 & 1 & h & h^2/2 & 1 & 2h & 2h^2 \\ 0 & 1 & 0 & 0 & 1 & h & 0 & 1 & 2h \\ 0 & 0 & 1 & 0 & 0 & 1 & 0 & 0 & 1 \end{bmatrix}$$

and

$$N_{CA} = \begin{bmatrix} C \\ CA \\ CA^2 \end{bmatrix} = \begin{bmatrix} 1 & 0 & 0 \\ 1 & h & h^2/2 \\ 1 & 2h & 2h^2 \end{bmatrix}$$

are both of full rank, so that the system (6.43) is both completely controllable and observable, it follows from Theorem 6.1 that there exists a positive definite symmetric matrix P such that

$$\lim_{k \to \infty} P_{k+1,k} = P,$$

where

$$P_{k+1,k} = A(I - G_k C^\top)P_{k,k-1}A^\top + \begin{bmatrix} \sigma_p & 0 & 0 \\ 0 & \sigma_v & 0 \\ 0 & 0 & \sigma_a \end{bmatrix}$$

with

$$G_k = P_{k,k-1}C(C^\top P_{k,k-1}C + \sigma_m)^{-1}.$$

Hence, substituting G_k into the expression for $P_{k+1,k}$ above and then taking the limit, we arrive at the following matrix Riccati equation:

$$P = A[P - PC(C^\top PC + \sigma_m)^{-1}C^\top P]A^\top + \begin{bmatrix} \sigma_p & 0 & 0 \\ 0 & \sigma_v & 0 \\ 0 & 0 & \sigma_a \end{bmatrix}. \qquad (6.44)$$

Now, solving this matrix Riccati equation for the positive definite matrix P, we obtain the limiting Kalman gain

$$G = PC/(C^\top PC + \sigma_m)$$

and the limiting (or steady-state) Kalman filtering equations:

$$\begin{cases} \vec{\mathbf{x}}_{k+1} = A\vec{\mathbf{x}}_k + G(v_k - CA\vec{\mathbf{x}}_k) \\ \vec{\mathbf{x}}_0 = E(\mathbf{x}_0). \end{cases} \qquad (6.45)$$

Since the matrix Riccati equation (6.44) may be solved before the filtering process is being performed, this limiting Kalman filter gives rise to an extremely efficient real-time tracker. Of course, in view of Theorem 6.3, the estimate $\vec{\mathbf{x}}_k$ and the optimal estimate $\hat{\mathbf{x}}_k$ are exponentially close to each other.

Exercises

6.1. Prove that the estimate $\tilde{\mathbf{x}}_{k-1}$ in (6.7) is an unbiased estimate of $\mathbf{x}_{k-1}$ in the sense that $E(\tilde{\mathbf{x}}_{k-1}) = E(\mathbf{x}_{k-1})$.

6.2. Verify that

$$\frac{d}{ds}A^{-1}(s) = -A^{-1}(s)\left[\frac{d}{ds}A(s)\right]A^{-1}(s).$$

6.3. Show that if λ_{min} is the smallest eigenvalue of P, then $P \geq \lambda_{min}I$. Similarly, if λ_{max} is the largest eigenvalue of P then $P \leq \lambda_{max}I$.

6.4. Let F be an $n \times n$ matrix. Suppose that all the eigenvalues of F are of absolute value less than 1. Show that $F^k \to 0$ as $k \to \infty$.

6.5. Prove that for any $n \times n$ matrices A and B,

$$(A + B)\,(A + B)^\top \leq 2(AA^\top + BB^\top).$$

6.6. Let $\{\underline{\xi}_k\}$ and $\{\underline{\eta}_k\}$ be sequences of zero-mean Gaussian white system and measurement noise processes, respectively, and $\tilde{\mathbf{x}}_k$ be defined by (6.4). Show that

$$\langle \mathbf{x}_{k-1} - \tilde{\mathbf{x}}_{k-1},\ \underline{\xi}_{k-1} \rangle = 0$$

and

$$\langle \mathbf{x}_{k-1} - \tilde{\mathbf{x}}_{k-1},\ \underline{\eta}_k \rangle = 0.$$

6.7. Verify that for the Kalman gain G_k, we have

$$-(I - G_kC)P_{k,k-1}C^\top G_k^\top + G_kR_kG_k^\top = 0.$$

Using this formula, show that

$$P_{k,k} = (I - G_kC)AP_{k-1,k-1}A^\top(I - G_kC)^\top$$
$$+ (I - G_kC)\Gamma Q_k\Gamma^\top(I - G_kC)^\top + G_kRG_k^\top.$$

6.8. By imitating the proof of Lemma 6.8, show that all the eigenvalues of $(I - GC)A$ are of absolute value less than 1.

6.9. Let $\underline{\epsilon}_k = \hat{\mathbf{x}}_k - \tilde{\mathbf{x}}_k$ where $\tilde{\mathbf{x}}_k$ is defined by (6.4), and let $\underline{\delta}_k = \mathbf{x}_k - \hat{\mathbf{x}}_k$. Show that

$$\langle \underline{\epsilon}_{k-1},\ \underline{\xi}_{k-1} \rangle = 0, \qquad \langle \underline{\epsilon}_{k-1},\ \underline{\eta}_k \rangle = 0,$$

$$\langle \underline{\delta}_{k-1},\ \underline{\xi}_{k-1} \rangle = 0, \qquad \langle \underline{\delta}_{k-1},\ \underline{\eta}_k \rangle = 0,$$

where $\{\underline{\xi}_k\}$ and $\{\underline{\eta}_k\}$ are zero-mean Gaussian white system and measurement noise processes, respectively.

6.10. Let

$$B_j = \langle \underline{\epsilon}_j, \, \underline{\delta}_j \rangle A^\top C^\top, \qquad j = 0, 1, \cdots,$$

where $\underline{\epsilon}_j = \hat{\mathbf{x}}_j - \vec{\mathbf{x}}_j$, $\underline{\delta}_j = \mathbf{x}_j - \hat{\mathbf{x}}_j$, and $\vec{\mathbf{x}}_j$ is defined by (6.4). Prove that B_j are componentwise uniformly bounded.

6.11. Derive formula (6.41).

6.12. Derive the limiting (or steady-state) Kalman filtering algorithm for the scalar system:

$$\begin{cases} x_{k+1} = ax_k + \gamma \xi_k \\ \quad v_k = cx_k + \eta_k, \end{cases}$$

where a, γ, and c are constants and $\{\xi_k\}$ and $\{\eta_k\}$ are zero-mean Gaussian white noise sequences with variances q and r, respectively.

7. Sequential and Square-Root Algorithms

It is now clear that the only time-consuming operation in the Kalman filtering process is the computation of the Kalman gain matrices:

$$G_k = P_{k,k-1}C_k^\top(C_k P_{k,k-1}C_k^\top + R_k)^{-1}.$$

Since the primary concern of the Kalman filter is its real-time capability, it is of utmost importance to be able to compute G_k preferably without directly inverting a matrix at each time instant, and/or to perform efficiently and accurately a modified operation, whether it would involve matrix inversions or not. The sequential algorithm, which we will first discuss, is designed to avoid a direct computation of the inverse of the matrix $(C_k P_{k,k-1}C_k^\top + R_k)$, while the square-root algorithm, which we will then study, only requires inversion of triangular matrices and improve the computational accuracy by working with the square-root of possibly very large or very small numbers. We also intend to combine these two algorithms to yield a fairly efficient computational scheme for real-time applications.

7.1 Sequential Algorithm

The sequential algorithm is especially efficient if the positive definite matrix R_k is a diagonal matrix, namely:

$$R_k = diag[\, r_k^1, \; \cdots \;, \; r_k^q \,],$$

where $r_k^1, \cdots, r_k^q > 0$. If R_k is not diagonal, then an orthogonal matrix T_k may be determined so that the transformation $T_k^\top R_k T_k$ is a diagonal matrix. In doing so, the observation equation

$$\mathbf{v}_k = C_k \mathbf{x}_k + \underline{\eta}_k$$

of the state-space description is changed to

$$\tilde{\mathbf{v}}_k = \tilde{C}_k \mathbf{x}_k + \tilde{\underline{\eta}}_k \,,$$

where $\tilde{\mathbf{v}}_k = T_k^\top \mathbf{v}_k$, $\tilde{C}_k = T_k^\top C_k$, and $\tilde{\underline{\eta}}_k = T_k^\top \underline{\eta}_k$, so that

$$Var(\tilde{\underline{\eta}}_k) = T_k^\top R_k T_k \,.$$

In the following discussion, we will assume that R_k is diagonal. Since we are only interested in computing the Kalman gain matrix G_k and the corresponding optimal estimate $\hat{\mathbf{x}}_{k|k}$ of the state vector $\mathbf{x}_k$ for a fixed k, we will simply drop the indices k whenever no chance of confusion arises. For instance, we write

$$\mathbf{v}_k = \begin{bmatrix} v^1 \\ \vdots \\ v^q \end{bmatrix}_{q \times 1}, \qquad C_k^\top = \begin{bmatrix} \mathbf{c}^1 & \cdots & \mathbf{c}^q \end{bmatrix}_{n \times q}$$

and

$$R_k = diag \begin{bmatrix} r^1, & \cdots, & r^q \end{bmatrix}.$$

The sequential algorithm can be described as follows.

Theorem 7.1. *Let k be fixed and set*

$$P^0 = P_{k,k-1} \qquad and \qquad \hat{\mathbf{x}}^0 = \hat{\mathbf{x}}_{k|k-1} \,. \tag{7.1}$$

For $i = 1, \cdots, q$, compute

$$\begin{cases} \mathbf{g}^i = \dfrac{1}{(\mathbf{c}^i)^\top P^{i-1} \mathbf{c}^i + r^i} P^{i-1} \mathbf{c}^i \\ \hat{\mathbf{x}}^i = \hat{\mathbf{x}}^{i-1} + (v^i - (\mathbf{c}^i)^\top \hat{\mathbf{x}}^{i-1}) \mathbf{g}^i \\ P^i = P^{i-1} - \mathbf{g}^i (\mathbf{c}^i)^\top P^{i-1} \,. \end{cases} \tag{7.2}$$

Then we have

$$G_k = P^q C_k^\top R_k^{-1} \tag{7.3}$$

and

$$\hat{\mathbf{x}}_{k|k} = \hat{\mathbf{x}}^q \,, \tag{7.4}$$

(cf. Fig.7.1).

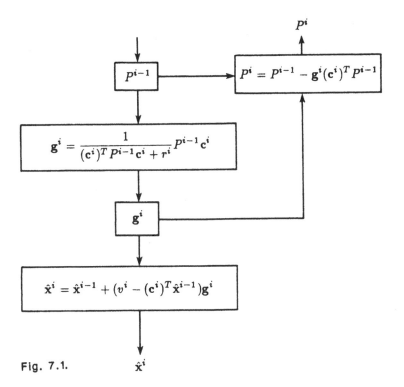

Fig. 7.1. $\hat{\mathbf{x}}^i$

To prove (7.3), we first verify that

$$G_k = P_{k,k} C_k^\top R_k^{-1}.\tag{7.5}$$

This can be seen by returning to the filtering equations. We have

$$P_{k,k} = (I - G_k C_k) P_{k,k-1}$$
$$= P_{k,k-1} - G_k C_k P_{k,k-1},$$

so that

$$G_k = P_{k,k-1} C_k^\top (C_k P_{k,k-1} C_k^\top + R_k)^{-1}$$
$$= (P_{k,k} + G_k C_k P_{k,k-1}) C_k^\top (C_k P_{k,k-1} C_k^\top + R_k)^{-1}$$

or

$$G_k (C_k P_{k,k-1} C_k^\top + R_k) = (P_{k,k} + G_k C_k P_{k,k-1}) C_k^\top,$$

which yields (7.5). Hence, to prove (7.3), it is sufficient to show that

$$P_{k,k} = P^q.\tag{7.6}$$

A direct proof of this identity does not seem to be available. Hence, we appeal to the matrix inversion lemma (Lemma 1.2). Let $\epsilon > 0$ be given, and set $P_\epsilon{}^0 = P_{k,k-1} + \epsilon I$, which is now positive definite. Also, set

$$P_\epsilon{}^i = P_\epsilon{}^{i-1} - \mathbf{g}_\epsilon^i (\mathbf{c}^i)^\top P_\epsilon{}^{i-1}$$

and

$$\mathbf{g}_\epsilon^i = \frac{1}{(\mathbf{c}^i)^\top P_\epsilon{}^{i-1} \mathbf{c}^i + r^i} P_\epsilon{}^{i-1} \mathbf{c}^i .$$

Then by an inductive argument starting with $i = 1$ and using (1.3) of the matrix inversion lemma, it can be seen that the matrix

$$P_\epsilon{}^i = P_\epsilon{}^{i-1} - P_\epsilon{}^{i-1} \mathbf{c}^i \left[(\mathbf{c}^i)^\top P_\epsilon{}^{i-1} \mathbf{c}^i + r^i \right]^{-1} (\mathbf{c}^i)^\top P_\epsilon{}^{i-1}$$

is invertible and

$$(P_\epsilon{}^i)^{-1} = (P_\epsilon{}^{i-1})^{-1} + \mathbf{c}^i (r^i)^{-1} (\mathbf{c}^i)^\top .$$

Hence, using all these equations for $i = q, q-1, \cdots, 1$, consecutively, we have

$$(P_\epsilon{}^q)^{-1} = (P_\epsilon{}^{q-1})^{-1} + \mathbf{c}^q (r^q)^{-1} (\mathbf{c}^q)^\top$$

$$= \cdots \cdots$$

$$= (P_\epsilon{}^0)^{-1} + \sum_{i=1}^{q} \mathbf{c}^i (r^i)^{-1} (\mathbf{c}^i)^\top$$

$$= (P_\epsilon{}^0)^{-1} + C_k^\top R_k^{-1} C_k .$$

On the other hand, again by the matrix inversion lemma, the matrix

$$\tilde{P}_\epsilon := P_\epsilon{}^0 - P_\epsilon{}^0 C_k (C_k P_\epsilon{}^0 C_k^\top + R_k)^{-1} P_\epsilon{}^0$$

is also invertible with

$$\tilde{P}_\epsilon{}^{-1} = (P_\epsilon{}^0)^{-1} + C_k^\top R_k^{-1} C_k .$$

Hence, we have $\tilde{P}_\epsilon{}^{-1} = (P_\epsilon{}^q)^{-1}$, so that

$$\tilde{P}_\epsilon = P_\epsilon{}^q .$$

From the Kalman filtering equations, we have

$$\tilde{P}_\epsilon \to P^0 - P^0 C_k (C_k P^0 C_k^\top + R_k)^{-1} P^0$$
$$= (I - G_k C_k) P_{k,k-1} = P_{k,k}$$

as $\epsilon \to 0$; while from the definition, we have

$$P_\epsilon^q \to P^q$$

as $\epsilon \to 0$. This means that (7.6) holds so that (7.3) is verified.
To establish (7.4), we first observe that

$$\mathbf{g}^i = P^i \mathbf{c}^i (r^i)^{-1}. \tag{7.7}$$

Indeed, since

$$P^{i-1} = P^i + \mathbf{g}^i (\mathbf{c}^i)^\top P^{i-1}$$

which follows from the third equation in (7.2), we have, using the first equation in (7.2),

$$\mathbf{g}^i = \frac{1}{(\mathbf{c}^i)^\top P^{i-1} \mathbf{c}^i + r^i} (P^i + \mathbf{g}^i (\mathbf{c}^i)^\top P^{i-1}) \mathbf{c}^i.$$

This, upon simplification, is (7.7). Now, from the third equation in (7.2) again, we obtain

$$P^q = (I - \mathbf{g}^q (\mathbf{c}^q)^\top) P^{q-1}$$
$$= \cdots \cdots$$
$$= (I - \mathbf{g}^q (\mathbf{c}^q)^\top) \cdots (I - \mathbf{g}^{i+1} (\mathbf{c}^{i+1})^\top) P^i \tag{7.8}$$

for any i, $0 \le i \le q - 1$. Hence, by consecutive applications of the correction equation of the Kalman filter, (7.3), (7.1), (7.8), and (7.7), we have

$$\hat{\mathbf{x}}_{k|k} = \hat{\mathbf{x}}_{k|k-1} + G_k(\mathbf{v}_k - C_k\hat{\mathbf{x}}_{k|k-1})$$

$$= (I - G_kC_k)\hat{\mathbf{x}}_{k|k-1} + G_k\mathbf{v}_k$$

$$= (I - P^qC_k^\top R_k^{-1}C_k)\hat{\mathbf{x}}_{k|k-1} + P^qC_k^\top R_k^{-1}\mathbf{v}_k$$

$$= \left(I - \sum_{i=1}^{q} P^q\mathbf{c}^i(r^i)^{-1}(\mathbf{c}^i)^\top\right)\hat{\mathbf{x}}^0 + \sum_{i=1}^{q} P^q\mathbf{c}^i(r^i)^{-1}v^i$$

$$= \left[\left(I - P^q\mathbf{c}^q(r^q)^{-1}(\mathbf{c}^q)^\top\right) - \sum_{i=1}^{q-1}(I - \mathbf{g}^q(\mathbf{c}^q)^\top)\right.$$

$$\left. \cdots (I - \mathbf{g}^{i+1}(\mathbf{c}^{i+1})^\top)P^i\mathbf{c}^i(r^i)^{-1}(\mathbf{c}^i)^\top\right]\hat{\mathbf{x}}^0 + \sum_{i=1}^{q-1}(I - \mathbf{g}^q(\mathbf{c}^q)^\top)$$

$$\cdots (I - \mathbf{g}^{i+1}(\mathbf{c}^{i+1})^\top)P^i\mathbf{c}^i(r^i)^{-1}v^i + P^q\mathbf{c}^q(r^q)^{-1}v^q$$

$$= \left[\left(I - \mathbf{g}^q(\mathbf{c}^q)^\top\right) - \sum_{i=1}^{q-1}(I - \mathbf{g}^q(\mathbf{c}^q)^\top)\right.$$

$$\left. \cdots (I - \mathbf{g}^{i+1}(\mathbf{c}^{i+1})^\top)\mathbf{g}^i(\mathbf{c}^i)^\top\right]\hat{\mathbf{x}}^0 + \sum_{i=1}^{q-1}(I - \mathbf{g}^q(\mathbf{c}^q)^\top)$$

$$\cdots (I - \mathbf{g}^{i+1}(\mathbf{c}^{i+1})^\top)\mathbf{g}^iv^i + \mathbf{g}^qv^q$$

$$= (I - \mathbf{g}^q(\mathbf{c}^q)^\top) \cdots (I - \mathbf{g}^1(\mathbf{c}^1)^\top)\hat{\mathbf{x}}_0$$

$$+ \sum_{i=1}^{q-1}(I - \mathbf{g}^q(\mathbf{c}^q)^\top) \cdots (I - \mathbf{g}^{i+1}(\mathbf{c}^{i+1})^\top)\mathbf{g}^iv^i + \mathbf{g}^qv^q .$$

On the other hand, from the second equation in (7.2), we also have

$$\hat{\mathbf{x}}^q = (I - \mathbf{g}^q(\mathbf{c}^q)^\top)\hat{\mathbf{x}}^{q-1} + \mathbf{g}^qv^q$$

$$= (I - \mathbf{g}^q(\mathbf{c}^q)^\top)(I - \mathbf{g}^{q-1}(\mathbf{c}^{q-1})^\top)\hat{\mathbf{x}}^{q-2}$$

$$+ (I - \mathbf{g}^q(\mathbf{c}^q)^\top)\mathbf{g}^{q-1}v^{q-1} + \mathbf{g}^qv^q$$

$$= \cdots \cdots$$

$$= (I - \mathbf{g}^q(\mathbf{c}^q)^\top) \cdots (I - \mathbf{g}^1(\mathbf{c}^1)^\top)\hat{\mathbf{x}}^0$$

$$+ \sum_{i=1}^{q-1}(I - \mathbf{g}^q(\mathbf{c}^q)^\top) \cdots (I - \mathbf{g}^{i+1}(\mathbf{c}^{i+1})^\top)\mathbf{g}^iv^i + \mathbf{g}^qv^q$$

which is the same as the above expression. That is, we have proved that $\hat{\mathbf{x}}_{k|k} = \hat{\mathbf{x}}^q$, completing the proof of Theorem 7.1.

7.2 Square-Root Algorithm

We now turn to the square-root algorithm. The following result from linear algebra is important for this consideration. Its proof is left to the reader (cf. Exercise 7.1).

Lemma 7.1. *To any positive definite symmetric matrix A, there is a unique lower triangular matrix A^c such that $A = A^c(A^c)^\top$. More generally, to any $n \times (n + p)$ matrix A, there is an $n \times n$ matrix $\tilde{A}$ such that $\tilde{A}\tilde{A}^\top = AA^\top$.*

A^c has the property of being a "square-root" of A, and since it is lower triangular, its inverse can be computed more efficiently (cf. Exercise 7.3). Note also that in going to the square-root, very small numbers become larger and very large numbers become smaller, so that computation is done more accurately. The factorization of a matrix into the product of a lower triangular matrix and its transpose is usually done by a Gauss elimination scheme known as Cholesky factorization, and this explains why the superscript c is being used. For the general case, $\tilde{A}$ is also called a "square-root" of $AA^\top$.

In the square-root algorithm to be discussed below, the inverse of the lower triangular factor

$$H_k := \left(C_k P_{k,k-1} C_k^\top + R_k \right)^c \tag{7.9}$$

will be taken. To improve the accuracy of the algorithm, we will also use R_k^c instead of the positive definite square-root $R_k^{1/2}$. Of course, if R_k is a diagonal matrix, then $R_k^c = R_k^{1/2}$.

We first consider the following recursive scheme: Let

$$J_{0,0} = \left(Var(\mathbf{x}_0) \right)^{1/2},$$

$J_{k,k-1}$ be a square-root of the matrix

$$\left[A_{k-1}J_{k-1,k-1} \quad \Gamma_{k-1}Q_{k-1}^{1/2} \right]_{n \times (n+p)} \left[A_{k-1}J_{k-1,k-1} \quad \Gamma_{k-1}Q_{k-1}^{1/2} \right]_{n \times (n+p)}^\top,$$

and

$$J_{k,k} = J_{k,k-1}\left[I - J_{k,k-1}^\top C_k^\top (H_k^\top)^{-1}(H_k + R_k^c)^{-1}C_k J_{k,k-1} \right]$$

for $k = 1, 2, \cdots$, where $(Var(\mathbf{x}_0))^{1/2}$ and $Q_{k-1}^{1/2}$ are arbitrary square-roots of $Var(\mathbf{x}_0)$ and Q_{k-1}, respectively. The auxiliary matrices $J_{k,k-1}$ and $J_{k,k}$ are also square-roots (of $P_{k,k-1}$ and $P_{k,k}$, respectively), although they are not necessarily lower triangular nor positive definite, as in the following:

Theorem 7.2. $J_{0,0}J_{0,0}^\top = P_{0,0}$, and for $k = 1, 2, \cdots$,

$$J_{k,k-1}J_{k,k-1}^\top = P_{k,k-1} \tag{7.10}$$

$$J_{k,k}J_{k,k}^\top = P_{k,k}. \tag{7.11}$$

The first statement is trivial since $P_{0,0} = Var(\mathbf{x}_0)$. We can prove (7.10) and (7.11) by mathematical induction. Suppose that (7.11) holds for $k - 1$; then (7.10) follows immediately by using the relation between $P_{k,k-1}$ and $P_{k-1,k-1}$ in the Kalman filtering process. Now we can verify (7.11) for k using (7.10) for the same k. Indeed, since

$$C_k P_{k,k-1} C_k^\top = H_k H_k^\top - R_k$$

so that

$$
\begin{aligned}
&(H_k^\top)^{-1}(H_k + R_k^c)^{-1} + [(H_k + R_k^c)^\top]^{-1}H_k^{-1} \\
&\quad - (H_k^\top)^{-1}(H_k + R_k^c)^{-1}C_k P_{k,k-1} C_k^\top [(H_k + R_k^c)^\top]^{-1}H_k^{-1} \\
&= (H_k^\top)^{-1}(H_k + R_k^c)^{-1}\{H_k(H_k + R_k^c)^\top + (H_k + R_k^c)H_k^\top \\
&\quad - H_k H_k^\top + R_k\}[(H_k + R_k^c)^\top]^{-1}H_k^{-1} \\
&= (H_k^\top)^{-1}(H_k + R_k^c)^{-1}\{H_k H_k^\top + H_k(R_k^c)^\top \\
&\quad + R_k^c H_k^\top + R_k\}[(H_k + R_k^c)^\top]^{-1}H_k^{-1} \\
&= (H_k^\top)^{-1}(H_k + R_k^c)^{-1}(H_k + R_k^c)(H_k + R_k^c)^\top[(H_k + R_k^c)^\top]^{-1}H_k^{-1} \\
&= (H_k^\top)^{-1}H_k^{-1} \\
&= (H_k H_k^\top)^{-1},
\end{aligned}
$$

it follows from (7.10) that

$$
\begin{aligned}
&J_{k,k}J_{k,k}^\top \\
&= J_{k,k-1}\big[I - J_{k,k-1}^\top C_k^\top (H_k^\top)^{-1}(H_k + R_k^c)^{-1}C_k J_{k,k-1}\big] \\
&\quad \cdot \big[I - J_{k,k-1}^\top C_k^\top [(H_k + R_k^c)^\top]^{-1}H_k^{-1}C_k J_{k,k-1}\big]J_{k,k-1}^\top \\
&= J_{k,k-1}\big\{ I - J_{k,k-1}^\top C_k^\top (H_k^\top)^{-1}(H_k + R_k^c)^{-1}C_k J_{k,k-1} \\
&\qquad - J_{k,k-1}^\top C_k^\top [(H_k + R_k^c)^\top]^{-1}H_k^{-1}C_k J_{k,k-1} \\
&\qquad + J_{k,k-1}^\top C_k^\top (H_k^\top)^{-1}(H_k + R_k^c)^{-1}C_k J_{k,k-1}J_{k,k-1}^\top C_k^\top \\
&\qquad \cdot [(H_k + R_k^c)^\top]^{-1}H_k^{-1}C_k J_{k,k-1}\big\}J_{k,k-1}^\top \\
&= P_{k,k-1} - P_{k,k-1}C_k^\top \{(H_k^\top)^{-1}(H_k + R_k^c)^{-1} + [(H_k + R_k^c)^\top]^{-1}H_k^{-1} \\
&\quad - (H_k^\top)^{-1}(H_k + R_k^c)^{-1}C_k P_{k,k-1} C_k^\top[(H_k + R_k^c)^\top]^{-1}H_k^{-1}\}C_k P_{k,k-1} \\
&= P_{k,k-1} - P_{k,k-1}C_k^\top (H_k H_k^\top)^{-1}C_k P_{k,k-1} \\
&= P_{k,k}.
\end{aligned}
$$

This completes the induction process.

In summary, the square-root Kalman filtering algorithm can be stated as follows:

(i) Compute $J_{0,0} = (Var(\mathbf{x}_0))^{1/2}$.

(ii) For $k = 1, 2, \cdots$, compute $J_{k,k-1}$, a square-root of the matrix

$$[A_{k-1}J_{k-1,k-1} \ \ \Gamma_{k-1}Q_{k-1}^{1/2}]_{n \times (n+p)}[A_{k-1}J_{k-1,k-1} \ \ \Gamma_{k-1}Q_{k-1}^{1/2}]_{n \times (n+p)}^{\mathsf{T}} \,,$$

and the matrix

$$H_k = (C_k J_{k,k-1} J_{k,k-1}^{\mathsf{T}} C_k^{\mathsf{T}} + R_k)^c \,,$$

and then compute

$$J_{k,k} = J_{k,k-1}\big[\, I - J_{k,k-1}^{\mathsf{T}} C_k^{\mathsf{T}} (H_k^{\mathsf{T}})^{-1}(H_k + R_k^c)^{-1} C_k J_{k,k-1}\big]\,.$$

(iii) Compute $\hat{\mathbf{x}}_{0|0} = E(\mathbf{x}_0)$, and for $k = 1, 2, \cdots$, using the information from (ii), compute

$$G_k = J_{k,k-1}J_{k,k-1}^{\mathsf{T}} C_k^{\mathsf{T}} (H_k^{\mathsf{T}})^{-1} H_k^{-1}$$

and

$$\hat{\mathbf{x}}_{k|k} = A_{k-1}\hat{\mathbf{x}}_{k-1|k-1} + G_k(\mathbf{v}_k - C_k A_{k-1}\hat{\mathbf{x}}_{k-1|k-1})\,,$$

(cf. Fig.7.2).

We again remark that we only have to invert triangular matrices, and in addition, these matrices are square-root of the ones which might have very small or very large entries.

7.3 An Algorithm for Real-Time Applications

In the particular case when

$$R_k = diag[\, r_k^1, \ \cdots \, , \ r_k^q \,]$$

is a diagonal matrix, the sequential and square-root algorithms can be combined to yield the following algorithm which does not require direct matrix inversions:

(i) Compute $J_{0,0} = (Var(\mathbf{x}_0))^{1/2}$.

(ii) For each fixed $k = 1, 2, \cdots$, compute
    (a) a square-root $J_{k,k-1}$ of the matrix

$$[A_{k-1}J_{k-1,k-1} \ \ \Gamma_{k-1}Q_{k-1}^{1/2}]_{n \times (n+p)}[A_{k-1}J_{k-1,k-1} \ \ \Gamma_{k-1}Q_{k-1}^{1/2}]_{n \times (n+p)}^{\mathsf{T}} \,,$$

    and

(b) for $i = 1, \cdots, k,$

$$\mathbf{g}_k^i = \frac{1}{(\mathbf{c}_k^i)^\top J_{k,k-1}^{i-1} (J_{k,k-1}^{i-1})^\top \mathbf{c}_k^i + r_k^i} J_{k,k-1}^{i-1} (J_{k,k-1}^{i-1})^\top \mathbf{c}_k^i,$$

$$J_{k,k-1}^i = \left(J_{k,k-1}^{i-1} (J_{k,k-1}^{i-1})^\top - \mathbf{g}_k^i (\mathbf{c}_k^i)^\top J_{k,k-1}^{i-1} (J_{k,k-1}^{i-1})^\top \right)^c,$$

where

$$J_{k,k-1}^0 := J_{k,k-1}, \quad J_{k,k-1}^q = J_{k,k} \quad and \quad C_k^\top := [\, \mathbf{c}_k^1 \ \cdots \ \mathbf{c}_k^q \,].$$

(iii) Compute $\hat{\mathbf{x}}_{0|0} = E(\mathbf{x}_0)$.
(iv) For each fixed $k = 1, 2, \cdots$, compute
   (a) $\hat{\mathbf{x}}_{k|k-1} = A_{k-1}\hat{\mathbf{x}}_{k-1|k-1}$,
   (b) for $i = 1, \cdots, q$, with $\hat{\mathbf{x}}_k^0 := \hat{\mathbf{x}}_{k|k-1}$, and using information
     from (ii)(b), compute

$$\hat{\mathbf{x}}_k^i = \hat{\mathbf{x}}_k^{i-1} + (v_k^i - (c_k^i)^\top \hat{\mathbf{x}}_k^{i-1})\mathbf{g}_k^i,$$

where $\mathbf{v}_k := [\, v_k^1 \ \cdots \ v_k^q \,]^\top$, so that

$$\hat{\mathbf{x}}_{k|k} = \hat{\mathbf{x}}_k^q.$$

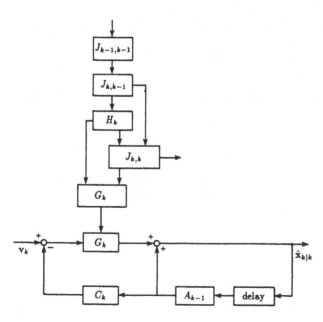

Fig. 7.2.

Exercises

7.1. Give a proof of Lemma 7.1.

7.2. Find the lower triangular matrix L that satisfies:

$$(a) \qquad LL^\mathsf{T} = \begin{bmatrix} 1 & 2 & 3 \\ 2 & 8 & 2 \\ 3 & 2 & 14 \end{bmatrix}.$$

$$(b) \qquad LL^\mathsf{T} = \begin{bmatrix} 1 & 1 & 1 \\ 1 & 3 & 2 \\ 1 & 2 & 4 \end{bmatrix}.$$

7.3. (a) Derive a formula to find the inverse of the matrix

$$L = \begin{bmatrix} \ell_{11} & 0 & 0 \\ \ell_{21} & \ell_{22} & 0 \\ \ell_{31} & \ell_{32} & \ell_{33} \end{bmatrix},$$

where ℓ_{11}, ℓ_{22}, and ℓ_{33} are nonzero.

(b) Formulate the inverse of

$$L = \begin{bmatrix} \ell_{11} & 0 & 0 & \cdots & 0 \\ \ell_{21} & \ell_{22} & 0 & \cdots & 0 \\ \vdots & \vdots & \ddots & \ddots & \vdots \\ \vdots & \vdots & & \ddots & 0 \\ \ell_{n1} & \ell_{n2} & \cdots & \cdots & \ell_{nn} \end{bmatrix},$$

where $\ell_{11}, \cdots, \ell_{nn}$ are nonzero.

7.4. Consider the following computer simulation of the Kalman filtering process. Let $\epsilon \ll 1$ be a small positive number such that

$$1 - \epsilon \not\simeq 1$$
$$1 - \epsilon^2 \simeq 1$$

where "$\simeq$" denotes equality after rounding in the computer. Suppose that we have

$$P_{k,k} = \begin{bmatrix} \frac{\epsilon^2}{1-\epsilon^2} & 0 \\ 0 & 1 \end{bmatrix}.$$

Compare the standard Kalman filter with the square-root filter for this example. Note that this example illustrates the improved numerical characteristics of the square-root filter.

7.5. Prove that to any positive definite symmetric matrix A, there is a unique upper triangular matrix A^u such that $A = A^u(A^u)^\mathsf{T}$.

7.6. Using the upper triangular decompositions instead of the lower triangular ones, derive a new square-root Kalman filter.

7.7. Combine the sequential algorithm and the square-root scheme with upper triangular decompositions to derive a new filtering algorithm.

8. Extended Kalman Filter and System Identification

The Kalman filtering process has been designed to estimate the state vector in a linear model. If the model turns out to be nonlinear, a linearization procedure is usually performed in deriving the filtering equations. We will consider a real-time linear Taylor approximation of the system function at the previous state estimate and that of the observation function at the corresponding predicted position. The Kalman filter so obtained will be called the *extended Kalman filter*. This idea to handle a nonlinear model is quite natural, and the filtering procedure is fairly simple and efficient. Furthermore, it has found many important real-time applications. One such application is *adaptive system identification* which we will also discuss briefly in this chapter. Finally, by improving the linearization procedure of the extended Kalman filtering algorithm, we will introduce a modified extended Kalman filtering scheme which has a parallel computational structure. We then give two numerical examples to demonstrate the advantage of the modified Kalman filter over the standard one in both state estimation and system parameter identification.

8.1 Extended Kalman Filter

A state-space description of a system which is not necessarily linear will be called a *nonlinear model* of the system. In this chapter, we will consider a nonlinear model of the form

$$\begin{cases} \mathbf{x}_{k+1} = \mathbf{f}_k(\mathbf{x}_k) + H_k(\mathbf{x}_k)\underline{\xi}_k \\ \mathbf{v}_k = \mathbf{g}_k(\mathbf{x}_k) + \underline{\eta}_k, \end{cases} \tag{8.1}$$

where $\mathbf{f}_k$ and $\mathbf{g}_k$ are vector-valued functions with ranges in $\mathbf{R}^n$ and $\mathbf{R}^q$, respectively, $1 \le q \le n$, and H_k a matrix-valued function with range in $\mathbf{R}^n \times \mathbf{R}^q$, such that for each k the first order partial derivatives of $\mathbf{f}_k(\mathbf{x}_k)$ and $\mathbf{g}_k(\mathbf{x}_k)$ with respect to all the components of $\mathbf{x}_k$ are continuous. As usual, we simply consider zero-mean Gaussian white noise sequences $\{\underline{\xi}_k\}$ and $\{\underline{\eta}_k\}$ with ranges in $\mathbf{R}^p$ and $\mathbf{R}^q$, respectively, $1 \leq p, q \leq n$, and

$$E(\underline{\xi}_k \underline{\xi}_\ell^\top) = Q_k \delta_{k\ell}, \qquad E(\underline{\eta}_k \underline{\eta}_\ell^\top) = R_k \delta_{k\ell},$$

$$E(\underline{\xi}_k \underline{\eta}_\ell^\top) = 0, \quad E(\underline{\xi}_k \mathbf{x}_0^\top) = 0, \quad E(\underline{\eta}_k \mathbf{x}_0^\top) = 0,$$

for all k and ℓ. The real-time linearization process is carried out as follows.

In order to be consistent with the linear model, the initial estimate $\hat{\mathbf{x}}_0 = \hat{\mathbf{x}}_{0|0}$ and predicted position $\hat{\mathbf{x}}_{1|0}$ are chosen to be

$$\hat{\mathbf{x}}_0 = E(\mathbf{x}_0), \qquad \hat{\mathbf{x}}_{1|0} = \mathbf{f}_0(\hat{\mathbf{x}}_0).$$

We will then formulate $\hat{\mathbf{x}}_k = \hat{\mathbf{x}}_{k|k}$, consecutively, for $k = 1, 2, \cdots$, using the predicted positions

$$\hat{\mathbf{x}}_{k+1|k} = \mathbf{f}_k(\hat{\mathbf{x}}_k) \tag{8.2}$$

and the linear state-space description

$$\begin{cases} \mathbf{x}_{k+1} = A_k \mathbf{x}_k + \mathbf{u}_k + \Gamma_k \underline{\xi}_k \\ \mathbf{w}_k = C_k \mathbf{x}_k + \underline{\eta}_k, \end{cases} \tag{8.3}$$

where $A_k, \mathbf{u}_k, \Gamma_k, \mathbf{w}_k$ and C_k are to be determined in real-time as follows. Suppose that $\hat{\mathbf{x}}_j$ has been determined so that $\hat{\mathbf{x}}_{j+1|j}$ is also defined using (8.2), for $j = 0, 1, \cdots, k$. We consider the linear Taylor approximation of $\mathbf{f}_k(\mathbf{x}_k)$ at $\hat{\mathbf{x}}_k$ and that of $\mathbf{g}_k(\mathbf{x}_k)$ at $\hat{\mathbf{x}}_{k|k-1}$; that is,

$$\begin{cases} \mathbf{f}_k(\mathbf{x}_k) \simeq \mathbf{f}_k(\hat{\mathbf{x}}_k) + A_k(\mathbf{x}_k - \hat{\mathbf{x}}_k) \\ \mathbf{g}_k(\mathbf{x}_k) \simeq \mathbf{g}_k(\hat{\mathbf{x}}_{k|k-1}) + C_k(\mathbf{x}_k - \hat{\mathbf{x}}_{k|k-1}), \end{cases} \tag{8.4}$$

where

$$A_k = \left[\frac{\partial \mathbf{f}_k}{\partial \mathbf{x}_k}(\hat{\mathbf{x}}_k) \right] \qquad and \qquad C_k = \left[\frac{\partial \mathbf{g}_k}{\partial \mathbf{x}_k}(\hat{\mathbf{x}}_{k|k-1}) \right]. \tag{8.5}$$

Here and throughout, for any vector-valued function

$$\mathbf{h}(\mathbf{x}_k) = \begin{bmatrix} h_1(\mathbf{x}_k) \\ \vdots \\ h_m(\mathbf{x}_k) \end{bmatrix}$$

where

$$\mathbf{x}_k = \begin{bmatrix} x_k^1 \\ \vdots \\ x_k^n \end{bmatrix},$$

we denote, as usual,

$$\left[\frac{\partial \mathbf{h}}{\partial \mathbf{x}_k}(\mathbf{x}_k^*) \right] = \begin{bmatrix} \frac{\partial h_1}{\partial x_k^1}(\mathbf{x}_k^*) & \cdots & \frac{\partial h_1}{\partial x_k^n}(\mathbf{x}_k^*) \\ \vdots & & \vdots \\ \frac{\partial h_m}{\partial x_k^1}(\mathbf{x}_k^*) & \cdots & \frac{\partial h_m}{\partial x_k^n}(\mathbf{x}_k^*) \end{bmatrix}. \tag{8.6}$$

Hence, by setting

$$\begin{cases} \mathbf{u}_k = \mathbf{f}_k(\hat{\mathbf{x}}_k) - A_k \hat{\mathbf{x}}_k \\ \Gamma_k = H_k(\hat{\mathbf{x}}_k) \\ \mathbf{w}_k = \mathbf{v}_k - \mathbf{g}_k(\hat{\mathbf{x}}_{k|k-1}) + C_k \hat{\mathbf{x}}_{k|k-1}, \end{cases} \tag{8.7}$$

the nonlinear model (8.1) is approximated by the linear model (8.3) using the matrices and vectors defined in (8.5) and (8.7) at the kth instant (cf. Exercise 8.2). Of course, this linearization is possible only if $\hat{\mathbf{x}}_k$ has been determined. We already have $\hat{\mathbf{x}}_0$, so that the system equation in (8.3) is valid for $k = 0$. From this, we define $\hat{\mathbf{x}}_1 = \hat{\mathbf{x}}_{1|1}$ as the optimal unbiased estimate (with optimal weight) of $\mathbf{x}_1$ in the linear model (8.3), using the data $[\mathbf{v}_0^\top \ \mathbf{w}_1^\top]^\top$. Now, by applying (8.2), (8.3) is established for $k = 1$, so that $\hat{\mathbf{x}}_2 = \hat{\mathbf{x}}_{2|2}$ can be determined analogously, using the data $[\mathbf{v}_0^\top \ \mathbf{w}_1^\top \ \mathbf{w}_2^\top]^\top$, etc. From the Kalman filtering results for linear deterministic/stochastic state-space descriptions in Chapters 2 and 3 (cf. (2.18) and Exercise 3.6), we arrive at the "correction" formula

$$\begin{aligned} \hat{\mathbf{x}}_k &= \hat{\mathbf{x}}_{k|k-1} + G_k(\mathbf{w}_k - C_k \hat{\mathbf{x}}_{k|k-1}) \\ &= \hat{\mathbf{x}}_{k|k-1} + G_k((\mathbf{v}_k - \mathbf{g}_k(\hat{\mathbf{x}}_{k|k-1}) + C_k \hat{\mathbf{x}}_{k|k-1}) - C_k \hat{\mathbf{x}}_{k|k-1}) \\ &= \hat{\mathbf{x}}_{k|k-1} + G_k(\mathbf{v}_k - \mathbf{g}_k(\hat{\mathbf{x}}_{k|k-1})), \end{aligned}$$

where G_k is the Kalman gain matrix for the linear model (8.3) at the kth instant.

The resulting filtering process is called the *extended Kalman filter*. The filtering algorithm may be summarized as follows (cf.

Exercise 8.3):

$$
\begin{cases}
P_{0,0} = Var(\mathbf{x}_0), \quad \hat{\mathbf{x}}_0 = E(\mathbf{x}_0). \\[4pt]
For \quad k = 1, 2, \cdots, \\[4pt]
P_{k,k-1} = \left[\dfrac{\partial \mathbf{f}_{k-1}}{\partial \mathbf{x}_{k-1}} (\hat{\mathbf{x}}_{k-1}) \right] P_{k-1,k-1} \left[\dfrac{\partial \mathbf{f}_{k-1}}{\partial \mathbf{x}_{k-1}} (\hat{\mathbf{x}}_{k-1}) \right]^{\mathsf{T}} \\[6pt]
\qquad\quad + H_{k-1}(\hat{\mathbf{x}}_{k-1}) Q_{k-1} H_{k-1}^{\mathsf{T}}(\hat{\mathbf{x}}_{k-1}) \\[6pt]
\hat{\mathbf{x}}_{k|k-1} = \mathbf{f}_{k-1}(\hat{\mathbf{x}}_{k-1}) \\[4pt]
G_k = P_{k,k-1} \left[\dfrac{\partial \mathbf{g}_k}{\partial \mathbf{x}_k} (\hat{\mathbf{x}}_{k|k-1}) \right]^{\mathsf{T}} \\[8pt]
\qquad \cdot \left[\left[\dfrac{\partial \mathbf{g}_k}{\partial \mathbf{x}_k} (\hat{\mathbf{x}}_{k|k-1}) \right] P_{k,k-1} \left[\dfrac{\partial \mathbf{g}_k}{\partial \mathbf{x}_k} (\hat{\mathbf{x}}_{k|k-1}) \right]^{\mathsf{T}} + R_k \right]^{-1} \\[8pt]
P_{k,k} = \left[I - G_k \left[\dfrac{\partial \mathbf{g}_k}{\partial \mathbf{x}_k} (\hat{\mathbf{x}}_{k|k-1}) \right] \right] P_{k,k-1} \\[6pt]
\hat{\mathbf{x}}_{k|k} = \hat{\mathbf{x}}_{k|k-1} + G_k (\mathbf{v}_k - \mathbf{g}_k(\hat{\mathbf{x}}_{k|k-1}))
\end{cases} \tag{8.8}
$$

(cf. Fig.8.1).

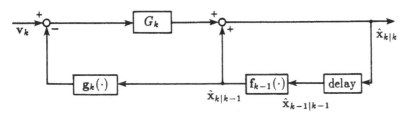

Fig. 8.1.

8.2 Satellite Orbit Estimation

Estimating the planar orbit of a satellite provides an interesting example for the extended Kalman filter. Let the governing equations of motion of the satellite in the planar orbit be given by

$$
\begin{cases}
\ddot{r} = r\dot{\theta}^2 - mgr^{-2} + \xi_r \\
\ddot{\theta} = -2r^{-1}\dot{r}\dot{\theta} + r^{-1}\xi_\theta,
\end{cases} \tag{8.9}
$$

where r is the (radial) distance (called the *range*) of the satellite from the center of the earth (called the *attracting point*), θ the angle measured from some reference axis (called the *angular displacement*), and m, g are constants, being the mass of the earth and the universal gravitational constant, respectively (cf. Fig.8.2). In addition, ξ_r and ξ_θ are assumed to be continuous-time uncorrelated zero-mean Gaussian white noise processes. By setting

$$\mathbf{x} = [r \ \ \dot{r} \ \ \theta \ \ \dot{\theta}]^\top = [x[1] \ \ x[2] \ \ x[3] \ \ x[4]]^\top,$$

the equations in (8.9) become:

$$\dot{\mathbf{x}} = \begin{bmatrix} x[2] \\ x[1]x[4]^2 - mg/x[1]^2 \\ x[4] \\ -2x[2]x[4]/x[1] \end{bmatrix} + \begin{bmatrix} 1 & 0 & 0 & 0 \\ 0 & 1 & 0 & 0 \\ 0 & 0 & 1 & 0 \\ 0 & 0 & 0 & 1/x[1] \end{bmatrix} \begin{bmatrix} 0 \\ \xi_r \\ 0 \\ \xi_\theta \end{bmatrix}.$$

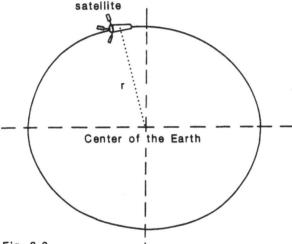

Fig. 8.2.

Hence, if we replace $\mathbf{x}$ by $\mathbf{x}_k$ and $\dot{\mathbf{x}}$ by $(\mathbf{x}_{k+1} - \mathbf{x}_k)h^{-1}$, where $h > 0$ denotes the sampling time, we obtain the discretized nonlinear model

$$\mathbf{x}_{k+1} = \mathbf{f}(\mathbf{x}_k) + H(\mathbf{x}_k)\underline{\xi}_k,$$

where

$$\mathbf{f}(\mathbf{x}_k) = \begin{bmatrix} x_k[1] + h \ x_k[2] \\ x_k[2] + h \ x_k[1] x_k[4]^2 - hmg/x_k[1]^2 \\ x_k[3] + h \ x_k[4] \\ x_k[4] - 2h \ x_k[2]x_k[4]/x_k[1] \end{bmatrix},$$

$$H(\mathbf{x}_k) = \begin{bmatrix} h & 0 & 0 & 0 \\ 0 & h & 0 & 0 \\ 0 & 0 & h & 0 \\ 0 & 0 & 0 & h/x_k[1] \end{bmatrix},$$

and $\underline{\xi}_k := [0 \ \ \xi_r(k) \ \ 0 \ \ \xi_\theta(k)]^\top$. Now, suppose that the range r is measured giving the data information

$$\mathbf{v}_k = \begin{bmatrix} 1 & 0 & 0 & 0 \\ 0 & 0 & 1 & 0 \end{bmatrix} \mathbf{x}_k + \eta_k \,,$$

where $\{\eta_k\}$ is a zero-mean Gaussian white noise sequence, independent of $\{\xi_r(k)\}$ and $\{\xi_\theta(k)\}$. It is clear that

$$\left[\frac{\partial \mathbf{f}}{\partial \mathbf{x}_{k-1}} (\hat{\mathbf{x}}_{k-1}) \right] =$$

$$\begin{bmatrix} 1 & 2hmg/\hat{x}_{k-1}[1]^3 + h\,\hat{x}_{k-1}[4]^2 & 0 & 2h\hat{x}_{k-1}[2]\hat{x}_{k-1}[4]/\hat{x}_{k-1}[1]^2 \\ h & 1 & 0 & -2h\hat{x}_{k-1}[4]/\hat{x}_{k-1}[1] \\ 0 & 0 & 1 & 0 \\ 0 & 2h\,\hat{x}_{k-1}[1]\,\hat{x}_{k-1}[4] & h & 1 - 2h\hat{x}_{k-1}[2]/\hat{x}_{k-1}[1] \end{bmatrix}^\top,$$

$$H(\hat{\mathbf{x}}_{k-1|k-1}) = \begin{bmatrix} h & 0 & 0 & 0 \\ 0 & h & 0 & 0 \\ 0 & 0 & h & 0 \\ 0 & 0 & 0 & h/\hat{x}_{k-1}[1] \end{bmatrix},$$

$$\left[\frac{\partial g}{\partial \mathbf{x}_k} (\hat{\mathbf{x}}_{k|k-1}) \right] = \begin{bmatrix} 1 & 0 & 0 & 0 \\ 0 & 0 & 1 & 0 \end{bmatrix}$$

and

$$g(\hat{\mathbf{x}}_{k|k-1}) = \hat{\mathbf{x}}_{k|k-1}[1] \,.$$

By using these quantities in the extended Kalman filtering equations (8.8), we have an algorithm to determine $\hat{\mathbf{x}}_k$ which gives an estimate of the planar orbit $(\hat{x}_k[1], \hat{x}_k[3])$ of the satellite.

8.3 Adaptive System Identification

As an application of the extended Kalman filter, let us discuss the following problem of adaptive system identification. Suppose that a linear system with state-space description

$$\begin{cases} \mathbf{x}_{k+1} = A_k(\underline{\theta})\mathbf{x}_k + \Gamma_k(\underline{\theta})\underline{\xi}_k \\ \mathbf{v}_k = C_k(\underline{\theta})\mathbf{x}_k + \underline{\eta}_k \end{cases} \tag{8.10}$$

is being considered, where, as usual, $\mathbf{x}_k \epsilon \mathbf{R}^n$, $\underline{\xi}_k \epsilon \mathbf{R}^p$, $\underline{\eta}_k \epsilon \mathbf{R}^q$, $1 \leq p, q \leq n$, and $\{\underline{\xi}_k\}$ and $\{\underline{\eta}_k\}$ are uncorrelated Gaussian white noise sequences. In this application, we assume that $A_k(\underline{\theta})$, $\Gamma_k(\underline{\theta})$, and $C_k(\underline{\theta})$ are known matrix-valued functions of some unknown constant vector $\underline{\theta}$. The objective is to "identify" $\underline{\theta}$.

It seems quite natural to set $\underline{\theta}_{k+1} = \underline{\theta}_k = \underline{\theta}$, since $\underline{\theta}$ is a constant vector. Surprisingly, this assumption does not lead us anywhere as we will see below and from the simple example in the next section. In fact, $\underline{\theta}$ must be treated as a random constant vector such as

$$\underline{\theta}_{k+1} = \underline{\theta}_k + \underline{\zeta}_k, \tag{8.11}$$

where $\{\underline{\zeta}_k\}$ is any zero-mean Gaussian white noise sequence uncorrelated with $\{\underline{\eta}_k\}$ and with preassigned positive definite variances $Var(\underline{\zeta}_k) = S_k$. In applications, we may choose $S_k = S > 0$ for all k (see Section 8.4). Now, the system (8.10) together with the assumption (8.11) can be reformulated as the nonlinear model:

$$\begin{cases} \begin{bmatrix} \mathbf{x}_{k+1} \\ \underline{\theta}_{k+1} \end{bmatrix} = \begin{bmatrix} A_k(\underline{\theta}_k)\mathbf{x}_k \\ \underline{\theta}_k \end{bmatrix} + \begin{bmatrix} \Gamma_k(\underline{\theta}_k)\underline{\xi}_k \\ \underline{\zeta}_k \end{bmatrix} \\ \mathbf{v}_k = [C_k(\underline{\theta}_k) \quad 0] \begin{bmatrix} \mathbf{x}_k \\ \underline{\theta}_k \end{bmatrix} + \underline{\eta}_k, \end{cases} \tag{8.12}$$

and the extended Kalman filtering procedure can be applied to estimate the state vector which contains $\underline{\theta}_k$ as its components. That is, $\underline{\theta}_k$ is estimated optimally in an adaptive way. However, in order to apply the extended Kalman filtering process (8.8), we still need an initial estimate $\hat{\underline{\theta}}_0 := \hat{\underline{\theta}}_{0|0}$. One method is to appeal to the state-space description (8.10). For instance, since $E(\mathbf{v}_0) = C_0(\underline{\theta})E(\mathbf{x}_0)$ so that $\mathbf{v}_0 - C_0(\underline{\theta})E(\mathbf{x}_0)$ is of zero-mean, we could start from $k = 0$, take the variances of both sides of the modified "observation equation"

$$\mathbf{v}_0 - C_0(\underline{\theta})E(\mathbf{x}_0) = C_0(\underline{\theta})\mathbf{x}_0 - C_0(\underline{\theta})E(\mathbf{x}_0) + \underline{\eta}_0,$$

and use the estimate $[\mathbf{v}_0 - C_0(\underline{\theta})E(\mathbf{x}_0)][\mathbf{v}_0 - C_0(\underline{\theta})E(\mathbf{x}_0)]^\top$ for $Var(\mathbf{v}_0 - C_0(\underline{\theta})E(\mathbf{x}_0))$ (cf. Exercise 2.12) to obtain approximately

$$\mathbf{v}_0\mathbf{v}_0^\top - C_0(\underline{\theta})E(\mathbf{x}_0)\mathbf{v}_0^\top - \mathbf{v}_0(C_0(\underline{\theta})E(\mathbf{x}_0))^\top$$
$$+ C_0(\underline{\theta})(E(\mathbf{x}_0)E(\mathbf{x}_0^\top) - Var(\mathbf{x}_0))C_0^\top(\underline{\theta}) - R_0 = 0 \tag{8.13}$$

(cf. Exercise 8.4). Now, solve for $\underline{\theta}$ and set one of the "most appropriate" solutions as the initial estimate $\hat{\underline{\theta}}_0$. If there is no solution of $\underline{\theta}$ in (8.13), we could use the equation

$$\mathbf{v}_1 = C_1(\underline{\theta})\mathbf{x}_1 + \underline{\eta}_1$$
$$= C_1(\underline{\theta})(A_0(\underline{\theta})\mathbf{x}_0 + \Gamma_0(\underline{\theta})\underline{\xi}_0) + \underline{\eta}_1$$

and apply the same procedure, yielding approximately

$$
\begin{aligned}
&\mathbf{v}_1\mathbf{v}_1^\top - C_1(\underline{\theta})A_0(\underline{\theta})E(\mathbf{x}_0)\mathbf{v}_1^\top - \mathbf{v}_1(C_1(\underline{\theta})A_0(\underline{\theta})E(\mathbf{x}_0))^\top \\
&- C_1(\underline{\theta})\Gamma_0(\underline{\theta})Q_0(C_1(\underline{\theta})\Gamma_0(\underline{\theta}))^\top + C_1(\underline{\theta})A_0(\underline{\theta})[E(\mathbf{x}_0)E(\mathbf{x}_0^\top) \\
&- Var(\mathbf{x}_0)]A_0^\top(\underline{\theta})C_1^\top(\underline{\theta}) - R_1 = 0
\end{aligned}
\tag{8.14}
$$

(cf. Exercise 8.5), etc. Once $\hat{\underline{\theta}}_0$ has been chosen, we can apply the extended Kalman filtering process (8.8) and obtain the following algorithm:

$$
\left\{
\begin{aligned}
&\begin{bmatrix} \hat{\mathbf{x}}_0 \\ \hat{\underline{\theta}}_0 \end{bmatrix} = \begin{bmatrix} E(\mathbf{x}_0) \\ \hat{\underline{\theta}}_0 \end{bmatrix}, \qquad
P_{0,0} = \begin{bmatrix} Var(\mathbf{x}_0) & 0 \\ 0 & S_0 \end{bmatrix}, \\[2mm]
&For \;\; k = 1, 2, \cdots, \\[1mm]
&\begin{bmatrix} \hat{\mathbf{x}}_{k|k-1} \\ \hat{\underline{\theta}}_{k|k-1} \end{bmatrix} = \begin{bmatrix} A_{k-1}(\hat{\underline{\theta}}_{k-1})\hat{\mathbf{x}}_{k-1} \\ \hat{\underline{\theta}}_{k-1} \end{bmatrix} \\[2mm]
&P_{k,k-1} = \begin{bmatrix} A_{k-1}(\hat{\underline{\theta}}_{k-1}) & \frac{\partial}{\partial\underline{\theta}}\left[A_{k-1}(\hat{\underline{\theta}}_{k-1})\hat{\mathbf{x}}_{k-1}\right] \\ 0 & I \end{bmatrix} P_{k-1,k-1} \\[2mm]
&\qquad\quad \cdot \begin{bmatrix} A_{k-1}(\hat{\underline{\theta}}_{k-1}) & \frac{\partial}{\partial\underline{\theta}}\left[A_{k-1}(\hat{\underline{\theta}}_{k-1})\hat{\mathbf{x}}_{k-1}\right] \\ 0 & I \end{bmatrix}^\top \\[2mm]
&\qquad\quad + \begin{bmatrix} \Gamma_{k-1}(\hat{\underline{\theta}}_{k-1})Q_{k-1}\Gamma_{k-1}^\top(\hat{\underline{\theta}}_{k-1}) & 0 \\ 0 & S_{k-1} \end{bmatrix} \\[2mm]
&G_k = P_{k,k-1}\left[C_k(\hat{\underline{\theta}}_{k|k-1}) \;\; 0\right]^\top \\[1mm]
&\qquad\quad \cdot \left[[C_k(\hat{\underline{\theta}}_{k|k-1}) \;\; 0]P_{k,k-1}[C_k(\hat{\underline{\theta}}_{k|k-1}) \;\; 0]^\top + R_k\right]^{-1} \\[2mm]
&P_{k,k} = [I - G_k[C_k(\hat{\underline{\theta}}_{k|k-1}) \;\; 0]]P_{k,k-1} \\[2mm]
&\begin{bmatrix} \hat{\mathbf{x}}_k \\ \hat{\underline{\theta}}_k \end{bmatrix} = \begin{bmatrix} \hat{\mathbf{x}}_{k|k-1} \\ \hat{\underline{\theta}}_{k|k-1} \end{bmatrix} + G_k(\mathbf{v}_k - C_k(\hat{\underline{\theta}}_{k|k-1})\hat{\mathbf{x}}_{k|k-1})
\end{aligned}
\right.
\tag{8.15}
$$

(cf. Exercise 8.6).

We remark that if the unknown constant vector $\underline{\theta}$ is considered to be deterministic; that is, $\underline{\theta}_{k+1} = \underline{\theta}_k = \underline{\theta}$ so that $S_k = 0$, then the procedure (8.15) only yields $\hat{\underline{\theta}}_k = \hat{\underline{\theta}}_{k-1}$ for all k, independent of the observation data (cf. Exercise 8.7), and this does not give us any information on $\hat{\underline{\theta}}_k$. In other words, by using $S_k = 0$, the unknown system parameter vector $\underline{\theta}$ cannot be identified via the extended Kalman filtering technique.

8.4 An Example of Constant Parameter Identification

The following simple example will demonstrate how well the extended Kalman filter performs for the purpose of adaptive system identification, even with an arbitrary choice of the initial estimate $\hat{\underline{\theta}}_0$.

Consider a linear system with the state-space description

$$\begin{cases} x_{k+1} = a \; x_k \\ \quad v_k = x_k + \eta_k \, , \end{cases}$$

where a is the unknown parameter that we must identify. Now, we treat a as a random variable; that is, we consider

$$a_{k+1} = a_k + \zeta_k \, ,$$

where a_k is the value of a at the kth instant and $E(\zeta_k) = 0$, $Var(\zeta_k) = 0.01$, say. Suppose that $E(x_0) = 1, Var(x_0) = 0.01$, and $\{\eta_k\}$ is a zero-mean Gaussian white noise sequence with $Var(\eta_k) = 0.01$. The objective is to estimate the unknown parameters a_k while performing the Kalman filtering procedure. By replacing a with a_k in the system equation, the above equations become the following nonlinear state-space description:

$$\begin{cases} \begin{bmatrix} x_{k+1} \\ a_{k+1} \end{bmatrix} = \begin{bmatrix} a_k x_k \\ a_k \end{bmatrix} + \begin{bmatrix} 0 \\ \zeta_k \end{bmatrix} \\ v_k = [1 \;\; 0] \begin{bmatrix} x_k \\ a_k \end{bmatrix} + \eta_k \, . \end{cases} \tag{8.16}$$

An application of (8.15) to this model yields

$$\begin{cases} P_{k,k-1} = \begin{bmatrix} \hat{a}_{k-1} & \hat{x}_{k-1} \\ 0 & 1 \end{bmatrix} P_{k,k-1} \begin{bmatrix} \hat{a}_{k-1} & 0 \\ \hat{x}_{k-1} & 1 \end{bmatrix} + \begin{bmatrix} 0 & 0 \\ 0 & 0.01 \end{bmatrix} \\ G_k = P_{k,k-1} \begin{bmatrix} 1 \\ 0 \end{bmatrix} \left[[1 \;\; 0] P_{k,k-1} \begin{bmatrix} 1 \\ 0 \end{bmatrix} + 0.01 \right]^{-1} \\ P_{k,k} = \begin{bmatrix} \begin{bmatrix} 1 & 0 \\ 0 & 1 \end{bmatrix} - G_k [1 \;\; 0] \end{bmatrix} P_{k,k-1} \\ \begin{bmatrix} \hat{x}_k \\ \hat{a}_k \end{bmatrix} = \begin{bmatrix} \hat{a}_{k-1} \hat{x}_{k-1} \\ \hat{a}_{k-1} \end{bmatrix} + G_k (v_k - \hat{a}_{k-1} \hat{x}_{k-1}) \, , \end{cases} \tag{8.17}$$

where the initial estimate of x_0 is $\hat{x}_0 = E(x_0) = 1$ but $\hat{a}_0$ is unknown.

To test this adaptive parameter identification algorithm, we create two pseudo-random sequences $\{\eta_k\}$ and $\{\zeta_k\}$ with zero mean and the above specified values of variances. Let us also (secretly) assign the value of a to be -1 in order to generate the data $\{v_k\}$.

To apply the algorithm described in (8.17), we need an initial estimate $\hat{a}_0$ of a. This can be done by using (8.14) with the first bit of data $v_1 = -1.1$ that we generate. In other words, we obtain

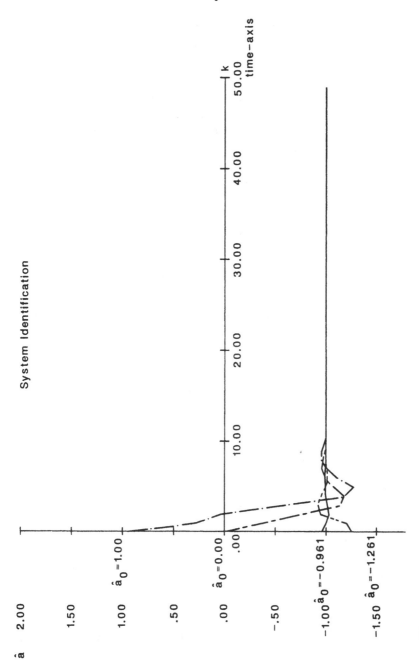

Fig. 8.3.

$$0.99a^2 + 2.2a + 1.2 = 0$$

or

$$\hat{a}_0 = -1.261, \quad -0.961.$$

In fact, the initial estimate $\hat{a}_0$ can be chosen arbitrarily. In Fig. 8.3. we have plots of $\hat{a}_k$ for four different choices of $\hat{a}_0$, namely: $-1.261, -0.961, 0$, and 1. Observe that for $k \geq 10$, all estimates $\hat{a}_k$ with any of these initial conditions are already very close to the actual value $a = -1$.

Suppose that a is considered to be deterministic, so that $a_{k+1} = a_k$ and (8.16) becomes

$$\begin{cases} \begin{bmatrix} x_{k+1} \\ a_{k+1} \end{bmatrix} = \begin{bmatrix} a_k x_k \\ a_k \end{bmatrix} \\ v_k = \begin{bmatrix} 1 & 0 \end{bmatrix} \begin{bmatrix} x_k \\ a_k \end{bmatrix} + \eta_k . \end{cases}$$

Then it is not difficult to see that the Kalman gain matrix becomes

$$G_k = \begin{bmatrix} g_k \\ 0 \end{bmatrix}$$

for some constant g_k and the "correlation" formula in the filtering algorithm is

$$\begin{bmatrix} \hat{x}_k \\ \hat{a}_k \end{bmatrix} = \begin{bmatrix} \hat{a}_{k-1}\hat{x}_{k-1} \\ \hat{a}_{k-1} \end{bmatrix} + \begin{bmatrix} g_k \\ 0 \end{bmatrix} (v_k - \hat{a}_{k-1}\hat{x}_{k-1}) .$$

Note that since $\hat{a}_k = \hat{a}_{k-1} = \hat{a}_0$ is independent of the data $\{v_k\}$, the value of a cannot be identified.

8.5 Modified Extended Kalman Filter

In this section, we modify the extended Kalman filtering algorithm discussed in the previous sections and introduce a more efficient parallel computational scheme for system parameters identification. The modification is achieved by an improved linearization procedure. As a result, the modified Kalman filtering algorithm can be applied to real-time system parameter identification even for time-varying stochastic systems. We will also give two numerical examples with computer simulations to demonstrate the effectiveness of this modified filtering scheme over the original extended Kalman filtering algorithm.

The nonlinear stochastic system under consideration is the following:

$$\begin{cases} \begin{bmatrix} \mathbf{x}_{k+1} \\ \mathbf{y}_{k+1} \end{bmatrix} = \begin{bmatrix} F_k(\mathbf{y}_k)\mathbf{x}_k \\ H_k(\mathbf{x}_k,\mathbf{y}_k) \end{bmatrix} + \begin{bmatrix} \Gamma_k^1(\mathbf{x}_k,\mathbf{y}_k) & 0 \\ \Gamma_k^2(\mathbf{x}_k,\mathbf{y}_k) & \Gamma_k^3(\mathbf{x}_k,\mathbf{y}_k) \end{bmatrix} \begin{bmatrix} \underline{\xi}_k^1 \\ \underline{\xi}_k^2 \end{bmatrix} \\ \\ \mathbf{v}_k = [C_k(\mathbf{x}_k,\mathbf{y}_k) \quad 0] \begin{bmatrix} \mathbf{x}_k \\ \mathbf{y}_k \end{bmatrix} + \underline{\eta}_k, \end{cases} \tag{8.18}$$

where, $\mathbf{x}_k$ and $\mathbf{y}_k$ are n- and m-dimensional vectors, respectively, $\left\{ \begin{bmatrix} \underline{\xi}_k^1 \\ \underline{\xi}_k^2 \end{bmatrix} \right\}$ and $\{\underline{\eta}_k\}$ uncorrelated zero-mean Gaussian white noise sequences with variance matrices

$$Q_k = Var \begin{bmatrix} \underline{\xi}_k^1 \\ \underline{\xi}_k^2 \end{bmatrix} \qquad and \qquad R_k = Var\,(\underline{\eta}_k),$$

respectively, and $F_k, H_k, \Gamma_k^1, \Gamma_k^2, \Gamma_k^3$ and C_k nonlinear matrix-valued functions. Assume that F_k and C_k are differentiable.

The modified Kalman filtering algorithm that we will derive consists of two sub-algorithms. Algorithm I shown below is a modification of the extended Kalman filter discussed previously. It differs from the previous one in that the real-time linear Taylor approximation is not taken at the previous estimate. Instead, in order to improve the performance, it is taken at the optimal estimate of $\mathbf{x}_k$ given by a standard Kalman filtering algorithm (which will be called Algorithm II below), from the subsystem

$$\begin{cases} \mathbf{x}_{k+1} = F_k(\tilde{\mathbf{y}}_k)\mathbf{x}_k + \Gamma_k^1(\tilde{\mathbf{x}}_k,\tilde{\mathbf{y}}_k)\underline{\xi}_k^1 \\ \mathbf{v}_k = C(\tilde{\mathbf{x}}_k,\tilde{\mathbf{y}}_k)\mathbf{x}_k + \underline{\eta}_k \end{cases} \tag{8.19}$$

of (8.18), evaluated at the estimate $(\tilde{\mathbf{x}}_k, \tilde{\mathbf{y}}_k)$ from Algorithm I. In other words, the two algorithms are applied in parallel starting with the same initial estimate as shown in Figure 8.4, where Algorithm I (namely, the modified Kalman filtering algorithm) is used for yielding the estimate $\begin{bmatrix} \tilde{\mathbf{x}}_k \\ \tilde{\mathbf{y}}_k \end{bmatrix}$ with the input $\hat{\mathbf{x}}_{k-1}$ obtained from Algorithm II (namely: the standard Kalman algorithm for the linear system (8.19)); and Algorithm II is used for yielding the estimate $\hat{\mathbf{x}}_k$ with the input $\begin{bmatrix} \tilde{\mathbf{x}}_{k-1} \\ \tilde{\mathbf{y}}_{k-1} \end{bmatrix}$ obtained from Algorithm I. The two algorithms together will be called the *parallel algorithm* (I and II) later.

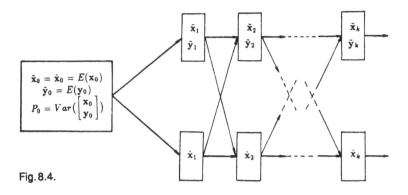

Fig. 8.4.

Algorithm I: Set

$$\begin{bmatrix} \tilde{\mathbf{x}}_0 \\ \tilde{\mathbf{y}}_0 \end{bmatrix} = \begin{bmatrix} E(\mathbf{x}_0) \\ E(\mathbf{y}_0) \end{bmatrix} \quad and \quad P_0 = Var(\begin{bmatrix} \mathbf{x}_0 \\ \mathbf{y}_0 \end{bmatrix}).$$

For $k = 1, 2, \cdots$, compute

$$P_{k,k-1} = \left[\frac{\partial}{\partial \begin{bmatrix} \mathbf{x}_{k-1} \\ \mathbf{y}_{k-1} \end{bmatrix}} \begin{bmatrix} F_{k-1}(\tilde{\mathbf{y}}_{k-1})\hat{\mathbf{x}}_{k-1} \\ H_{k-1}(\hat{\mathbf{x}}_{k-1}, \tilde{\mathbf{y}}_{k-1}) \end{bmatrix} \right] P_{k-1}$$

$$\cdot \left[\frac{\partial}{\partial \begin{bmatrix} \mathbf{x}_{k-1} \\ \mathbf{y}_{k-1} \end{bmatrix}} \begin{bmatrix} F_{k-1}(\tilde{\mathbf{y}}_{k-1})\hat{\mathbf{x}}_{k-1} \\ H_{k-1}(\hat{\mathbf{x}}_{k-1}, \tilde{\mathbf{y}}_{k-1}) \end{bmatrix} \right]^{\top}$$

$$+ \begin{bmatrix} \Gamma^1_{k-1}(\tilde{\mathbf{x}}_{k-1}, \tilde{\mathbf{y}}_{k-1}) & 0 \\ \Gamma^2_{k-1}(\tilde{\mathbf{x}}_{k-1}, \tilde{\mathbf{y}}_{k-1}) & \Gamma^3_{k-1}(\tilde{\mathbf{x}}_{k-1}, \tilde{\mathbf{y}}_{k-1}) \end{bmatrix} Q_{k-1}$$

$$\cdot \begin{bmatrix} \Gamma^1_{k-1}(\tilde{\mathbf{x}}_{k-1}, \tilde{\mathbf{y}}_{k-1}) & 0 \\ \Gamma^2_{k-1}(\tilde{\mathbf{x}}_{k-1}, \tilde{\mathbf{y}}_{k-1}) & \Gamma^3_{k-1}(\tilde{\mathbf{x}}_{k-1}, \tilde{\mathbf{y}}_{k-1}) \end{bmatrix}^{\top}$$

$$\begin{bmatrix} \tilde{\mathbf{x}}_{k|k-1} \\ \tilde{\mathbf{y}}_{k|k-1} \end{bmatrix} = \begin{bmatrix} F_{k-1}(\tilde{\mathbf{y}}_{k-1})\hat{\mathbf{x}}_{k-1} \\ H_{k-1}(\hat{\mathbf{x}}_{k-1}, \tilde{\mathbf{y}}_{k-1}) \end{bmatrix}$$

$$G_k = P_{k,k-1} \left[\frac{\partial}{\partial \begin{bmatrix} \mathbf{x}_k \\ \mathbf{y}_k \end{bmatrix}} C_k(\tilde{\mathbf{x}}_{k|k-1}, \tilde{\mathbf{y}}_{k|k-1}) \right]^{\top} \left\{ \left[\frac{\partial}{\partial \begin{bmatrix} \mathbf{x}_k \\ \mathbf{y}_k \end{bmatrix}} C_k(\tilde{\mathbf{x}}_{k|k-1}, \tilde{\mathbf{y}}_{k|k-1}) \right] \right.$$

$$\left. \cdot P_{k,k-1} \left[\frac{\partial}{\partial \begin{bmatrix} \mathbf{x}_k \\ \mathbf{y}_k \end{bmatrix}} C_k(\tilde{\mathbf{x}}_{k|k-1}, \tilde{\mathbf{y}}_{k|k-1}) \right]^{\top} + R_k \right\}^{-1}$$

$$P_k = \left[I - G_k \left[\frac{\partial}{\partial \begin{bmatrix} \mathbf{x}_k \\ \mathbf{y}_k \end{bmatrix}} C_k(\tilde{\mathbf{x}}_{k|k-1}, \tilde{\mathbf{y}}_{k|k-1}) \right] \right] P_{k,k-1}$$

$$\begin{bmatrix} \tilde{\mathbf{x}}_k \\ \tilde{\mathbf{y}}_k \end{bmatrix} = \begin{bmatrix} \tilde{\mathbf{x}}_{k|k-1} \\ \tilde{\mathbf{y}}_{k|k-1} \end{bmatrix} + G_k(\mathbf{v}_k - C_k(\tilde{\mathbf{x}}_{k|k-1}, \tilde{\mathbf{y}}_{k|k-1})\tilde{\mathbf{x}}_{k|k-1}),$$

where $Q_k = Var\left(\begin{bmatrix} \xi_k^1 \\ \xi_k^2 \end{bmatrix} \right)$ and $R_k = Var(\underline{\eta}_k)$; and $\hat{\mathbf{x}}_{k-1}$ is obtained by using the following algorithm.

Algorithm II: Set

$$\hat{\mathbf{x}}_0 = E(\mathbf{x}_0) \quad and \quad P_0 = Var(\mathbf{x}_0).$$

For $k = 1, 2, \cdots$, compute

$$\begin{aligned}
P_{k,k-1} =& [F_{k-1}(\tilde{\mathbf{y}}_{k-1})]P_{k-1}[F_{k-1}(\tilde{\mathbf{y}}_{k-1})]^\top \\
&+ [\Gamma_{k-1}^1(\tilde{\mathbf{x}}_{k-1}, \tilde{\mathbf{y}}_{k-1})]Q_{k-1}[\Gamma_{k-1}^1(\tilde{\mathbf{x}}_{k-1}, \tilde{\mathbf{y}}_{k-1})]^\top \\
\hat{\mathbf{x}}_{k|k-1} =& [F_{k-1}(\tilde{\mathbf{y}}_{k-1})]\hat{\mathbf{x}}_{k-1} \\
G_k =& P_{k,k-1}[C_k(\tilde{\mathbf{x}}_{k-1}, \tilde{\mathbf{y}}_{k-1})]^\top \\
&\cdot [[C_k(\tilde{\mathbf{x}}_{k-1}, \tilde{\mathbf{y}}_{k-1})]P_{k,k-1}[C_k(\tilde{\mathbf{x}}_{k-1}, \tilde{\mathbf{y}}_{k-1})]^\top + R_k]^{-1} \\
P_k =& [I - G_k[C_k(\tilde{\mathbf{x}}_{k-1}, \tilde{\mathbf{y}}_{k-1})]]P_{k,k-1} \\
\hat{\mathbf{x}}_k =& \hat{\mathbf{x}}_{k|k-1} + G_k(\mathbf{v}_k - [C_k(\tilde{\mathbf{x}}_{k-1}, \tilde{\mathbf{y}}_{k-1})]\hat{\mathbf{x}}_{k|k-1}),
\end{aligned}$$

where $Q_k = Var(\underline{\xi}_k^1)$, $R_k = Var(\underline{\eta}_k)$, and $(\tilde{\mathbf{x}}_{k-1}, \tilde{\mathbf{y}}_{k-1})$ is obtained from Algorithm I.

Here, the following notation is used:

$$\begin{bmatrix} \frac{\partial}{\partial \begin{bmatrix} \mathbf{x}_{k-1} \\ \mathbf{y}_{k-1} \end{bmatrix}} \begin{bmatrix} F_{k-1}(\tilde{\mathbf{y}}_{k-1})\hat{\mathbf{x}}_{k-1} \\ H_{k-1}(\hat{\mathbf{x}}_{k-1}, \tilde{\mathbf{y}}_{k-1}) \end{bmatrix} \end{bmatrix}$$

$$= \begin{bmatrix} \frac{\partial}{\partial \begin{bmatrix} \mathbf{x}_{k-1} \\ \mathbf{y}_{k-1} \end{bmatrix}} \begin{bmatrix} F_k(\mathbf{y}_{k-1})\mathbf{x}_{k-1} \\ H_{k-1}(\mathbf{x}_{k-1}, \mathbf{y}_{k-1}) \end{bmatrix} \end{bmatrix}_{\substack{\mathbf{x}_{k-1} = \hat{\mathbf{x}}_{k-1} \\ \mathbf{y}_{k-1} = \tilde{\mathbf{y}}_{k-1}}} .$$

We remark that the modified Kalman filtering algorithm (namely: Algorithm I) is different from the original extended Kalman filtering scheme in that the Jacobian matrix of the (non-linear) vector-valued function $F_{k-1}(\mathbf{y}_{k-1})\mathbf{x}_{k-1}$ and the prediction term $\begin{bmatrix} \tilde{\mathbf{x}}_{k|k-1} \\ \tilde{\mathbf{y}}_{k|k-1} \end{bmatrix}$ are both evaluated at the optimal position $\underline{\hat{\mathbf{x}}}_{k-1}$

at each time instant, where $\hat{\mathbf{x}}_{k-1}$ is determined by the standard Kalman filtering algorithm (namely, Algorithm II).

We next give a derivation of the modified extended Kalman filtering algorithm. Let $\hat{\mathbf{x}}_k$ be the optimal estimate of the state vector $\mathbf{x}_k$ in the linear system (8.19), in the sense that

$$Var(\hat{\mathbf{x}}_k - \mathbf{x}_k) \leq Var(\mathbf{z}_k - \mathbf{x}_k) \qquad (8.20)$$

among all linear and unbiased estimates $\mathbf{z}_k$ of $\mathbf{x}_k$ using all data information $\mathbf{v}_1, \cdots, \mathbf{v}_k$. Since (8.19) is a subsystem of the original system evaluated at $(\tilde{\mathbf{x}}_k, \tilde{\mathbf{y}}_k)$ just as the extended Kalman filter is derived, it is clear that

$$Var(\hat{X}_k - X_k) \leq Var(\widetilde{X}_k - X_k), \qquad (8.21)$$

where

$$X_k = \begin{bmatrix} \mathbf{x}_k \\ \mathbf{y}_k \end{bmatrix}, \widehat{X}_k = \begin{bmatrix} \hat{\mathbf{x}}_k \\ \tilde{\mathbf{y}}_k \end{bmatrix} \quad and \quad \widetilde{X}_k = \begin{bmatrix} \tilde{\mathbf{x}}_k \\ \tilde{\mathbf{y}}_k \end{bmatrix}.$$

Now, consider an $(n + m)$-dimensional nonlinear differentiable vector-valued function

$$Z = \mathbf{f}(X) \qquad (8.22)$$

defined on $\mathbf{R}^{n+m}$. Since the purpose is to estimate

$$Z_k = \mathbf{f}(X_k)$$

from some (optimal) estimate $\widehat{X}_k$ of X_k, we use $\widehat{X}_k$ as the center of the linear Taylor approximation. In doing so, choosing a better estimate of X_k as the center should yield a better estimate for Z_k. In other words, if $\hat{X}_k$ is used in place of $\widetilde{X}_k$ as the center for the linear Taylor approximation of Z_k, we should obtain a better estimate of Z_k as shown in the illustrative diagram shown in Figure 8.5. Here, $\tilde{X}_k$ is used as the center for the linear Taylor approximation in the standard extended Kalman filter.

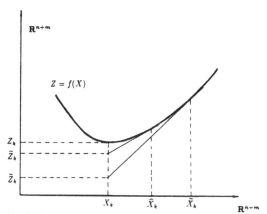

Fig. 8.5.

Now, if $\mathbf{f}'(X)$ denotes the Jacobian matrix of $\mathbf{f}(X)$ at X, then the linear Taylor estimate $\widehat{Z}_k$ of Z_k with center at $\widehat{X}_k$ is given by

$$\widehat{Z}_k = \mathbf{f}(\widehat{X}_k) + \mathbf{f}'(\widehat{X}_k)(X_k - \widehat{X}_k)$$

or, equivalently,

$$\widehat{Z}_k = \mathbf{f}'(\widehat{X}_k)X_k + (\mathbf{f}(\widehat{X}_k) - \mathbf{f}'(\widehat{X}_k)\widehat{X}_k). \tag{8.23}$$

We next return to the nonlinear model (8.18) and apply the linearization formula (8.23) to $\mathbf{f}_k$ defined by

$$\mathbf{f}_k \left(\begin{bmatrix} \mathbf{x}_k \\ \mathbf{y}_k \end{bmatrix} \right) = \begin{bmatrix} F_k(\mathbf{y}_k)\mathbf{x}_k \\ H_k(\mathbf{x}_k, \mathbf{y}_k) \end{bmatrix}. \tag{8.24}$$

Suppose that $(\tilde{\mathbf{x}}_k, \tilde{\mathbf{y}}_k)$ and $\hat{\mathbf{x}}_k$ have already been determined by using the parallel algorithm (I and II) at the kth instant. Going the $(k+1)$st instant, we apply (8.23) with center $\begin{bmatrix} \hat{\mathbf{x}}_k \\ \tilde{\mathbf{y}}_k \end{bmatrix}$ (instead of $\begin{bmatrix} \tilde{\mathbf{x}}_k \\ \tilde{\mathbf{y}}_k \end{bmatrix}$), using the Jacobian matrix

$$\mathbf{f}'_k \left(\begin{bmatrix} \hat{\mathbf{x}}_k \\ \tilde{\mathbf{y}}_k \end{bmatrix} \right) = \begin{bmatrix} \dfrac{\partial}{\partial \begin{bmatrix} \mathbf{x}_k \\ \mathbf{y}_k \end{bmatrix}} \begin{bmatrix} F_k(\tilde{\mathbf{y}}_k)\hat{\mathbf{x}}_k \\ H_k(\hat{\mathbf{x}}_k, \tilde{\mathbf{y}}_k) \end{bmatrix} \end{bmatrix},$$

to linearize the model (8.18). Moreover, as is usually done in the standard extended Kalman filter, we use the zeroth order Taylor approximations $C_k(\tilde{\mathbf{x}}_k, \tilde{\mathbf{y}}_k)$ and $\Gamma^i_k(\tilde{\mathbf{x}}_k, \tilde{\mathbf{y}}_k)$ for the matrices $C_k(\mathbf{x}_k, \mathbf{y}_k)$ and $\Gamma^i_k(\mathbf{x}_k, \mathbf{y}_k), i = 1, 2, 3$, respectively, in the linearization of the model (8.18). This leads to the following linear state-space description:

$$
\begin{cases}
\begin{bmatrix} \mathbf{x}_{k+1} \\ \mathbf{y}_{k+1} \end{bmatrix} = \left[\dfrac{\partial}{\partial \begin{bmatrix} \mathbf{x}_k \\ \mathbf{y}_k \end{bmatrix}} \begin{bmatrix} F_k(\tilde{\mathbf{y}}_k)\hat{\mathbf{x}}_k \\ H_k(\hat{\mathbf{x}}_k, \tilde{\mathbf{y}}_k) \end{bmatrix} \right] \begin{bmatrix} \mathbf{x}_k \\ \mathbf{y}_k \end{bmatrix} + \mathbf{u}_k \\
\qquad + \begin{bmatrix} \Gamma_k^1(\tilde{\mathbf{x}}_k, \tilde{\mathbf{y}}_k) & 0 \\ \Gamma_k^2(\tilde{\mathbf{x}}_k, \tilde{\mathbf{y}}_k) & \Gamma_k^3(\tilde{\mathbf{x}}_k, \tilde{\mathbf{y}}_k) \end{bmatrix} \begin{bmatrix} \underline{\xi}_k^1 \\ \underline{\xi}_k^2 \end{bmatrix} \\
\mathbf{v}_k = [C_k(\tilde{\mathbf{x}}_k, \tilde{\mathbf{y}}_k) \quad 0] \begin{bmatrix} \mathbf{x}_k \\ \mathbf{y}_k \end{bmatrix} + \underline{\eta}_k,
\end{cases} \tag{8.25}
$$

in which the constant vector

$$
\mathbf{u}_k = \begin{bmatrix} F_k(\tilde{\mathbf{y}}_k)\hat{\mathbf{x}}_k \\ H_k(\hat{\mathbf{x}}_k, \tilde{\mathbf{y}}_k) \end{bmatrix} - \left[\dfrac{\partial}{\partial \begin{bmatrix} \mathbf{x}_k \\ \mathbf{y}_k \end{bmatrix}} \begin{bmatrix} F_k(\tilde{\mathbf{y}}_k)\hat{\mathbf{x}}_k \\ H_k(\hat{\mathbf{x}}_k, \tilde{\mathbf{y}}_k) \end{bmatrix} \right] \begin{bmatrix} \hat{\mathbf{x}}_k \\ \tilde{\mathbf{y}}_k \end{bmatrix}
$$

can be considered as a deterministic control input. Hence, similar to the derivation of the standard Kalman filter, we obtain Algorithm I for the linear system (8.25). Here, it should be noted that the prediction term $\begin{bmatrix} \tilde{\mathbf{x}}_{k|k-1} \\ \tilde{\mathbf{y}}_{k|k-1} \end{bmatrix}$ in Algorithm I is the only term that contains $\mathbf{u}_k$ as formulated below:

$$
\begin{bmatrix} \tilde{\mathbf{x}}_{k|k-1} \\ \tilde{\mathbf{y}}_{k|k-1} \end{bmatrix} = \left[\dfrac{\partial}{\partial \begin{bmatrix} \mathbf{x}_{k-1} \\ \mathbf{y}_{k-1} \end{bmatrix}} \begin{bmatrix} F_{k-1}(\tilde{\mathbf{y}}_{k-1})\hat{\mathbf{x}}_{k-1} \\ H_{k-1}(\hat{\mathbf{x}}_{k-1}, \tilde{\mathbf{y}}_{k-1}) \end{bmatrix} \right] \begin{bmatrix} \hat{\mathbf{x}}_{k-1} \\ \tilde{\mathbf{y}}_{k-1} \end{bmatrix} + \mathbf{u}_{k-1}
$$

$$
= \left[\dfrac{\partial}{\partial \begin{bmatrix} \mathbf{x}_{k-1} \\ \mathbf{y}_{k-1} \end{bmatrix}} \begin{bmatrix} F_{k-1}(\tilde{\mathbf{y}}_{k-1})\hat{\mathbf{x}}_{k-1} \\ H_{k-1}(\hat{\mathbf{x}}_{k-1}, \tilde{\mathbf{y}}_{k-1}) \end{bmatrix} \right] \begin{bmatrix} \hat{\mathbf{x}}_{k-1} \\ \tilde{\mathbf{y}}_{k-1} \end{bmatrix} + \begin{bmatrix} F_{k-1}(\tilde{\mathbf{y}}_{k-1})\hat{\mathbf{x}}_{k-1} \\ H_{k-1}(\hat{\mathbf{x}}_{k-1}, \tilde{\mathbf{y}}_{k-1}) \end{bmatrix}
$$

$$
- \left[\dfrac{\partial}{\partial \begin{bmatrix} \mathbf{x}_{k-1} \\ \mathbf{y}_{k-1} \end{bmatrix}} \begin{bmatrix} F_{k-1}(\tilde{\mathbf{y}}_{k-1})\hat{\mathbf{x}}_{k-1} \\ H_{k-1}(\hat{\mathbf{x}}_{k-1}, \tilde{\mathbf{y}}_{k-1}) \end{bmatrix} \right] \begin{bmatrix} \hat{\mathbf{x}}_{k-1} \\ \tilde{\mathbf{y}}_{k-1} \end{bmatrix}
$$

$$
= \begin{bmatrix} F_{k-1}(\tilde{\mathbf{y}}_{k-1})\hat{\mathbf{x}}_{k-1} \\ H_{k-1}(\hat{\mathbf{x}}_{k-1}, \tilde{\mathbf{y}}_{k-1}) \end{bmatrix}.
$$

8.6 Time-Varying Parameter Identification

In this section, we give two numerical examples to demonstrate the advantage of the modified extended Kalman filter over the standard one, for both state estimation and system parameter identification.

For this purpose, we first consider the nonlinear system

$$\begin{cases} \begin{bmatrix} x_{k+1} \\ y_{k+1} \\ z_{k+1} \end{bmatrix} = \begin{bmatrix} \begin{bmatrix} 1 & z_k \\ -0.1 & 1 \\ & z_k \end{bmatrix} \begin{bmatrix} x_k \\ y_k \end{bmatrix} \end{bmatrix} + \begin{bmatrix} \xi_k^1 \\ \xi_k^2 \\ \xi_k^3 \end{bmatrix}, \quad \begin{bmatrix} x_0 \\ y_0 \\ z_0 \end{bmatrix} = \begin{bmatrix} 1.0 \\ 1.0 \\ 0.1 \end{bmatrix} \\[4em] v_k = x_k + \eta_k, \end{cases}$$

where $\underline{\xi}_k$ and η_k are uncorrelated zero-mean Gaussian white noise sequences with $Var(\underline{\xi}_k) = 0.1 I_3$ and $Var(\eta_k) = 0.01$ for all k. We then create two pseudo-random noise sequences and use

$$\begin{bmatrix} \hat{x}_0 \\ \hat{y}_0 \\ \hat{z}_0 \end{bmatrix} = \begin{bmatrix} \tilde{x}_0 \\ \tilde{y}_0 \\ \tilde{z}_0 \end{bmatrix} = \begin{bmatrix} 100 \\ 100 \\ 1.0 \end{bmatrix} \quad \text{and} \quad P_0 = I_3$$

as the initial estimates in the comparison of the extended Kalman filtering algorithm and the parallel algorithm (I and II).

Computer simulation shows that both methods give excellent estimates to the actual x_k component of the state vector. However, the parallel algorithm (I and II) provides much better estimates to the y_k and z_k components than the standard extended Kalman filtering algorithm.

In Figures 8.6 and 8.7, we plot the estimates of y_k against the actual graph of y_k by using the standard extended Kalman filtering scheme and the parallel algorithm, respectively. Analogous plots for the z_k component are shown in Figures 8.8 and 8.9, respectively.

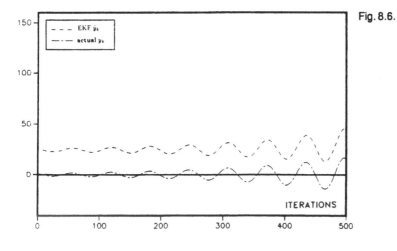

Fig. 8.6.

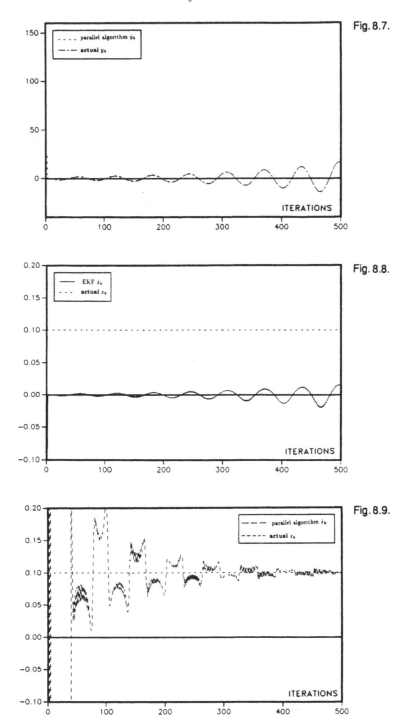

Fig. 8.7.

Fig. 8.8.

Fig. 8.9.

As a second example, we consider the following system parameter identification problem for the time-varying stochastic system (8.19) with an unknown time-varying parameter vector $\underline{\theta}_k$. As usual, we only assume that $\underline{\theta}_k$ is a zero-mean random vector. This leads to the nonlinear model (8.25) with $\mathbf{y}_k$ replaced by $\underline{\theta}_k$. Applying the modified extended Kalman filtering algorithm to this model, we have the following:

Algorithm I':   Let $\text{Var}(\underline{\theta}_0) > 0$ and set

$$\begin{bmatrix} \tilde{\mathbf{x}}_0 \\ \tilde{\underline{\theta}}_0 \end{bmatrix} = \begin{bmatrix} E(\mathbf{x}_0) \\ 0 \end{bmatrix}, \quad P_{0,0} = \begin{bmatrix} Var(\mathbf{x}_0) & 0 \\ 0 & Var(\underline{\theta}_0) \end{bmatrix}.$$

For $k = 1, 2, \cdots$, compute

$$\begin{bmatrix} \tilde{\mathbf{x}}_{k,|k-1} \\ \tilde{\underline{\theta}}_{k|k-1} \end{bmatrix} = \begin{bmatrix} A_{k-1}(\tilde{\underline{\theta}}_{k-1})\hat{\mathbf{x}}_{k-1} \\ \tilde{\underline{\theta}}_{k-1} \end{bmatrix}$$

$$P_{k,k-1} = \begin{bmatrix} A_{k-1}(\tilde{\underline{\theta}}_{k-1}) & \frac{\partial}{\partial \underline{\theta}} \left[A_{k-1}(\tilde{\underline{\theta}}_{k-1})\hat{\mathbf{x}}_{k-1} \right] \\ 0 & I \end{bmatrix} P_{k-1,k-1}$$

$$\cdot \begin{bmatrix} A_{k-1}(\tilde{\underline{\theta}}_{k-1}) & \frac{\partial}{\partial \underline{\theta}} \left[A_{k-1}(\tilde{\underline{\theta}}_{k-1})\hat{\mathbf{x}}_{k-1} \right] \\ 0 & I \end{bmatrix}^{\top}$$

$$+ \begin{bmatrix} \Gamma_{k-1}(\tilde{\underline{\theta}}_{k-1})Q_{k-1}\Gamma_{k-1}^{\top}(\tilde{\underline{\theta}}_{k-1}) & 0 \\ 0 & 0 \end{bmatrix}$$

$$G_k = P_{k,k-1}[C_k(\tilde{\underline{\theta}}_{k|k-1}) \quad 0]^{\top}$$
$$\cdot [[C_k(\tilde{\underline{\theta}}_{k|k-1}) \quad 0]P_{k,k-1}[C_k(\tilde{\underline{\theta}}_{k|k-1}) \quad 0]^{\top} + R_k]^{-1}$$

$$P_{k,k} = [I - G_k[C_k(\tilde{\underline{\theta}}_{k|k-1}) \quad 0]]P_{k,k-1}$$

$$\begin{bmatrix} \tilde{\mathbf{x}}_k \\ \tilde{\underline{\theta}}_k \end{bmatrix} = \begin{bmatrix} \tilde{\mathbf{x}}_{k|k-1} \\ \tilde{\underline{\theta}}_{k|k-1} \end{bmatrix} + G_k(v_k - C_k(\tilde{\underline{\theta}}_{k|k-1})\tilde{\mathbf{x}}_{k|k-1}),$$

where $\hat{\mathbf{x}}_k$ is obtained in parallel by applying Algorithm II with $\tilde{\mathbf{y}}_k$ replaced by $\tilde{\underline{\theta}}_k$.

To demonstrate the performance of the parallel algorithm (I' and II), we consider the following stochastic system with an unknown system parameter θ_k:

$$\begin{cases} \mathbf{x}_{k+1} = \begin{bmatrix} 1 & 1 \\ 0 & \theta_k \end{bmatrix} \mathbf{x}_k + \underline{\xi}_k, \quad \mathbf{x}_0 = \begin{bmatrix} 1 \\ 1 \end{bmatrix} \\ v_k = [1 \quad 0]\mathbf{x}_k + \eta_k. \end{cases}$$

For computer simulation, we use changing values of θ_k given by

$$\theta_k = \begin{cases} 1.0 & 0 \le k \le 20 \\ 1.2 & 20 < k \le 40 \\ 1.5 & 40 < k \le 60 \\ 2.0 & 60 < k \le 80 \\ 2.3 & 80 < k. \end{cases} \quad with \quad \hat{\theta}_0 = 5.0.$$

The problem now is to identify these (unknown) values. It turns out that the standard extended Kalman filtering algorithm does not give any reasonable estimate of θ_k, and in fact the estimates seem to grow exponentially with k. In Figure 8.10 the estimates of θ_k are given using the parallel algorithm (I' and II) with initial variances (i) Var(θ_0)=0.1, (ii) Var(θ_0)=1.0, and (iii) Var(θ_0)=50.

We finally remark that all existing extended Kalman filtering algorithms are *ad hoc* schemes since different linearizations have to be used to derive the results. Hence, there is no rigorous theory to guarantee the optimality of the extended or modified extended Kalman filtering algorithm in general.

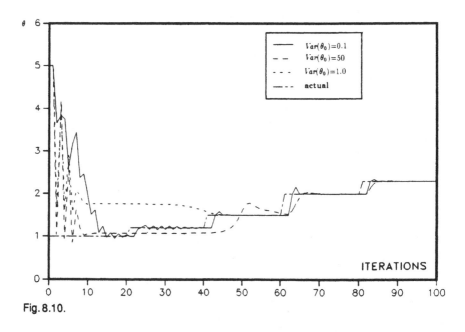

Fig. 8.10.

Exercises

8.1. Consider the two-dimensional radar tracking system shown
in Fig.8.11, where for simplicity the missile is assumed to
travel in the positive y-direction, so that $\dot{x} = 0$, $\dot{y} = v$, and
$\ddot{y} = a$, with v and a denoting the velocity and acceleration of
the missile, respectively.

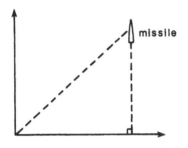

Fig. 8.11.

(a) Suppose that the radar is located at the origin and
measures the range r and angular displacement θ where
(r, θ) denotes the polar coordinates. Derive the nonlinear
equations

$$\begin{cases} \dot{r} = v sin\theta \\ \dot{\theta} = \dfrac{v}{r} cos\theta \end{cases}$$

and

$$\begin{cases} \ddot{r} = a\ sin\theta + \dfrac{v^2}{r^2} cos^2\theta \\ \ddot{\theta} = \left(\dfrac{ar - v^2 sin\theta}{r^2} \right) cos\theta - \dfrac{v^2}{r^2} sin\theta\ cos\theta \end{cases}$$

for this radar tracking model.

(b) By introducing the state vector

$$\mathbf{x} := \begin{bmatrix} r \\ \dot{r} \\ \theta \\ \dot{\theta} \end{bmatrix},$$

establish a vector-valued nonlinear differential equation
for this model.

(c) Assume that only the range r is observed and that both
the system and observation equations involve random
disturbances $\{\underline{\xi}\}$ and $\{\eta\}$. By replacing $\mathbf{x}$, $\dot{\mathbf{x}}$, $\underline{\xi}$, and η

by $\mathbf{x}_k$, $(\mathbf{x}_{k+1} - \mathbf{x}_k)h^{-1}$, $\underline{\xi}_k$, and η_k, respectively, where $h > 0$ denotes the sampling time, establish a discrete-time nonlinear system for the above model.

(d) Assume that $\{\underline{\xi}_k\}$ and $\{\eta_k\}$ in the above nonlinear system are zero-mean uncorrelated Gaussian white noise sequences. Describe the extended Kalman filtering algorithm for this nonlinear system.

8.2. Verify that the nonlinear model (8.1) can be approximated by the linear model (8.3) using the matrices and vectors defined in (8.5) and (8.7).

8.3. Verify that the extended Kalman filtering algorithm for the nonlinear model (8.1) can be obtained by applying the standard Kalman filtering equations (2.17) or (3.25) to the linear model (8.3) as shown in (8.8).

8.4. Verify equation (8.13).

8.5. Verify equation (8.14).

8.6. Verify the algorithm given in (8.15).

8.7. Prove that if the unknown constant vector $\underline{\theta}$ in the system (8.10) is deterministic (i.e., $\underline{\theta}_{k+1} = \underline{\theta}_k$ for all k), then the algorithm (8.15) fails to identify $\underline{\theta}$.

8.8. Consider the one-dimensional model

$$\begin{cases} x_{k+1} = x_k + \xi_k \\ v_k = c\, x_k + \eta_k\,, \end{cases}$$

where $E(x_0) = x^0$, $Var(x_0) = p_0$, $\{\xi_k\}$ and $\{\eta_k\}$ are both zero-mean Gaussian white noise sequences satisfying

$$E(\xi_k \xi_\ell) = q_k \delta_{k\ell}\,, \quad E(\eta_k \eta_\ell) = r_k \delta_{k\ell}\,,$$
$$E(\xi_k \eta_\ell) = E(\xi_k x_0) = E(\eta_k x_0) = 0\,.$$

Suppose that the unknown constant c is treated as a random constant:

$$c_{k+1} = c_k + \zeta_k\,,$$

where ζ_k is also a zero-mean Gaussian white noise sequence with known variances $Var(\zeta_k) = s_k$. Derive an algorithm to estimate c. Try the special case where $s_k = s > 0$.

9. Decoupling of Filtering Equations

The limiting (or steady-state) Kalman filter provides a very efficient method for estimating the state vector in a time-invariant linear system in real-time. However, if the state vector has a very high dimension n, and only a few components are of interest, this filter gives an abundance of useless information, an elimination of which should improve the efficiency of the filtering process. A decoupling method is introduced in this chapter for this purpose. It allows us to decompose an n-dimensional limiting Kalman filtering algorithm into n independent one-dimensional recursive formulas so that we may drop the ones that are of little interest.

9.1 Decoupling Formulas

Consider the time-invariant linear stochastic system

$$\begin{cases} \mathbf{x}_{k+1} = A\mathbf{x}_k + \Gamma \underline{\xi}_k \\ \mathbf{v}_k = C\mathbf{x}_k + \underline{\eta}_k , \end{cases} \tag{9.1}$$

where all items have been defined in Chapter 6 (cf. Section 6.1 for more details). Recall that the limiting Kalman gain matrix is given by

$$G = PC^\top (CPC^\top + R)^{-1} , \tag{9.2}$$

where P is the positive definite solution of the matrix Riccati equation

$$P = A[P - PC^\top (CPC^\top + R)^{-1} CP]A^\top + \Gamma Q \Gamma^\top , \tag{9.3}$$

and the steady-state estimate $\vec{\mathbf{x}}_k$ of $\mathbf{x}_k$ is given by

$$\begin{aligned} \vec{\mathbf{x}}_k &= A\vec{\mathbf{x}}_{k-1} + G(\mathbf{v}_k - CA\vec{\mathbf{x}}_{k-1}) \\ &= (I - GC)A\vec{\mathbf{x}}_{k-1} + G\mathbf{v}_k , \end{aligned} \tag{9.4}$$

(cf. (6.4)). Here, Q and R are variance matrices defined by

$$Q = E(\underline{\xi}_k \underline{\xi}_k^\top) \qquad and \qquad R = E(\underline{\eta}_k \underline{\eta}_k^\top)$$

for all k (cf. Section 6.1). Let $\Phi = (I - GC)A := [\phi_{ij}]_{n \times n}$, $G = [g_{ij}]_{n \times q}$, $1 \le q \le n$, and set

$$\vec{\mathbf{x}}_k = \begin{bmatrix} x_{k,1} \\ \vdots \\ x_{k,n} \end{bmatrix} \qquad and \qquad \mathbf{v}_k = \begin{bmatrix} v_{k,1} \\ \vdots \\ v_{k,q} \end{bmatrix}.$$

We now consider the z-transforms

$$\begin{cases} X_j = X_j(z) = \sum_{k=0}^{\infty} x_{k,j} z^{-k}, & j = 1, 2, \cdots, n, \\ V_j = V_j(z) = \sum_{k=0}^{\infty} v_{k,j} z^{-k}, & j = 1, 2, \cdots, n, \end{cases} \qquad (9.5)$$

of the jth components of $\{\vec{\mathbf{x}}_k\}$ and $\{\mathbf{v}_k\}$, respectively. Since (9.4) can be formulated as

$$x_{k+1,j} = \sum_{i=1}^{n} \phi_{ji} x_{k,i} + \sum_{i=1}^{q} g_{ji} v_{k+1,i}$$

for $k = 0, 1, \cdots$, we have

$$z X_j = \sum_{i=1}^{n} \phi_{ji} X_i + z \sum_{i=1}^{q} g_{ji} V_i.$$

Hence, by setting

$$\Lambda = \Lambda(z) = (zI - \Phi),$$

we arrive at

$$\Lambda \begin{bmatrix} X_1 \\ \vdots \\ X_n \end{bmatrix} = zG \begin{bmatrix} V_1 \\ \vdots \\ V_q \end{bmatrix}. \qquad (9.6)$$

Note that for large values of $|z|$, Λ is diagonal dominant and is therefore invertible. Hence, Cramer's rule can be used to solve for $X_1, \cdots, X_n$ in (9.6). Let Λ_i be obtained by replacing the ith column of Λ with

$$zG \begin{bmatrix} V_1 \\ \vdots \\ V_q \end{bmatrix}.$$

Then, $det\Lambda$ and $det\Lambda_i$ are both polynomials in z of degree n, and

$$(det\Lambda)X_i(z) = det\Lambda_i , \qquad (9.7)$$

$i = 1, \cdots, n$. In addition, we may write

$$det\Lambda = z^n + b_1 z^{n-1} + b_2 z^{n-2} + \cdots + b_{n-1}z + b_n , \qquad (9.8)$$

where

$$b_1 = -(\lambda_1 + \lambda_2 + \cdots + \lambda_n) ,$$
$$b_2 = (\lambda_1\lambda_2 + \lambda_1\lambda_3 + \cdots + \lambda_1\lambda_n + \lambda_2\lambda_3 + \cdots + \lambda_{n-1}\lambda_n) ,$$
$$\cdots\cdots$$
$$b_n = (-1)^n \lambda_1\lambda_2\cdots\lambda_n ,$$

with $\lambda_i, i = 1, 2, \cdots, n$, being the eigenvalues of matrix Φ. Similarly, we have

$$det\Lambda_i = \left(\sum_{\ell=0}^{n} c_\ell^1 z^{n-\ell}\right)V_1 + \cdots + \left(\sum_{\ell=0}^{n} c_\ell^q z^{n-\ell}\right)V_\ell , \qquad (9.9)$$

where c_ℓ^i, $\ell = 0, 1, \cdots, n$, $i = 1, 2, \cdots, q$, can also be computed explicitly. Now, by substituting (9.8) and (9.9) into (9.7) and then taking the inverse z-transforms on both sides, we obtain the following recursive (decoupling) formulas:

$$x_{k,i} = - b_1 x_{k-1,i} - b_2 x_{k-2,i} - \cdots - b_n x_{k-n,i}$$
$$+ c_0^1 v_{k,1} + c_1^1 v_{k-1,1} + \cdots + c_n^1 v_{k-n,1}$$
$$\cdots\cdots$$
$$+ c_0^q v_{k,q} + c_1^q v_{k-1,q} + \cdots + c_n^q v_{k-n,q} , \qquad (9.10)$$

$i = 1, 2, \cdots, n$. Note that the coefficients $b_1, \cdots, b_n$ and $c_0^i, \cdots, c_n^i, i = 1, \cdots, q$, can be computed before the filtering process is applied. We also remark that in the formula (9.10), each $x_{k,i}$ depends only on the previous state variables $x_{k-1,i}, \cdots, x_{k-n,i}$ and the data information, but not on any other state variables $x_{k-\ell,j}$ with $j \neq i$. This means that the filtering formula (9.4) has been decomposed into n one-dimensional recursive ones.

9.2 Real-Time Tracking

To illustrate the decoupling technique, let us return to the real-time tracking example studied in Section 3.5. As we have seen there and in Exercise 3.8, this real-time tracking model may be simplified to take on the formulation:

$$\begin{cases} \mathbf{x}_{k+1} = A\mathbf{x}_k + \underline{\xi}_k \\ v_k = C\mathbf{x}_k + \eta_k , \end{cases} \tag{9.11}$$

where

$$A = \begin{bmatrix} 1 & h & h^2/2 \\ 0 & 1 & h \\ 0 & 0 & 1 \end{bmatrix} , \qquad C = [\, 1 \ \ 0 \ \ 0 \,], \qquad h > 0 ,$$

and $\{\underline{\xi}_k\}$ and $\{\eta_k\}$ are both zero-mean Gaussian white noise sequences satisfying the assumption that

$$E(\underline{\xi}_k\underline{\xi}_\ell^{\top}) = \begin{bmatrix} \sigma_p & 0 & 0 \\ 0 & \sigma_v & 0 \\ 0 & 0 & \sigma_a \end{bmatrix} \delta_{k\ell} , \qquad E(\eta_k\eta_\ell) = \sigma_m\delta_{k\ell} ,$$

$$E(\underline{\xi}_k\eta_\ell) = 0 , \qquad E(\underline{\xi}_k\mathbf{x}_0^{\top}) = 0 , \qquad E(\eta_k\mathbf{x}_0) = 0 ,$$

with $\sigma_p, \sigma_v, \sigma_a \geq 0 , \sigma_p + \sigma_v + \sigma_a > 0$, and $\sigma_m > 0$.

As we have also shown in Chapter 6 that the limiting Kalman filter for this system is given by

$$\begin{cases} \vec{\mathbf{x}}_k = \Phi\vec{\mathbf{x}}_{k-1} + v_k G \\ \vec{\mathbf{x}}_0 = E(\mathbf{x}_0) , \end{cases}$$

where $\Phi = (I - GC)A$ with

$$G = \begin{bmatrix} g_1 \\ g_2 \\ g_3 \end{bmatrix} = PC/(C^{\top}PC + \sigma_m) = \frac{1}{P[1,1] + \sigma_m} \begin{bmatrix} P[1,1] \\ P[2,1] \\ P[3,1] \end{bmatrix}$$

and $P = [P[i,j]]_{3\times3}$ being the positive definite solution of the following matrix Riccati equation:

$$P = A[P - PC(C^{\top}PC + \sigma_m)^{-1}C^{\top}P]A^{\top} + \begin{bmatrix} \sigma_p & 0 & 0 \\ 0 & \sigma_v & 0 \\ 0 & 0 & \sigma_a \end{bmatrix}$$

or

$$P = A\left[P - \frac{1}{P[1,1] + \sigma_m}P\begin{bmatrix}1 & 0 & 0 \\ 0 & 0 & 0 \\ 0 & 0 & 0\end{bmatrix}P\right]A^{\top}$$

$$+ \begin{bmatrix}\sigma_p & 0 & 0 \\ 0 & \sigma_v & 0 \\ 0 & 0 & \sigma_a\end{bmatrix}. \tag{9.12}$$

Since

$$\Phi = (I - GC)A = \begin{bmatrix}1 - g_1 & (1-g_1)h & (1-g_1)h^2/2 \\ -g_2 & 1 - g_2h & h - g_2h^2/2 \\ -g_3 & -g_3h & 1 - g_3h^2/2\end{bmatrix},$$

Equation (9.6) now becomes

$$\begin{bmatrix}z - 1 + g_1 & -h + hg_1 & -h^2/2 + h^2g_1/2 \\ g_2 & z - 1 + hg_2 & -h + h^2g_2/2 \\ g_3 & hg_3 & z - 1 + h^2g_3/2\end{bmatrix}\begin{bmatrix}X_1 \\ X_2 \\ X_3\end{bmatrix} = z\begin{bmatrix}g_1 \\ g_2 \\ g_3\end{bmatrix}V,$$

and by Cramer's rule, we have

$$X_i = H_i V,$$

$i = 1, 2, 3$, where

$$\begin{aligned}H_1 =&\{g_1 + (g_3h^2/2 + g_2h - 2g_1)z^{-1} + (g_3h^2/2 - g_2h + g_1)z^{-2}\} \\ &\cdot \{1 + ((g_1 - 3) + g_2h + g_3h^2/2)z^{-1} \\ &+ ((3 - 2g_1) - g_2h + g_3h^2/2)z^{-2} + (g_1 - 1)z^{-3}\},\end{aligned}$$

$$\begin{aligned}H_2 =&\{g_2 + (hg_3 - 2g_2)z^{-1} + (g_2 - hg_3)z^{-2}\} \\ &\cdot \{1 + ((g_1 - 3) + g_2h + g_3h^2/2)z^{-1} \\ &+ ((3 - 2g_1) - g_2h + g_3h^2/2)z^{-2} + (g_1 - 1)z^{-3}\},\end{aligned}$$

$$\begin{aligned}H_3 =&\{g_3 - 2g_3z^{-1} + g_3z^{-2}\} \cdot \{1 + ((g_1 - 3) + g_2h + g_3h^2)z^{-1} \\ &+ ((3 - 2g_1) - g_2h + g_3h^2/2)z^{-2} + (g_1 - 1)z^{-3}\}.\end{aligned}$$

Thus, if we set

$$\vec{\mathbf{x}}_k = \begin{bmatrix}x_k \\ \dot{x}_k \\ \ddot{x}_k\end{bmatrix},$$

and take the inverse z-transforms, we obtain

$$
\begin{aligned}
x_k = &- ((g_1 - 3) + g_2 h + g_3 h^2/2)x_{k-1} - ((3 - 2g_1) - g_2 h + g_3 h^2/2)x_{k-2} \\
&- (g_1 - 1)x_{k-3} + g_1 v_k + (g_3 h^2/2 + g_2 h - 2g_1)v_{k-1} \\
&+ (g_3 h^2/2 - g_2 h + g_1)v_{k-2}, \\
\dot{x}_k = &- ((g_1 - 3) + g_2 h + g_3 h^2/2)\dot{x}_{k-1} - ((3 - 2g_1) - g_2 h + g_3 h^2/2)\dot{x}_{k-2} \\
&- (g_1 - 1)\dot{x}_{k-3} + g_2 v_k + (hg_3 - 2g_2)v_{k-1} + (g_2 - hg_3)v_{k-2}, \\
\ddot{x}_k = &- ((g_1 - 3) + g_2 h + g_3 h^2/2)\ddot{x}_{k-1} - ((3 - 2g_1) - g_2 h + g_3 h^2/2)\ddot{x}_{k-2} \\
&- (g_1 - 1)\ddot{x}_{k-3} + g_3 v_k - 2g_3 v_{k-1} + g_3 v_{k-2},
\end{aligned}
$$

$k = 0, 1, \cdots$, with initial conditions $x_{-1}, \dot{x}_{-1}$, and $\ddot{x}_{-1}$, where $v_k = 0$ for $k < 0$ and $x_k = \dot{x}_k = \ddot{x}_k = 0$ for $k < -1$ (cf. Exercise 9.2).

9.3 The $\alpha - \beta - \gamma$ Tracker

One of the most popular trackers is the so-called $\alpha - \beta - \gamma$ tracker. It is a "suboptimal" filter and can be described by

$$
\begin{cases}
\check{\mathbf{x}}_k = A\check{\mathbf{x}}_{k-1} + H(v_k - CA\check{\mathbf{x}}_{k-1}) \\
\check{\mathbf{x}}_0 = E(\mathbf{x}_0),
\end{cases}
\tag{9.13}
$$

where $H = [\alpha \quad \beta/h \quad \gamma/h^2]^\top$ for some constants α, β, and γ (cf. Fig.9.1). In practice, the α, β, γ values are chosen according to the physical model and depending on the user's experience. In this section, we only consider the example where

$$
A = \begin{bmatrix} 1 & h & h^2/2 \\ 0 & 1 & h \\ 0 & 0 & 1 \end{bmatrix} \qquad and \qquad C = [\, 1 \quad 0 \quad 0 \,].
$$

Hence, by setting

$$
g_1 = \alpha, \qquad g_2 = \beta/h, \qquad and \qquad g_3 = \gamma/h^2,
$$

the decoupled filtering formulas derived in Section 9.2 become a decoupled $\alpha - \beta - \gamma$ tracker. We will show that under certain conditions on the α, β, γ values, the $\alpha - \beta - \gamma$ tracker for the time-invariant system (9.11) is actually a limiting Kalman filter, so that these conditions will guarantee "near-optimal" performance of the tracker.

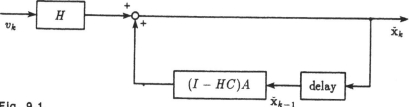

Fig. 9.1.

Since the matrix P in (9.12) is symmetric, we may write

$$P = \begin{bmatrix} p_{11} & p_{21} & p_{31} \\ p_{21} & p_{22} & p_{32} \\ p_{31} & p_{32} & p_{33} \end{bmatrix},$$

so that (9.12) becomes

$$P = A\left[P - \frac{1}{p_{11} + \sigma_m}P\begin{bmatrix} 1 & 0 & 0 \\ 0 & 0 & 0 \\ 0 & 0 & 0 \end{bmatrix}P\right]A^{\mathsf{T}} + \begin{bmatrix} \sigma_p & 0 & 0 \\ 0 & \sigma_v & 0 \\ 0 & 0 & \sigma_a \end{bmatrix} \qquad (9.14)$$

and

$$G = PC/(C^{\mathsf{T}}PC + \sigma_m)$$

$$= \frac{1}{p_{11} + \sigma_m}\begin{bmatrix} p_{11} \\ p_{21} \\ p_{31} \end{bmatrix}. \qquad (9.15)$$

A necessary condition for the $\alpha - \beta - \gamma$ tracker (9.13) to be a limiting Kalman filter is $H = G$, or equivalently,

$$\begin{bmatrix} \alpha \\ \beta/h \\ \gamma/h^2 \end{bmatrix} = \frac{1}{p_{11} + \sigma_m}\begin{bmatrix} p_{11} \\ p_{21} \\ p_{31} \end{bmatrix},$$

so that

$$\begin{bmatrix} p_{11} \\ p_{21} \\ p_{31} \end{bmatrix} = \frac{\sigma_m}{1 - \alpha}\begin{bmatrix} \alpha \\ \beta/h \\ \gamma/h^2 \end{bmatrix}. \qquad (9.16)$$

On the other hand, by simple algebra, it follows from (9.14),

(9.15), and (9.16) that

$$
\begin{bmatrix} p_{11} & p_{21} & p_{31} \\ p_{21} & p_{22} & p_{32} \\ p_{31} & p_{32} & p_{33} \end{bmatrix} =
$$

$$
\begin{bmatrix} \begin{array}{c} p_{11} + 2hp_{21} + h^2p_{31} \\ +h^2p_{22} + h^3p_{32} + h^4p_{33}/4 \end{array} & \begin{array}{c} p_{21} + hp_{31} + hp_{22} \\ +3h^2p_{32}/2 + h^3p_{33}/2 \end{array} & \begin{array}{c} p_{31} + hp_{32} \\ +h^2p_{33}/2 \end{array} \\[1.5em] \begin{array}{c} p_{21} + hp_{31} + hp_{22} \\ +3h^2p_{32}/2 + h^3p_{33}/2 \end{array} & p_{22} + 2hp_{32} + h^2p_{33} & p_{32} + hp_{33} \\[1.5em] p_{31} + hp_{32} + h^2p_{33}/2 & p_{32} + hp_{33} & p_{33} \end{array} \end{bmatrix}
$$

$$
- \frac{1}{p_{11} + \sigma_m} \begin{bmatrix} \begin{array}{c} (\alpha+\beta)^2 \\ +\gamma(\alpha+\beta+\gamma/4) \end{array} & \begin{array}{c} (\alpha\beta+\alpha\gamma+\beta^2 \\ +3\beta\gamma/2+\gamma^2/2)/h \end{array} & \begin{array}{c} \gamma(\alpha+\beta \\ +\gamma/2)/h^2 \end{array} \\[1.5em] \begin{array}{c} (\alpha\beta+\alpha\gamma+\beta^2 \\ +3\beta\gamma/2+\gamma^2/2)/h \end{array} & (\beta+\gamma)^2/h^2 & \gamma(\beta+\gamma)/h^3 \\[1.5em] \gamma(\alpha+\beta+\gamma/2)/h^2 & \gamma(\beta+\gamma)/h^3 & \gamma^2/h^4 \end{array} \end{bmatrix}
$$

$$
+ \begin{bmatrix} \sigma_p & 0 & 0 \\ 0 & \sigma_v & 0 \\ 0 & 0 & \sigma_a \end{bmatrix} .
$$

Substituting (9.16) into the above equation yields

$$
\begin{cases}
\dfrac{h^4}{\gamma^2}\sigma_a = p_{11} + \sigma_m \\[1em]
p_{11} = \dfrac{h^4(\beta+\gamma)^2}{\gamma^3(2\alpha+2\beta+\gamma)}\sigma_a - \dfrac{h}{\gamma(2\alpha+2\beta+\gamma)}\sigma_v \\[1em]
2hp_{21} + h^2p_{31} + h^2p_{22} \\[0.5em]
\quad = \dfrac{h^4}{4\gamma^2}(4\alpha^2 + 8\alpha\beta + 2\beta^2 + 4\alpha\gamma + \beta\gamma)\sigma_a + \dfrac{h^2}{2}\sigma_v - \sigma_p \\[1em]
p_{31} + p_{22} = \dfrac{h^4}{4\gamma^2}(4\alpha+\beta)(\beta+\gamma)\sigma_a + \dfrac{3}{4}\sigma_v ,
\end{cases} \tag{9.17}
$$

and

$$
\begin{cases}
p_{22} = \dfrac{\sigma_m}{(1-\alpha)h^2}[\beta(\alpha+\beta+\gamma/4) - \gamma(2+\alpha)/2] \\[1em]
p_{32} = \dfrac{\sigma_m}{(1-\alpha)h^3}\gamma(\alpha+\beta/2) \\[1em]
p_{33} = \dfrac{\sigma_m}{(1-\alpha)h^4}\gamma(\beta+\gamma) .
\end{cases} \tag{9.18}
$$

Hence, from (9.16), (9.17) , and (9.18) we have

$$
\begin{cases}
\dfrac{\sigma_p}{\sigma_m} = \dfrac{1}{1-\alpha}(\alpha^2 + \alpha\beta + \alpha\gamma/2 - 2\beta) \\[2mm]
\dfrac{\sigma_v}{\sigma_m} = \dfrac{1}{1-\alpha}(\beta^2 - 2\alpha\gamma)h^{-2} \\[2mm]
\dfrac{\sigma_a}{\sigma_m} = \dfrac{1}{1-\alpha}\gamma^2 h^{-4}
\end{cases}
\tag{9.19}
$$

and

$$
P = \frac{\sigma_m}{1-\alpha}
\begin{bmatrix}
\alpha & \beta/h & \gamma/h^2 \\
\beta/h & (\beta(\alpha+\beta+\gamma/4) - \gamma(2+\alpha)/2)/h^2 & \gamma(\alpha+\beta/2)/h^3 \\
\gamma/h^2 & \gamma(\alpha+\beta/2)/h^3 & \gamma(\beta+\gamma)/h^4
\end{bmatrix}
\tag{9.20}
$$

(cf. Exercise 9.4). Since P must be positive definite (cf. Theorem 6.1), the α, β, γ values can be characterized as follows (cf. Exercise 9.5):

Theorem 9.1. *Let the α, β, γ values satisfy the conditions in (9.19) and suppose that $\sigma_m > 0$. Then the $\alpha - \beta - \gamma$ tracker is a limiting Kalman filter if and only if the following conditions are satisfied:*
(i) $0 < \alpha < 1,$ $\gamma > 0$,
(ii) $\sqrt{2\alpha\gamma} \leq \beta \leq \frac{\alpha}{2-\alpha}(\alpha+\gamma/2)$, and
(iii) the matrix

$$
\tilde{P} =
\begin{bmatrix}
\alpha & \beta & \gamma \\
\beta & \beta(\alpha+\beta+\gamma/4) - \gamma(2+\gamma)/2 & \gamma(\alpha+\beta/2) \\
\gamma & \gamma(\alpha+\beta/2) & \gamma(\beta+\gamma)
\end{bmatrix}
$$

is non-negative definite.

9.4 An Example

Let us now consider the special case of the real-time tracking system (9.11) where $\sigma_p = \sigma_v = 0$ and $\sigma_a, \sigma_m > 0$. It can be verified that (9.16-18) together yield

$$
\begin{cases}
\alpha = 1 - s\gamma^2 \\
\beta^2 = 2\alpha\gamma \\
\alpha^2 + \alpha\beta + \alpha\gamma/2 - 2\beta = 0
\end{cases}
\tag{9.21}
$$

where

$$
s = \frac{\sigma_m}{\sigma_a}h^{-4}
$$

(cf. Exercise 9.6). By simple algebra, (9.21) gives

$$f(\gamma) := s^3\gamma^6 + s^2\gamma^5 - 3s(s - 1/12)\gamma^4$$
$$+ 6s\gamma^3 + 3(s - 1/12)\gamma^2 + \gamma - 1$$
$$= 0 \tag{9.22}$$

(cf. Exercise 9.7), and in order to satisfy condition (i) in Theorem 9.1, we must solve (9.22) for a positive γ. To do so, we note that since $f(0) = -1$ and $f(+\infty) = +\infty$, there is at least one positive root γ. In addition, by the Descartes rule of signs, there are at most 3 real roots. In the following, we give the values of γ for different choices of s:

s	0.09	0.08	0.07	0.06	0.05	0.04	0.03	0.02	0.01
γ	0.755	0.778	0.804	0.835	0.873	0.919	0.979	1.065	1.211

Exercises

9.1. Consider the two-dimensional real-time tracking system

$$\begin{cases} \mathbf{x}_{k+1} = \begin{bmatrix} 1 & h \\ 0 & 1 \end{bmatrix} \mathbf{x}_k + \underline{\xi}_k \\ v_k = [\,1 \quad 0\,]\mathbf{x}_k + \eta_k, \end{cases}$$

where $h > 0$, and $\{\underline{\xi}_k\}$, $\{\eta_k\}$ are both uncorrelated zero-mean Gaussian white noise sequences. The $\alpha - \beta$ tracker associated with this system is defined by

$$\begin{cases} \check{\mathbf{x}}_k = \begin{bmatrix} 1 & h \\ 0 & 1 \end{bmatrix} \check{\mathbf{x}}_{k-1} + \begin{bmatrix} \alpha \\ \beta/h \end{bmatrix} (v_k - [\,1 \quad 0\,]\begin{bmatrix} 1 & h \\ 0 & 1 \end{bmatrix}\check{\mathbf{x}}_{k-1}) \\ \check{\mathbf{x}}_0 = E(\mathbf{x}_0). \end{cases}$$

(a) Derive the decoupled Kalman filtering algorithm for this $\alpha - \beta$ tracker.

(b) Give the conditions under which this $\alpha - \beta$ tracker is a limiting Kalman filter.

9.2. Verify the decoupled formulas of $x_k, \dot{x}_k$, and $\ddot{x}_k$ given in Section 9.2 for the real-time tracking system (9.11).

9.3. Consider the three-dimensional radar-tracking system

$$\begin{cases} \mathbf{x}_{k+1} = \begin{bmatrix} 1 & h & h^2/2 \\ 0 & 1 & h \\ 0 & 0 & 1 \end{bmatrix} \mathbf{x}_k + \underline{\xi}_k \\ v_k = [\, 1 \;\; 0 \;\; 0 \,]\mathbf{x}_k + w_k \,, \end{cases}$$

where $\{w_k\}$ is a sequence of colored noise defined by

$$w_k = s w_{k-1} + \eta_k$$

and $\{\underline{\xi}_k\}$, $\{\eta_k\}$ are both uncorrelated zero-mean Gaussian white noise sequences, as described in Chapter 5. The associated $\alpha - \beta - \gamma - \theta$ *tracker* for this system is defined by the algorithm:

$$\begin{cases} \check{X}_k = \begin{bmatrix} A & 0 \\ 0 & s \end{bmatrix} \check{X}_{k-1} + \begin{bmatrix} \alpha \\ \beta/h \\ \gamma/h^2 \\ \theta \end{bmatrix} \{ v_k - [\, 1 \;\; 0 \;\; 0 \,] \begin{bmatrix} A & 0 \\ 0 & s \end{bmatrix} \check{X}_{k-1} \} \\ \check{X}_0 = \begin{bmatrix} E(\mathbf{x}_0) \\ 0 \end{bmatrix} \,, \end{cases}$$

where α, β, γ, and θ are constants (cf. Fig.9.2).

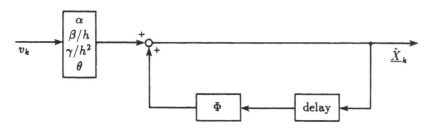

Fig. 9.2.

(a) Compute the matrix

$$
\Phi = \{I - \begin{bmatrix} \alpha \\ \beta/h \\ \gamma/h^2 \\ \theta \end{bmatrix} [\,1\ \ 0\ \ 0\,]\}\begin{bmatrix} A & 0 \\ 0 & s \end{bmatrix}.
$$

(b) Use Cramer's rule to solve the system

$$
[zI - \Phi]\begin{bmatrix} \tilde{X}_1 \\ \tilde{X}_2 \\ \tilde{X}_3 \\ W \end{bmatrix} = z\begin{bmatrix} \alpha \\ \beta/h \\ \gamma/h^2 \\ \theta \end{bmatrix}V
$$

for $\tilde{X}_1, \tilde{X}_2, \tilde{X}_3$ and W. (The above system is obtained when the z-transform of the $\alpha - \beta - \gamma - \theta$ filter is taken.)

(c) By taking the inverse z-transforms of $\tilde{X}_1$, $\tilde{X}_2$, $\tilde{X}_3$, and W, give the decoupled filtering equations for the $\alpha - \beta - \gamma - \theta$ filter.

(d) Verify that when the colored noise sequence $\{\eta_k\}$ becomes white; namely, $s = 0$ and θ is chosen to be zero, the decoupled filtering equations obtained in part (c) reduce to those obtained in Section 9.2 with $g_1 = \alpha$, $g_2 = \beta/h$, and $g_3 = \gamma/h^2$.

9.4. Verify equations (9.17-20).

9.5. Prove Theorem 9.1 and observe that conditions (i)-(iii) are independent of the sampling time h.

9.6. Verify the equations in (9.21).

9.7. Verify the equation in (9.22).

10. Kalman Filtering for Interval Systems

If some system parameters such as certain elements of the system matrix are not precisely known or gradually change with time, then the Kalman filtering algorithm cannot be directly applied. In this case, robust Kalman filtering that has the ability of handling uncertainty is needed. In this chapter we introduce one of such robust Kalman filtering algorithms.

Consider the nominal system

$$\begin{cases} \mathbf{x}_{k+1} = A_k \mathbf{x}_k + \Gamma_k \underline{\xi}_k, \\ \mathbf{v}_k = C_k \mathbf{x}_k + \underline{\eta}_k, \end{cases} \tag{10.1}$$

where A_k, Γ_k and C_k are known $n \times n$, $n \times p$ and $q \times n$ matrices, respectively, with $1 \leq p, q \leq n$, and where

$$E(\underline{\xi}_k) = 0, \qquad E(\underline{\xi}_k \underline{\xi}_\ell^\top) = Q_k \delta_{k\ell},$$
$$E(\underline{\eta}_k) = 0, \qquad E(\underline{\eta}_k \underline{\eta}_\ell^\top) = R_k \delta_{k\ell},$$
$$E(\underline{\xi}_k \underline{\eta}_\ell^\top) = 0, \qquad E(\mathbf{x}_0 \underline{\xi}_k) = 0, \qquad E(\mathbf{x}_0 \underline{\eta}_k) = 0,$$

for all $k, \ell = 0, 1, \cdots$, with Q_k and R_k being positive definite and symmetric matrices.

If all the constant matrices, A_k, Γ_k, and C_k, are known, then the Kalman filter can be applied to the nominal system (10.1), which yields optimal estimates $\{\hat{\mathbf{x}}_k\}$ of the unknown state vectors $\{\mathbf{x}_k\}$ using the measurement data $\{\mathbf{v}_k\}$ in a recursive scheme. However, if some of the elements of these system matrices are unknown or uncertain, modification of the entire setting for filtering is necessary. Suppose that the uncertain parameters are only known to be bounded. Then we can write

$$A_k^I = A_k + \Delta A_k = \left[A_k - |\Delta A_k|, A_k + |\Delta A_k| \right],$$
$$\Gamma_k^I = \Gamma_k + \Delta \Gamma_k = \left[\Gamma_k - |\Delta \Gamma_k|, \Gamma_k + |\Delta \Gamma_k| \right],$$
$$C_k^I = C_k + \Delta C_k = \left[C_k - |\Delta C_k|, C_k + |\Delta C_k| \right],$$

$k = 0, 1, \cdots$, where $|\Delta A_k|$, $|\Delta \Gamma_k|$, and $|\Delta C_k|$ are constant bounds for the unknowns. The corresponding system

$$\begin{cases} \mathbf{x}_{k+1} = A_k^I \mathbf{x}_k + \Gamma_k^I \underline{\xi}_k \,, \\ \mathbf{v}_k = C_k^I \mathbf{x}_k + \underline{\eta}_k \,, \end{cases} \tag{10.2}$$

$k = 0, 1, \cdots$, is then called an *interval system*.

Under this framework, how is the original Kalman filtering algorithm modified and applied to the interval system (10.2)? This question is to be addressed in this chapter.

10.1 Interval Mathematics

In this section, we first provide some preliminary results on interval arithmetic and interval analysis that are needed throughout the chapter.

10.1.1 Intervals and Their Properties

A closed and bounded subset $[\underline{x}, \overline{x}]$ in $R = (-\infty, \infty)$ is referred to as an interval. In particular, a single point $x \in R$ is considered as a degenerate interval with $\underline{x} = \overline{x} = x$.

Some useful concepts and properties of intervals are:
(a) *Equality:* Two intervals, $[\underline{x}_1, \overline{x}_1]$ and $[\underline{x}_2, \overline{x}_2]$, are said to be *equal*, and denoted by

$$[\underline{x}_1, \overline{x}_1] = [\underline{x}_2, \overline{x}_2] \,,$$

if and only if $\underline{x}_1 = \underline{x}_2$ and $\overline{x}_1 = \overline{x}_2$.
(b) *Intersection:* The intersection of two intervals, $[\underline{x}_1, \overline{x}_1]$ and $[\underline{x}_2, \overline{x}_2]$, is defined to be

$$[\underline{x}_1, \overline{x}_1] \cap [\underline{x}_2, \overline{x}_2] = [max\{\underline{x}_1, \underline{x}_2\}, min\{\overline{x}_1, \overline{x}_2\}] \,.$$

Furthermore, these two intervals are said to be *disjoint*, and denoted by

$$[\underline{x}_1, \overline{x}_1] \cap [\underline{x}_2, \overline{x}_2] = \phi \,,$$

if and only if $\underline{x}_1 > \overline{x}_2$ or $\underline{x}_2 > \overline{x}_1$.
(c) *Union:* The union of two non-disjoint intervals, $[\underline{x}_1, \overline{x}_1]$ and $[\underline{x}_2, \overline{x}_2]$, is defined to be

$$[\underline{x}_1, \overline{x}_1] \cup [\underline{x}_2, \overline{x}_2] = [min\{\underline{x}_1, \underline{x}_2\}, max\{\overline{x}_1, \overline{x}_2\}] \,.$$

Note that the union is defined only if the two intervals are not disjoint, i.e.,

$$[\underline{x}_1, \overline{x}_1] \cap [\underline{x}_2, \overline{x}_2] \neq \phi \, ;$$

otherwise, it is undefined since the result is not an interval.

(d) *Inequality*: The interval $[\underline{x}_1, \overline{x}_1]$ is said to be *less than* (resp., *greater than*) the interval $[\underline{x}_2, \overline{x}_2]$, denoted by

$$[\underline{x}_1, \overline{x}_1] < [\underline{x}_2, \overline{x}_2] \qquad (resp., \quad [\underline{x}_1, \overline{x}_1] > [\underline{x}_2, \overline{x}_2])$$

if and only if $\overline{x}_1 < \underline{x}_2$ (resp., $\underline{x}_1 > \overline{x}_2$); otherwise, they cannot be compared. Note that the relations $\leq$ and $\geq$ are not defined for intervals.

(e) *Inclusion*: The interval $[\underline{x}_1, \overline{x}_1]$ is said to be *included* in $[\underline{x}_2, \overline{x}_2]$, denoted

$$[\underline{x}_1, \overline{x}_1] \subseteq [\underline{x}_2, \overline{x}_2]$$

if and only if $\underline{x}_2 \leq \underline{x}_1$ and $\overline{x}_1 \leq \overline{x}_2$; namely, if and only if $[\underline{x}_1, \overline{x}_1]$ is a subset (subinterval) of $[\underline{x}_2, \overline{x}_2]$.

For example, for three given intervals, $X_1 = [-1, 0]$, $X_2 = [-1, 2]$, and $X_3 = [2, 10]$, we have

$$X_1 \cap X_2 = [-1, 0] \cap [-1, 2] = [-1, 0] \, ,$$
$$X_1 \cap X_3 = [-1, 0] \cap [2, 10] = \phi \, ,$$
$$X_2 \cap X_3 = [-1, 2] \cap [2, 10] = [2, 2] = 2 \, ,$$
$$X_1 \cup X_2 = [-1, 0] \cup [-1, 2] = [-1, 2] \, ,$$
$$X_1 \cup X_3 = [-1, 0] \cup [2, 10] \;\; is \; undefined \, ,$$
$$X_2 \cup X_3 = [-1, 2] \cup [2, 10] = [-1, 10] \, ,$$
$$X_1 = [-1, 0] < [2, 10] = X_3 \, ,$$
$$X_1 = [-1, 0] \subset [-1, 10] = X_2 \, .$$

10.1.2 Interval Arithmetic

Let $[\underline{x}, \overline{x}]$, $[\underline{x}_1, \overline{x}_1]$, and $[\underline{x}_2, \overline{x}_2]$ be intervals. The basic arithmetic operations of intervals are defined as follows:

(a) *Addition*:

$$[\underline{x}_1, \overline{x}_1] + [\underline{x}_2, \overline{x}_2] = [\underline{x}_1 + \underline{x}_2, \overline{x}_1 + \overline{x}_2] \, .$$

(b) *Subtraction*:
$$[\underline{x}_1, \overline{x}_1] - [\underline{x}_2, \overline{x}_2] = [\underline{x}_1 - \overline{x}_2, \overline{x}_1 - \underline{x}_2].$$

(c) *Reciprocal operation*:

$$If \ \ 0 \notin [\underline{x}, \overline{x}] \ \ then \ \ [\underline{x}, \overline{x}]^{-1} = [1/\overline{x}, 1/\underline{x}];$$
$$If \ \ 0 \in [\underline{x}, \overline{x}] \ \ then \ \ [\underline{x}, \overline{x}]^{-1} \ \ is \ undefined.$$

(d) *Multiplication*:
$$[\underline{x}_1, \overline{x}_1] \cdot [\underline{x}_2, \overline{x}_2] = [\underline{y}, \overline{y}],$$

where
$$\underline{y} = min \{\underline{x}_1 \underline{x}_2, \underline{x}_1 \overline{x}_2, \overline{x}_1 \underline{x}_2, \overline{x}_1 \overline{x}_2\},$$
$$\overline{y} = max \{\underline{x}_1 \underline{x}_2, \underline{x}_1 \overline{x}_2, \overline{x}_1 \underline{x}_2, \overline{x}_1 \overline{x}_2\}.$$

(e) *Division*:
$$[\underline{x}_1, \overline{x}_1] / [\underline{x}_2, \overline{x}_2] = [\underline{x}_1, \overline{x}_1] \cdot [\underline{x}_2, \overline{x}_2]^{-1},$$

provided that $0 \notin [\underline{x}_2, \overline{x}_2]$; otherwise, it is undefined.

For three intervals, $X = [\underline{x}, \overline{x}]$, $Y = [\underline{y}, \overline{y}]$, and $Z = [\underline{z}, \overline{z}]$, consider the interval operations of addition $(+)$, subtraction $(-)$, multiplication $(\cdot)$, and division $(/)$, namely,

$$Z = X * Y, \qquad * \in \{+, -, \cdot, /\}.$$

It is clear that $X * Y$ is also an interval. In other words, the family of intervals under the four operations $\{+, -, \cdot, /\}$ is algebraically closed. It is also clear that the real numbers $x, y, z, \cdots$ are isomorphic to degenerate intervals $[x, x]$, $[y, y]$, $[z, z]$, $\cdots$, so we will simply denote the point-interval operation $[x, x] * Y$ as $x * Y$. Moreover, the multiplication symbol "$\cdot$" will often be dropped for notational convenience.

Similar to conventional arithmetic, the interval arithmetic has the following basic algebraic properties (cf. Exercise 10.1):

$X + Y = Y + X,$

$Z + (X + Y) = (Z + X) + Y,$

$XY = YX,$

$Z(XY) = (ZX)Y,$

$X + 0 = 0 + X = X \quad and \quad X0 = 0X = 0, \ where \ 0 = [0, 0],$

$XI = IX = X, \ where \ I = [1, 1],$

$Z(X + Y) \subseteq ZX + ZY, \ where \ = \ holds \ only \ if \ either$

   (a) $Z = [z, z],$

   (b) $X = Y = 0,$ or

   (c) $xy \geq 0 \ for \ all \ x \in X \ and \ y \in Y.$

In addition, the following is an important property of interval operations, called the *monotonic inclusion* property.

Theorem 10.1. *Let X_1, X_2, Y_1, and Y_2 be intervals, with*

$$X_1 \subseteq Y_1 \quad and \quad X_2 \subseteq Y_2.$$

Then for any operation $* \in \{+, -, \cdot, /\}$, it follows that

$$X_1 * X_2 \subseteq Y_1 * Y_2.$$

This property is an immediate consequence of the relations $X_1 \subseteq Y_1$ and $X_2 \subseteq Y_2$, namely,

$$
\begin{aligned}
X_1 * X_2 &= \{x_1 * x_2 \,|\, x_1 \in X_1, x_2 \in X_2\} \\
&\subseteq \{y_1 * y_2 \,|\, y_1 \in Y_1, y_2 \in Y_2\} \\
&= Y_1 * Y_2.
\end{aligned}
$$

Corollary 10.1. *Let X and Y be intervals and let $x \in X$ and $y \in Y$. Then,*

$$x * y \subseteq X * Y, \quad for\ all\ * \in \{+, -, \cdot, /\}.$$

What is seemingly counter-intuitive in the above theorem and corollary is that some operations such as reciprocal, subtraction, and division do not seem to satisfy such a monotonic inclusion property. However, despite the above proof, let us consider a simple example of two intervals, $X = [0.2, 0.4]$ and $Y = [0.1, 0.5]$. Clearly, $X \subseteq Y$. We first show that $I/X \subseteq I/Y$, where $I = [1.0, 1.0]$. Indeed,

$$\frac{I}{X} = \frac{[1.0, 1.0]}{[0.2, 0.4]} = [2.5, 5.0] \quad and \quad \frac{I}{Y} = \frac{[1.0, 1.0]}{[0.1, 0.5]} = [2.0, 10.0].$$

We also observe that $I - X \subseteq I - Y$, by noting that

$$I - X = [1.0, 1.0] - [0.2, 0.4] = [0.6, 0.8]$$

and

$$I - Y = [1.0, 1.0] - [0.1, 0.5] = [0.5, 0.9].$$

Moreover, as a composition of these two operations, we again have $\frac{I}{I-X} \subseteq \frac{I}{I-Y}$, since

$$\frac{I}{I-X} = [5/4, 5/3] \quad and \quad \frac{I}{I-Y} = [10/9, 2].$$

We next extend the notion of intervals and interval arithmetic to include interval vectors and interval matrices. Interval vectors and interval matrices are similarly defined. For example,

$$A^I = \begin{bmatrix} [2,3] & [0,1] \\ [1,2] & [2,3] \end{bmatrix} \quad and \quad \mathbf{b}^I = \begin{bmatrix} [0,10] \\ [-6,1] \end{bmatrix}$$

are an interval matrix and an interval vector, respectively.

Let $A^I = [a_{ij}^I]$ and $B^I = [b_{ij}^I]$ be $n \times m$ interval matrices. Then, A^I and B^I are said to be *equal* if $a_{ij}^I = b_{ij}^I$ for all $i = 1, \cdots, n$ and $j = 1, \cdots, m$; A^I is said to be *contained* in B^I, denoted $A^I \subseteq B^I$, if $a_{ij}^I \subseteq b_{ij}^I$ for all $i = 1, \cdots, n$ and $j = 1, \cdots, m$, where, in particular, if $A^I = A$ is an ordinary constant matrix, we write $A \in B^I$.

Fundamental operations of interval matrices include:

(a) *Addition* and *Subtraction*:

$$A^I \pm B^I = [a_{ij}^I \pm b_{ij}^I] .$$

(b) *Multiplication*: For two $n \times r$ and $r \times m$ interval matrices A^I and B^I,

$$A^I B^I = \left[\sum_{k=1}^{r} a_{ik}^I b_{kj}^I \right] .$$

(c) *Inversion*: For an $n \times n$ interval matrix A^I with $det\,[A^I] \neq 0$,

$$[A^I]^{-1} = \frac{adj\,[A^I]}{det\,[A^I]} .$$

For instance, if $A^I = \begin{bmatrix} [2,3] & [0,1] \\ [1,2] & [2,3] \end{bmatrix}$, then

$$[A^I]^{-1} = \frac{adj\,[A^I]}{det\,[A^I]} = \frac{\begin{bmatrix} [2,3] & -[0,1] \\ -[1,2] & [2,3] \end{bmatrix}}{[2,3]\,[2,3] - [0,1]\,[1,2]}$$

$$= \begin{bmatrix} [2/9, 3/2] & [-1/2, 0] \\ [-1, -1/9] & [2/9, 3/2] \end{bmatrix} .$$

Interval matrices (including vectors) obey many algebraic operational rules that are similar to those for intervals (cf. Exercise 10.2).

10.1.3 Rational Interval Functions

Let S_1 and S_2 be intervals in R and $f : S_1 \rightarrow S_2$ be an ordinary one-variable real-valued (i.e., point-to-point) function. Denote by Σ_{S_1} and Σ_{S_2} families of all subintervals of S_1 and S_2, respectively. The interval-to-interval function, $f^I : \Sigma_{S_1} \rightarrow \Sigma_{S_2}$, defined by

$$f^I(X) = \left\{ f(x) \in S_2 : \quad x \in X, \ X \in \Sigma_{S_1} \right\}$$

is called the *united extension* of the point-to-point function f on S_1. Obviously, its range is

$$f^I(X) = \bigcup_{x \in X} \{f(x)\},$$

which is the union of all the subintervals of S_2 that contain the single point $f(x)$ for some $x \in X$.

The following property of the united extension $f^I : \Sigma_{S_1} \rightarrow \Sigma_{S_2}$ follows immediately from definition, namely,

$$X, Y \in \Sigma_{S_1} \quad and \quad X \subseteq Y \quad \Longrightarrow \quad f^I(X) \subseteq f^I(Y).$$

In general, an interval-to-interval function F of n-variables, $X_1, \cdots, X_n$, is said to have the *monotonic inclusion property*, if

$$X_i \subseteq Y_i, \quad \forall \ i = 1, \cdots, n \quad \Longrightarrow \quad F(X_1, \cdots, X_n) \subseteq F(Y_1, \cdots, Y_n).$$

Note that not all interval-to-interval functions have this property.

However, all united extensions have the monotonic inclusion property. Since interval arithmetic functions are united extensions of the real arithmetic functions: addition, subtraction, multiplication and division $(+, -, \cdot, /)$, interval arithmetic has the monotonic inclusion property, as previously discussed (cf. Theorem 10.1 and Corollary 10.1).

An interval-to-interval function will be called an *interval function* for simplicity. *Interval vectors* and *interval matrices* are similarly defined. An interval function is said to be *rational*, and so is called a *rational interval function*, if its values are defined by a finite sequence of interval arithmetic operations. Examples of rational interval functions include $X + Y^2 + Z^3$ and $(X^2 + Y^2)/Z$, etc., for intervals X, Y and Z, provided that $0 \notin Z$ for the latter.

It follows from the transitivity of the partially ordered relation $\subseteq$ that all the rational interval functions have the monotonic inclusion property. This can be verified by mathematical induction.

Next, let $f = f(x_1, \cdots, x_n)$ be an ordinary n-variable real-valued function, and $X_1, \cdots, X_n$ be intervals. An interval function, $F = F(X_1, \cdots, X_n)$, is said to be an *interval extension* of f if

$$F(x_1, \cdots, x_n) = f(x_1, \cdots, x_n), \qquad \forall \ x_i \in X_i, \ i = 1, \cdots, n.$$

Note also that not all the interval extensions have the monotonic inclusion property.

The following result can be established (cf. Exercise 10.3):

Theorem 10.2. *If F is an interval extension of f with the monotonic inclusion property, then the united extension f^I of f satisfies*

$$f^I(X_1, \cdots, X_n) \subseteq F(X_1, \cdots, X_n).$$

Since rational interval functions have the monotonic inclusion property, we have the following

Corollary 10.2. *If F is a rational interval function and is an interval extension of f, then*

$$f^I(X_1, \cdots, X_n) \subseteq F(X_1, \cdots, X_n).$$

This corollary provides a means of finite evaluation of upper and lower bounds on the value-range of an ordinary rational function over an n-dimensional rectangular domain in R^n.

As an example of the monotonic inclusion property of rational interval functions, consider calculating the function

$$f^I(X, A) = \frac{AX}{I - X}$$

for two cases: $X_1 = [2, 3]$ with $A_1 = [0, 2]$, and $X_2 = [2, 4]$ with $A_2 = [0, 3]$, respectively. Here, $X_1 \subset X_2$ and $A_1 \subset A_2$. A direct calculation yields

$$f_1^I(X_1, A_1) = \frac{[0, 2] \cdot [2, 3]}{[1, 1] - [2, 3]} = [-6, 0]$$

and

$$f_2^I(X_2, A_2) = \frac{[0, 3] \cdot [2, 4]}{[1, 1] - [2, 4]} = [-12, 0].$$

Here, we do have $f_1^I(X_1, A_1) \subset f_2^I(X_2, A_2)$, as expected.

We finally note that based on Corollary 10.2, when we have interval division of the type X^I/X^I where X^I does not contain zero, we can first examine its corresponding ordinary function and operation to obtain $x/x = 1$, and then return to the interval setting for the final answer. Thus, symbolically, we may write $X^I/X^I = 1$ for an interval X^I not containing zero. This is indeed a convention in interval calculations.

10.1.4 Interval Expectation and Variance

Let $f(x)$ be an ordinary function defined on an interval X. If f satisfies the ordinary Lipschitz condition

$$|f(x) - f(y)| \leq L\,|x - y|$$

for some positive constant L which is independent of $x, y \in X$, then the united extension f^I of f is said to be a *Lipschitz interval extension* of f over X.

Let $B(X)$ be a class of functions defined on X that are most commonly used in numerical computation, such as the four arithmetic functions $(+, -, \cdot, /)$ and the elementary type of functions like $e^{(\cdot)}$, $ln(\cdot)$, $\sqrt{\cdot}$, etc. We will only use some of these commonly used functions throughout this chapter, so $B(X)$ is introduced only for notational convenience.

Let N be a positive integer. Subdivide an interval $[a, b] \subseteq X$ into N subintervals, $X_1 = [\underline{X}_1, \overline{X}_1], \cdots, X_N = [\underline{X}_N, \overline{X}_N]$, such that

$$a = \underline{X}_1 < \overline{X}_1 = \underline{X}_2 < \overline{X}_2 = \cdots = \underline{X}_N < \overline{X}_N = b.$$

Moreover, for any $f \in B(X)$, let F be a Lipschitz interval extension of f defined on all X_i, $i = 1, \cdots, N$. Assume that F satisfies the monotonic inclusion property. Using the notation

$$S_N(F; [a, b]) = \frac{b - a}{N} \sum_{i=1}^{N} F(X_i),$$

we have

$$\int_a^b f(t)dt = \bigcap_{N=1}^{\infty} S_N(F; [a, b]) = \lim_{N \to \infty} S_N(F; [a, b]).$$

Note that if we recursively define

$$\begin{cases} Y_1 = S_1, \\ Y_{k+1} = S_{k+1} \cap Y_k, \quad k = 1, 2, \cdots, \end{cases}$$

where $S_k = S_k(F; [a, b])$, then $\{Y_k\}$ is a nested sequence of intervals that converges to the exact value of the integral $\int_a^b f(t)dt$.

Note also that a Lipschitz interval extension F used here has the property that $F(x)$ is a real number for any real number $x \in R$. However, for other interval functions that have the monotonic inclusion property but are not Lipschitz, the corresponding function $F(x)$ may have interval coefficients even if x is a real number.

Next, based on the interval mathematics introduced above, we introduce the following important concept.

Let X be an interval of real-valued random variables of interest, and let

$$f(x) = \frac{1}{\sqrt{2\pi}\sigma_x} exp\left\{\frac{-(x - \mu_x)^2}{2\sigma_x^2}\right\}, \qquad x \in X,$$

be an ordinary Gaussian density function with known μ_x and $\sigma_x > 0$. Then $f(x)$ has a Lipschitz interval extension, so that the *interval expectation*

$$E(X) = \int_{-\infty}^{\infty} x f(x)dx$$

$$= \int_{-\infty}^{\infty} \frac{x}{\sqrt{2\pi}\sigma_x} exp\left\{\frac{-(x - \mu_x)^2}{2\sigma_x^2}\right\} dx, \quad x \in X, \qquad (10.3)$$

and the *interval variance*

$$Var(X) = E([X - E(X)]^2)$$

$$= \int_{-\infty}^{\infty} (x - \mu_x)^2 f(x)dx$$

$$= \int_{-\infty}^{\infty} \frac{(x - \mu_x)^2}{\sqrt{2\pi}\sigma_x} exp\left\{\frac{-(x - \mu_x)^2}{2\sigma_x^2}\right\} dx, \quad x \in X, \quad (10.4)$$

are both well defined. This can be easily verified based on the definite integral defined above, with $a \to -\infty$ and $b \to \infty$. Also, with respect to another real interval Y of real-valued random variables, the *conditional interval expectation*

$$E(X|y \in Y) = \int_{-\infty}^{\infty} x f(x|y)dx$$

$$= \int_{-\infty}^{\infty} x \frac{f(x, y)}{f(y)} dx$$

$$= \int_{-\infty}^{\infty} \frac{x}{\sqrt{2\pi}\sigma_{xy}} exp\left\{\frac{-(x - \mu_{xy})^2}{2\sigma_{xy}^2}\right\} dx, \quad x \in X, \quad (10.5)$$

and the *conditional variance*

$$
\begin{aligned}
&Var(X|\,y \in Y) \\
&= E\big((x - \mu_x)^2\big|\ y \in Y\big) \\
&= \int_{-\infty}^{\infty} \big[x - E(x|\ y \in Y)\big]^2 f(x|y)dx \\
&= \int_{-\infty}^{\infty} \big[x - E(x|\,y \in Y)\big]^2 \frac{f(x,y)}{f(y)} dx \\
&= \int_{-\infty}^{\infty} \frac{\big[x - E(x|\,y \in Y)\big]^2}{\sqrt{2\pi}\tilde{\sigma}}\, exp\left\{\frac{-(x - \tilde{\mu})^2}{2\tilde{\sigma}^2}\right\} dx\,, \quad x \in X, \quad (10.6)
\end{aligned}
$$

are both well defined. This can be verified based on the same reasoning and the well-defined interval division operation (note that zero is not contained in the denominator for a Gaussian density interval function). In the above,

$$
\tilde{\mu} = \mu_x + \sigma_{xy}^2(y - \mu_y)/\sigma_y^2 \qquad and \qquad \tilde{\sigma}^2 = \sigma_x^2 - \sigma_{xy}^2\sigma_{yx}^2/\sigma_y^2,
$$

with

$$
\sigma_{xy}^2 = \sigma_{yx}^2 = E(XY) - E(X)E(Y) = E(xy) - E(x)E(y)\,, \quad x \in X.
$$

Moreover, it can be verified (cf. Exercise 10.4) that

$$
E(X|\,y \in Y) = E(x) + \sigma_{xy}^2[y - E(y)]/\sigma_y^2\,, \qquad x \in X\,, \qquad (10.7)
$$

and

$$
Var(X|\,y \in Y) = Var(x) - \sigma_{xy}^2\sigma_{yx}^2/\sigma_y^2\,, \qquad x \in X\,. \qquad (10.8)
$$

Finally, we note that all these quantities are well-defined rational interval functions, so that Corollary 10.2 can be applied to them.

10.2 Interval Kalman Filtering

Now, return to the interval system (10.2). Observe that this system has an upper boundary system defined by all upper bounds of elements of its interval matrices:

$$\begin{cases} \mathbf{x}_{k+1} = [A_k + |\Delta A_k|]\mathbf{x}_k + [\Gamma_k + |\Delta \Gamma_k|]\underline{\xi}_k, \\ \mathbf{v}_k = [C_k + |\Delta C_k|]\mathbf{x}_k + \underline{\eta}_k \end{cases} \tag{10.9}$$

and a lower boundary system using all lower bounds of the elements of its interval matrices:

$$\begin{cases} \mathbf{x}_{k+1} = [A_k - |\Delta A_k|]\mathbf{x}_k + [\Gamma_k - |\Delta \Gamma_k|]\underline{\xi}_k, \\ \mathbf{v}_k = [C_k - |\Delta C_k|]\mathbf{x}_k + \underline{\eta}_k. \end{cases} \tag{10.10}$$

We first point out that by performing the standard Kalman filtering algorithm for these two boundary systems, the resulting two filtering trajectories do not encompass all possible optimal solutions of the interval system (10.2) (cf. Exercise 10.5). As a matter of fact, there is no specific relation between these two boundary trajectories and the entire family of optimal filtering solutions: the two boundary trajectories and their neighboring ones are generally intercrossing each other due to the noise perturbations. Therefore, a new filtering algorithm that can provide all-inclusive estimates for the interval system is needed. The interval Kalman filtering scheme derived below serves this purpose.

10.2.1 The Interval Kalman Filtering Scheme

Recall the derivation of the standard Kalman filtering algorithm given in Chapter 3, in which only matrix algebraic operations (additions, subtractions, multiplications, and inversions) and (conditional) expectations and variances are used. Since all these operations are well defined for interval matrices and rational interval functions, as discussed in the last section, the same derivation can be carried out for interval systems in exactly the same way to yield a Kalman filtering algorithm for the interval system (10.2). This interval Kalman filtering algorithm is simply summarized as follows:

The Interval Kalman Filtering Scheme

The main-process:

$$\widehat{\mathbf{x}}_0^I = E(\mathbf{x}_0^I)\,,$$
$$\widehat{\mathbf{x}}_k^I = A_{k-1}^I \widehat{\mathbf{x}}_{k-1}^I + G_k^I \Big[\mathbf{v}_k^I - C_k^I A_{k-1}^I \widehat{\mathbf{x}}_{k-1}^I \Big]\,,$$
$$k = 1, 2, \cdots.\tag{10.11}$$

The co-process:

$$P_0^I = Var\left(\mathbf{x}_0^I\right)\,,$$
$$M_{k-1}^I = A_{k-1}^I P_{k-1}^I \left[A_{k-1}^I\right]^\top + B_{k-1}^I Q_{k-1} \left[B_{k-1}^I\right]^\top\,,$$
$$G_k^I = M_{k-1}^I \left[C_k^I\right]^\top \left[\left[C_k^I\right] M_{k-1}^I \left[C_k^I\right]^\top + R_k\right]^{-1}\,,$$
$$P_k^I = \left[I - G_k^I C_k^I\right] M_{k-1}^I \left[I - G_k^I C_k^I\right]^\top + \left[G_k^I\right] R_k \left[G_k^I\right]^\top\,,$$
$$k = 1, 2, \cdots.\tag{10.12}$$

A comparison of this algorithm with the standard Kalman filtering scheme (3.25) reveals that they are exactly the same in form, except that all matrices and vectors in (10.11)–(10.12) are intervals. As a result, the interval estimate trajectory will diverge rather quickly. However, this is due to the conservative interval modeling but not the new filtering algorithm.

It should be noted that from the theory this interval Kalman filtering algorithm is optimal for the interval system (10.2), in the same sense as the standard Kalman filtering scheme, since no approximation is needed in its derivation. The filtering result produced by the interval Kalman filtering scheme is a sequence of interval estimates, $\{\widehat{\mathbf{x}}_k^I\}$, that encompasses all possible optimal estimates $\{\widehat{\mathbf{x}}_k\}$ of the state vectors $\{\mathbf{x}_k\}$ which the interval system may generate. Hence, the filtering result produced by this interval Kalman filtering scheme is inclusive but generally conservative in the sense that the range of interval estimates is often unnecessarily wide in order to include all possible optimal solutions.

It should also be remarked that just like the random vector (the measurement data) $\mathbf{v}_k$ in the ordinary case, the interval data vector $\mathbf{v}_k^I$ shown in the interval Kalman filtering scheme above is an uncertain interval vector before its realization (i.e., before the data actually being obtained), but will be an ordinary constant vector after it has been measured and obtained. This should avoid possible confusion in implementing the algorithm.

10.2.2 Suboptimal Interval Kalman Filter

To improve the computational efficiency, appropriate approximations of the interval Kalman filtering algorithm (10.11)–(10.12) may be applied. In this subsection, we suggest a suboptimal interval Kalman filtering scheme, by replacing its interval matrix inversion with its worst-case inversion, while keeping everything else unchanged.

To do so, let

$$C_k^I = C_k + \Delta C_k \qquad and \qquad M_{k-1}^I = M_{k-1} + \Delta M_{k-1} ,$$

where C_k is the center point of C_k^I and M_{k-1} is center point of M_{k-1}^I (i.e., the nominal values of the interval matrices). Write

$$\left[[C_k^I] M_{k-1}^I [C_k^I]^\top + R_k \right]^{-1}$$
$$= \left[[C_k + \Delta C_k] [M_{k-1} + \Delta M_{k-1}] [C_k + \Delta C_k]^\top + R_k \right]^{-1}$$
$$= \left[C_k M_{k-1} C_k^\top + \Delta R_k \right]^{-1} ,$$

where

$$\Delta R_k = C_k M_{k-1} [\Delta C_k]^\top + C_k [\Delta M_{k-1}] C_k^\top + C_k [\Delta M_{k-1}] [\Delta C_k]^\top$$
$$+ [\Delta C_k] M_{k-1} C_k^\top + [\Delta C_k] M_{k-1} [\Delta C_k]^\top + [\Delta C_k] [\Delta M_{k-1}] C_k^\top$$
$$+ [\Delta C_k] [\Delta M_{k-1}] [\Delta C_k]^\top + R_k .$$

Then, in the algorithm (10.11)–(10.12), replace ΔR_k by its upper bound matrix, $|\Delta R_k|$, which consists of all the upper bounds of the interval elements of $\Delta R_k = [[-r_k(i,j), r_k(i,j)]]$, namely,

$$|\Delta R_k| = [r_{ij}] , \qquad r_k(i,j) \geq 0 . \tag{10.13}$$

We should note that this $|\Delta R_k|$ is an ordinary (non-interval) matrix, so that when the ordinary inverse matrix $[C_k M_{k-1} C_k^\top + |\Delta R_k|]^{-1}$ is used to replace the interval matrix inverse $[[C_k^I] M_{k-1}^I [C_k^I]^\top + R_k]^{-1}$, the matrix inversion becomes much easier. More importantly, when the perturbation matrix $\Delta C_k = 0$ in (10.13), meaning that the measurement equation in system (10.2) is as accurate as the nominal system model (10.1), we have $|\Delta R_k| = R_k$.

Thus, by replacing ΔR_k with $|\Delta R_k|$, we obtain the following *suboptimal* interval Kalman filtering scheme.

A Suboptimal Interval Kalman Filtering Scheme

The main-process:

$$\widehat{\mathbf{x}}_0^I = E\big(\mathbf{x}_0^I\big),$$
$$\widehat{\mathbf{x}}_k^I = A_{k-1}^I \widehat{\mathbf{x}}_{k-1}^I + G_k^I\big[\mathbf{v}_k^I - C_k^I A_{k-1}^I \widehat{\mathbf{x}}_{k-1}^I\big],$$
$$k = 1, 2, \cdots. \tag{10.14}$$

The co-process:

$$P_0^I = Var\big(x_0^I\big),$$
$$M_{k-1}^I = A_{k-1}^I P_{k-1}^I \big[A_{k-1}^I\big]^{\mathsf{T}} + B_{k-1}^I Q_{k-1}\big[B_{k-1}^I\big]^{\mathsf{T}},$$
$$G_k^I = M_{k-1}^I \big[C_k^I\big]^{\mathsf{T}}\big[C_k M_{k-1} C_k^{\mathsf{T}} + |\Delta R_k|\big]^{-1},$$
$$P_k^I = \big[I - G_k^I C_k^I\big]M_{k-1}^I\big[I - G_k^I C_k^I\big]^{\mathsf{T}} + \big[G_k^I\big]R_k\big[G_k^I\big]^{\mathsf{T}},$$
$$k = 1, 2, \cdots. \tag{10.15}$$

Finally, we remark that the worst-case matrix $|\Delta R_k|$ given in (10.13) contains the largest possible perturbations and is in some sense the "best" matrix that yields a numerically stable inverse. Another possible approximation is, if ΔC_k is small, to simply use $|\Delta R_k| \approx R_k$. For some specific systems such as the radar tracking system to be discussed in the next subsection, special techniques are also possible to improve the speed and/or accuracy in performing suboptimal interval filtering.

10.2.3 An Example of Target Tracking

In this subsection, we show a computer simulation by comparing the interval Kalman filtering with the standard one, for a simplified version of the radar tracking system (3.26), see also (4.22), (5.22), and (6.43),

$$\begin{cases} \mathbf{x}_{k+1} = \begin{bmatrix} 1 & h^I \\ 0 & 1 \end{bmatrix}\mathbf{x}_k + \xi_k, \\ v_k = \begin{bmatrix} 1 & 0 \end{bmatrix}x_k + \eta_k, \end{cases} \tag{10.16}$$

where basic assumptions are the same as those stated for the system (10.2). Here, the system has uncertainty in an interval entry:

$$h^I = [h - \Delta h, h + \Delta h] = [0.01 - 0.001, 0.01 + 0.001] = [0.009, 0.011],$$

in which the modeling error Δh was taken to be 10% of the nominal value of $h = 0.01$. Suppose that the other given data are:

$$E(x_0) = \begin{bmatrix} x_{01} \\ x_{02} \end{bmatrix} = \begin{bmatrix} 1 \\ 1 \end{bmatrix}, \quad Var(x_0) = \begin{bmatrix} P_{00} & P_{01} \\ P_{10} & P_{11} \end{bmatrix} = \begin{bmatrix} 0.5 & 0.0 \\ 0.0 & 0.5 \end{bmatrix},$$

$$Q_k = \begin{bmatrix} q & 0 \\ 0 & q \end{bmatrix} = \begin{bmatrix} 0.1 & 0.0 \\ 0.0 & 0.1 \end{bmatrix}, \quad R_k = r = 0.1.$$

For this model, using the interval Kalman filtering algorithm (10.11)–(10.12), we have

$$M_{k-1}^I = \begin{bmatrix} h^I\left[2P_{k-1}^I(1,0) + h^I P_{k-1}^I(1,1)\right] & P_{k-1}^I(0,1) + h^I P_{k-1}^I(1,1) \\ +P_{k-1}^I(0,0) + q & \\ P_{k-1}^I(1,0) + h^I P_{k-1}^I(1,1) & P_{k-1}^I(1,1) + q \end{bmatrix}$$

$$:= \begin{bmatrix} M_{k-1}^I(0,0) & M_{k-1}^I(0,1) \\ M_{k-1}^I(1,0) & M_{k-1}^I(1,1) \end{bmatrix}$$

$$G_k^I = \begin{bmatrix} 1 - r/(M_{00}^I + r) \\ M_{10}^I/(M_{00}^I + r) \end{bmatrix} := \begin{bmatrix} G_{k,1}^I \\ G_{k,2}^I \end{bmatrix}$$

$$P_k^I = \begin{bmatrix} rG_{k,1}^I & rG_{k,2}^I \\ & q + \left[P_{k-1}^I(1,1)\left[P_{k-1}^I(0,0) + q + r\right]\right. \\ rG_{k,2}^I & \left. -[P_{k-1}^I(0,1)]^2\right]/(M_{k-1}^I(0,0) + r) \end{bmatrix}$$

$$:= \begin{bmatrix} P_k^I(0,0) & P_k^I(0,1) \\ P_k^I(1,0) & P_k^I(1,1) \end{bmatrix}.$$

In the above, the matrices M_{k-1}^I and P_k^I are both symmetrical. Hence, $M_{k-1}^I(0,1) = M_{k-1}^I(1,0)$ and $P_{k-1}^I(0,1) = P_{k-1}^I(1,0)$. It follows from the filtering algorithm that

$$\begin{bmatrix} \hat{x}_{k,1}^I \\ \hat{x}_{k,2}^I \end{bmatrix} = \begin{bmatrix} \left[r\left(\hat{x}_{k-1,1}^I + h^I \hat{x}_{k-1,2}^I\right) + M_{k-1}^I(0,0)y_k\right]/M_{k-1}^I(0,0) \\ \hat{x}_{k-1,2}^I + G_{k,1}^I\left(y_k - \hat{x}_{k-1,1}^I - h^I \hat{x}_{k-1,2}^I\right) \end{bmatrix}.$$

The simulation results for $\hat{x}_{k,1}$ of this interval Kalman filtering versus the standard Kalman filtering, where the latter used the nominal value of h, are shown and compared in both Figures 10.1 and 10.2. From these two figures we can see that the new scheme (10.11)–(10.12) produces the upper and lower boundaries of the single estimated curve obtained by the standard Kalman filtering algorithm (3.25), and these two boundaries encompass all possible optimal estimates of the interval system (10.2).

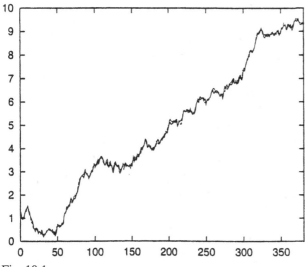

Fig. 10.1.

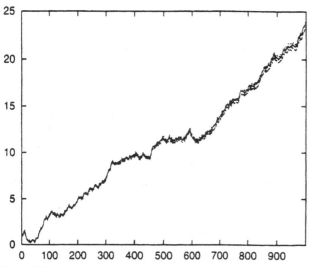

Fig. 10.2.

10.3 Weighted-Average Interval Kalman Filtering

As can be seen from Figures 10.1 and 10.2, the interval Kalman
filtering scheme produces the upper and lower boundaries for
all possible optimal trajectories obtained by using the standard
Kalman filtering algorithm. It can also be seen that as the iterations continue, the two boundaries are expanding. Here, it
should be emphasized once again that this seemingly divergent
result is not caused by the filtering algorithm, but rather, by the
iterations of the interval system model. That is, the upper and
lower trajectory boundaries of the interval system keeps expanding by themselves even if there is no noise in the model and no
filtering is performed. Hence, this phenomenon is inherent with
interval systems, although it is a natural and convenient way for
modeling uncertainties in dynamical systems.

To avoid this divergence while using interval system models,
a practical approach is to use a weighted average of all possible
optimal estimate trajectories encompassed by the two boundaries. An even more convenient way is to simply use a weighted
average of the two boundary estimates. For instance, taking a
certain weighted average of the two interval filtering trajectories
in Figures 10.1 and 10.2 gives the the results shown in Figures
10.3 and 10.4, respectively.

Finally, it is very important to remark that this averaging is
by nature different from the averaging of two standard Kalman
filtering trajectories produced by using the two (upper and lower)
boundary systems (10.9) and (10.10), respectively. The main reason is that the two boundaries of the filtering trajectories here,
as shown in Figures 10.1 and 10.2, encompass all possible optimal estimates, but the standard Kalman filtering trajectories
obtained from the two boundary systems do not cover all solutions (as pointed out above, cf. Exercise 10.5).

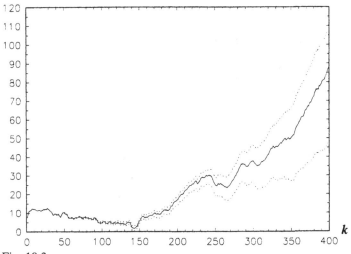

Fig. 10.3.

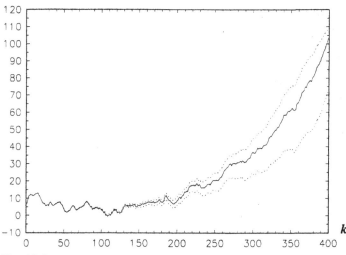

Fig. 10.4.

Exercises

10.1. For three intervals X, Y, and Z, verify that

$$X + Y = Y + X,$$
$$Z + (X + Y) = (Z + X) + Y,$$
$$XY = YX,$$
$$Z(XY) = (ZX)Y,$$
$$X + 0 = 0 + X = X \quad and \quad X0 = 0X = 0, \; where \; 0 = [0, 0],$$
$$XI = IX = X, \; where \; I = [1, 1],$$
$$Z(X + Y) \subseteq ZX + ZY, \; where \; = \; holds \; only \; if$$

    (a) $Z = [z, z]$;

    (b) $X = Y = 0$;

    (c) $xy \geq 0$ for all $x \in X$ and $y \in Y$.

10.2. Let A, B, C be ordinary constant matrices and A^I, B^I, C^I be interval matrices, respectively, of appropriate dimensions. Show that

(a) $A^I \pm B^I = \{A \pm B \mid A \in A^I, B \in B^I\}$;

(b) $A^I B = \{AB \mid A \in A^I\}$;

(c) $A^I + B^I = B^I + A^I$;

(d) $A^I + (B^I + C^I) = (A^I + B^I) + C^I$;

(e) $A^I + 0 = 0 + A^I = A^I$;

(f) $A^I I = I A^I = A^I$;

(g)   Subdistributive law:

(g.1) $(A^I + B^I)C^I \subseteq A^I C^I + B^I C^I$;

(g.2) $C^I(A^I + B^I) \subseteq C^I A^I + C^I B^I$;

(h) $(A^I + B^I)C = A^I C + B^I C$;

(i) $C(A^I + B^I) = CA^I + CB^I$;

(j)   Associative and Subassociative laws:

(j.1) $A^I(BC) \subseteq (A^I B)C$;

(j.2) $(AB^I)C^I \subseteq A(B^I C^I)$ if $C^I = -C^I$;

(j.3) $A(B^I C) = (AB^I)C$;

(j.4) $A^I(B^I C^I) = (A^I B^I)C^I$, if $B^I = -B^I$ and $C^I = -C^I$.

10.3. Prove Theorem 10.2.

10.4. Verify formulas (10.7) and (10.8).

10.5. Carry out a simple one-dimensional computer simulation to show that by performing the standard Kalman filtering algorithm for the two boundary systems (10.9)–(10.10) of the interval system (10.2), the two resulting filtering trajectories do not encompass all possible optimal estimation solutions for the interval system.

10.6. Consider the target tracking problem of tracking an uncertain incoming ballistic missile. This physical problem is described by the following simplified interval model:

$$\begin{cases} \mathbf{x}^I_{k+1} = A^I_k \mathbf{x}^I_k + \underline{\xi}_k \,, \\ \mathbf{v}^I_k = C^I_k \mathbf{x}^I_k + \underline{\eta}_k \,, \end{cases}$$

where $\mathbf{x}^I = \begin{bmatrix} x^I_1 & \cdots & x^I_7 \end{bmatrix}^{\mathsf{T}}$,

$$A_{11} = A_{22} = A_{33} = 1 \,, \quad A_{44} = -\frac{1}{2} g\, x_7 \frac{z^2 + x^2_4}{z} \,,$$

$$A_{45} = A_{54} = -\frac{1}{2} g\, \frac{x_7 x_4 x_5}{z} \,, \quad A_{46} = A_{64} = -\frac{1}{2} g\, \frac{x_7 x_4 x_6}{z} \,,$$

$$A_{47} = -\frac{1}{2} g\, x_4 z \,, \quad A_{55} = -\frac{1}{2} g\, x_7 \frac{z^2 + x^2_5}{z} \,,$$

$$A_{56} = A_{65} = -\frac{1}{2} g\, \frac{x_7 x_5 x_6}{z} \,, \quad A_{57} = -\frac{1}{2} g\, x_5 z \,,$$

$$A_{66} = -\frac{1}{2} g\, x_7 \frac{z^2 + x^2_6}{z} \,, \quad A_{67} = -\frac{1}{2} g\, x_6 z \,,$$

$$A_{76} = -K^I x_7 \,, \quad A_{77} = -K^I x_6 \,,$$

with all other $A_{ij} = 0$, where K^I is an uncertain system parameter, g is the gravety constant, $z = \sqrt{x^2_4 + x^2_5 + x^2_6}$, and

$$C = \begin{bmatrix} 1 & 0 & 0 & 0 & 0 & 0 & 0 \\ 0 & 1 & 0 & 0 & 0 & 0 & 0 \\ 0 & 0 & 1 & 0 & 0 & 0 & 0 \end{bmatrix}.$$

Both the dynamical and measurement noise sequences are zero-mean Gaussian, mutually independent, with covariances $\{Q_k\}$ and $\{R_k\}$, respectively. Perform interval Kalman filtering on this model, using the following data set:

$g = 0.981$,

$K^I = \begin{bmatrix} 2.3 \times 10^{-5}, 3.5 \times 10^{-5} \end{bmatrix}$,

$\mathbf{x}^I_0 = \big[3.2 \times 10^5, 3.2 \times 10^5, 2.1 \times 10^5, -1.5 \times 10^4, -1.5 \times 10^4,$
$\qquad -8.1 \times 10^3, 5 \times 10^{-10} \big]^{\mathsf{T}}$,

$P^I_0 = diag\big\{ 10^6, 10^6, 10^6, 10^6, 10^6, 1.286 \times 10^{-13} exp\{-23.616\} \big\}$,

$Q_k = \dfrac{1}{k+1}\, diag\big\{ 0, 0, 0, 100, 100, 100, 2.0 \times 10^{-18} \big\}$,

$R_k = \dfrac{1}{k+1}\, diag\big\{ 150, 150, 150 \big\}$.

11. Wavelet Kalman Filtering

In addition to the Kalman filtering algorithms discussed in the previous chapters, there are other computational schemes available for digital filtering performed in the time domain. Among them, perhaps the most exciting ones are wavelet algorithms, which are useful tools for multichannel signal processing (e.g., estimation or filtering) and multiresolution signal analysis. This chapter is devoted to introduce this effective technique of wavelet Kalman filtering by means of a specific application for illustration – simultaneous estimation and decomposition of random signals via a filter-bank-based Kalman filtering approach using wavelets.

11.1 Wavelet Preliminaries

The notion of wavelets was first introduced in the earlier 1980's, as a family of functions generated by a single function, called the "basic wavelet," by two simple operations: translation and scaling. Let $\psi(t)$ be such a basic wavelet. Then, with a scaling constant, a, and a translation constant, b, we obtain a family of wavelets of the form $\psi((t-b)/a)$. Using this family of wavelets as the integral kernel to define an integral transform, called the integral wavelet transform (IWT):

$$(W_\psi f)(b,a) = |a|^{-1/2} \int_{-\infty}^{\infty} f(t)\,\overline{\psi((t-b)/a))}\, dt\,, \quad f \in L^2\,, \qquad (11.1)$$

we can analyze the functions (or signals) $f(t)$ at different positions and at different scales according to the values of a and b. Note that the wavelet $\psi(t)$ acts as a time-window function whose "width" narrows as the value of the "scale" a in (11.1) decreases. Hence, if the frequency, ω in its Fourier transform $\hat{\psi}(\omega)$, is defined to be inversely proportional to the scale a, then the width of the time window induced by $\psi(t)$ narrows for studying high-frequency objects and widens for observing low-frequency situations. In addition, if the basic wavelet $\psi(t)$ is so chosen that its Fourier transform is also a window function, then the IWT, $(W_\psi f)(b,a)$, can be used for time-frequency localization and analysis of $f(t)$ around $t = b$ on a frequency band defined by the scale a.

11.1.1 Wavelet Fundamentals

An elegant approach to studying wavelets is via "multiresolution analysis." Let L^2 denote the space of real-valued finite-energy functions in the continuous-time domain $(-\infty, \infty)$, in which the inner product is defined by $\langle f, g \rangle = \int_{-\infty}^{\infty} f(t) \bar{g}(t) \, dt$ and the norm is defined by $\|f\|_{L^2} = \sqrt{|\langle f, f \rangle|}$. A nested sequence $\{V_k\}$ of closed subspaces of L^2 is said to form a *multiresolution analysis* of L^2 if there exists some window function (cf. Exercise 11.1), $\phi(t) \in L^2$, which satisfies the following properties:

(i)   for each integer k, the set

$$\{\phi_{kj}(t) := \phi(2^k t - j) : \quad j = \cdots, -1, 0, 1, \cdots \}$$

is an unconditional basis of V_k, namely, its linear span is dense in V_k and for each k,

$$\alpha \|\{c_j\}\|_{\ell^2}^2 \leq \left\| \sum_{j=-\infty}^{\infty} c_j \phi_{kj} \right\|_{L^2}^2 \leq \beta \|\{c_j\}\|_{\ell^2}^2 \qquad (11.2)$$

for all $\{c_j\} \in \ell^2$, where $\|\{c_j\}\|_{\ell^2} = \sqrt{\sum_{j=-\infty}^{\infty} |c_j|^2}$;

(ii)   the union of V_k is dense in L^2;

(iii)   the intersection of all V_k's is the zero function; and

(iv) $f(t) \in V_k$ if and only if $f(2t) \in V_{k+1}$.

Let W_k be the orthogonal complementary subspace of V_{k+1} relative to V_k, and we use the notation

$$V_{k+1} = V_k \bigoplus W_k . \qquad (11.3)$$

Then it is clear that $W_k \perp W_n$ for all $k \neq n$, and the entire space L^2 is an orthogonal sum of the spaces W_k, namely:

$$L^2 = \bigoplus_{k=-\infty}^{\infty} W_k . \qquad (11.4)$$

Suppose that there is a function, $\psi(t) \in W_0$, such that both $\psi(t)$ and its Fourier transform $\hat{\psi}(\omega)$ have sufficiently fast decay at $\pm\infty$ (cf. Exercise 11.2), and that for each integer k,

$$\{\psi_{kj}(t) := 2^{k/2} \psi(2^k t - j) : \quad j = \cdots, -1, 0, 1, \cdots \} \qquad (11.5)$$

is an unconditional basis of W_k. Then $\psi(t)$ is called a *wavelet* (cf. Exercise 11.3). Let $\tilde{\psi}(t) \in W_0$ be the *dual* of $\psi(t)$, in the sense that

$$\int_{-\infty}^{\infty} \tilde{\psi}(t-i)\,\overline{\psi(t-j)}\,dt = \delta_{ij}, \qquad i,j = \cdots, -1, 0, 1, \cdots.$$

Then both $\tilde{\psi}(t)$ and $\hat{\tilde{\psi}}(\omega)$ are window functions, in the time and frequency domains, respectively. If $\tilde{\psi}(t)$ is used as a basic wavelet in the definition of the IWT, then real-time algorithms are available to determine $(W_{\tilde{\psi}}f)(b,a)$ at the dyadic time instants $b = j/2^k$ on the kth frequency bands defined by the scale $a = 2^{-k}$. Also, $f(t)$ can be reconstructed in real-time from information of $(W_{\tilde{\psi}}f)(b,a)$ at these dyadic data values.

More precisely, by defining

$$\psi_{kj}(t) = 2^{k/2}\psi(2^k t - j), \tag{11.6}$$

any function $f(t) \in L^2$ can be expressed as a wavelet series

$$f(t) = \sum_{k=-\infty}^{\infty} \sum_{j=-\infty}^{\infty} d_j^k \psi_{k,j}(t) \tag{11.7}$$

with

$$d_j^k = \left(W_{\tilde{\psi}}f\right)\left(j2^{-k}, 2^{-k}\right). \tag{11.8}$$

According to (11.3), there exist two sequences $\{a_j\}$ and $\{b_j\}$ in ℓ^2 such that

$$\phi(2t - \ell) = \sum_{j=-\infty}^{\infty} \left[a_{\ell-2j}\phi(t-j) + b_{\ell-2j}\psi(t-j)\right] \tag{11.9}$$

for all integers ℓ; and it follows from property (i) and (11.3) that two sequences $\{p_j\}$ and $\{q_j\}$ in ℓ^2 are uniquely determined such that

$$\phi(t) = \sum_{j=-\infty}^{\infty} p_j\,\phi(2t - j) \tag{11.10}$$

and

$$\psi(t) = \sum_{j=-\infty}^{\infty} q_j\,\phi(2t - j). \tag{11.11}$$

The pair of sequences $(\{a_j\}, \{b_j\})$ yields a pyramid algorithm for finding the IWT values $\{d_j^k\}$; while the pair of sequences $(\{p_j\}, \{q_j\})$

yields a pyramid algorithm for reconstructing $f(t)$ from the IWT values $\{d_j^k\}$.

We remark that if $\phi(t)$ is chosen as a B-spline function, then a compactly supported wavelet $\psi(t)$ is obtained such that its dual $\tilde{\psi}(t)$ has exponential decay at $\pm\infty$ and the IWT with respect to $\tilde{\psi}(t)$ has linear phase. Moreover, both sequences $\{p_j\}$ and $\{q_j\}$ are finite, while $\{a_j\}$ and $\{b_j\}$ have very fast exponential decay. It should be noted that the spline wavelets $\psi(t)$ and $\tilde{\psi}(t)$ have explicit formulations and can be easily implemented.

11.1.2 Discrete Wavelet Transform and Filter Banks

For a given sequence of scalar *deterministic* signals, $\{x(i,n)\} \in \ell^2$, at a fixed resolution level i, a lower resolution signal can be derived by lowpass filtering with a halfband lowpass filter having an impulse response $\{h(n)\}$. More precisely, a sequence of the lower resolution signal (indicated by an index L) is obtained by downsampling the output of the lowpass filter by two, namely, $h(n) \to h(2n)$, so that

$$x_L(i-1, n) = \sum_{k=-\infty}^{\infty} h(2n-k)\, x(i, k). \qquad (11.12)$$

Here, (11.12) defines a mapping from ℓ^2 to itself. The wavelet coefficients, as a complement to $x_L(i-1, n)$, will be denoted by $\{x_H(i-1, n)\}$, which can be computed by first using a highpass filter with an impulse response $\{g(n)\}$ and then using downsampling the output of the highpass filtering by two. This yields

$$x_H(i-1, n) = \sum_{k=-\infty}^{\infty} g(2n-k)\, x(i, k). \qquad (11.13)$$

The original signal $\{x(i,n)\}$ can be recovered from the two filtered and downsampled (lower resolution) signals $\{x_L(i-1,n)\}$ and $\{x_H(i-1,n)\}$. Filters $\{h(n)\}$ and $\{g(n)\}$ must meet some constraints in order to produce a perfect reconstruction for the signal. The most important constraint is that the filter impulse responses form an orthonormal set. For this reason, (11.12) and (11.13) together can be considered as a decomposition of the original signal onto an orthonormal basis, and the reconstruction

$$x(i, n) = \sum_{k=-\infty}^{\infty} h(2k-n) x_L(i-1, k)$$

$$+ \sum_{k=-\infty}^{\infty} g(2k - n)x_H(i - 1, k) \qquad (11.14)$$

can be considered as a sum of orthogonal projections.

The operation defined by (11.12) and (11.13) is called the *discrete (forward) wavelet transform*, while the *discrete inverse wavelet transform* is defined by (11.14).

To be implementable, we use FIR (finite impulse response) filters for both $\{h(n)\}$ and $\{g(n)\}$ (namely, these two sequences have finite length, L), and we require that

$$g(n) = (-1)^n h(L - 1 - n), \qquad (11.15)$$

where L must be even under this relation. Clearly, once the lowpass filter $\{h(n)\}$ is determined, the highpass filter is also determined.

The discrete wavelet transform can be implemented by an octave-band filter bank, as shown in Figure 11.1 (b), where only three levels are depicted. The dimensions of different decomposed signals at different levels are shown in Figure 11.1 (a).

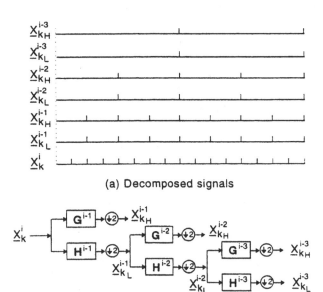

(a) Decomposed signals

(b) A two-channel filter bank

Fig. 11.1.

For a sequence of deterministic signals with finite length, it is more convenient to describe the wavelet transform in an operator form. Consider a sequence of signals at resolution level i with length M:

$$\underline{X}_k^i = \left[x(i, k - M + 1), x(i, k - M + 2), \cdots, x(i, k)\right]^\top.$$

Formulas (11.12) and (11.13) can be written in the following operator form:

$$\underline{X}_{k_L}^{i-1} = \mathbf{H}^{i-1}\underline{X}_k^i \qquad and \qquad \underline{X}_{k_H}^{i-1} = \mathbf{G}^{i-1}\underline{X}_k^i,$$

where operators $\mathbf{H}^{i-1}$ and $\mathbf{G}^{i-1}$ are composed of lowpass and highpass filter responses [the $\{h(n)\}$ and $\{g(n)\}$ in (11.12) and (11.13)], mapping from level i to level $i-1$. Similarly, when mapping from level $i-1$ to level i, (11.14) can be written in operator form as (cf. Exercise 11.4)

$$\underline{X}_k^i = (\mathbf{H}^{i-1})^\top \underline{X}_{k_L}^{i-1} + (\mathbf{G}^{i-1})^\top \underline{X}_{k_H}^{i-1}. \tag{11.16}$$

On the other hand, the orthogonality constraint can also be expressed in operator form as

$$(\mathbf{H}^{i-1})^\top \mathbf{H}^{i-1} + (\mathbf{G}^{i-1})^\top \mathbf{G}^{i-1} = \mathbf{I}$$

and

$$\begin{bmatrix} \mathbf{H}^{i-1}(\mathbf{H}^{i-1})^\top & \mathbf{H}^{i-1}(\mathbf{G}^{i-1})^\top \\ \mathbf{G}^{i-1}(\mathbf{H}^{i-1})^\top & \mathbf{G}^{i-1}(\mathbf{G}^{i-1})^\top \end{bmatrix} = \begin{bmatrix} \mathbf{I} & 0 \\ 0 & \mathbf{I} \end{bmatrix}.$$

A simultaneous multilevel signal decomposition can be carried out by a filter bank. For instance, to decompose $\underline{X}_k^i$ into three levels, as shown in Figure 11.1, the following composite transform can be applied:

$$\begin{bmatrix} \underline{X}_{k_L}^{i-3} \\ \underline{X}_{k_H}^{i-3} \\ \underline{X}_{k_H}^{i-2} \\ \underline{X}_{k_H}^{i-1} \end{bmatrix} = \mathbf{T}^{i-3|i}\,\underline{X}_k^i,$$

where

$$\mathbf{T}^{i-3|i} = \begin{bmatrix} \mathbf{H}^{i-3}\mathbf{H}^{i-2}\mathbf{H}^{i-1} \\ \mathbf{G}^{i-3}\mathbf{H}^{i-2}\mathbf{H}^{i-1} \\ \mathbf{G}^{i-2}\mathbf{H}^{i-1} \\ \mathbf{G}^{i-1} \end{bmatrix}$$

is an orthogonal matrix, simultaneously mapping $\underline{X}_k^i$ onto the three levels of the filter bank.

11.2 Singal Estimation and Decomposition

In random signal estimation and decomposition, a general approach is first to estimate the unknown signal using its measurement data and then to decompose the estimated signal according to the resolution requirement. Such a two-step approach is off-line in nature and is often not desirable for real-time applications.

In this section, a technique for simultaneous optimal estimation and multiresolutional decomposition of a random signal is developed. This is used as an example for illustrating wavelet Kalman filtering. An algorithm that simultaneously performs estimation and decomposition is derived based on the discrete wavelet transform and is implemented by a Kalman filter bank. The algorithm preserves the merits of the Kalman filtering scheme for estimation, in the sense that it produces an optimal (linear, unbiased, and minimum error-variance) estimate of the unknown signal, in a recursive manner using sampling data obtained from the noisy signal.

The approach to be developed has the following special features: First, instead of a two-step approach, it determines in one step the estimated signal such that the resulting signal naturally possesses the desired decomposition. Second, the recursive Kalman filtering scheme is employed in the algorithm, so as to achieve the estimation and decomposition of the unknown but noisy signal, not only simultaneously but also optimally. Finally, the entire signal processing is performed on-line, which is a real-time process in the sense that as a block of new measurements flows in, a block of estimates of the signal flows out in the required decomposition form. In this procedure, the signal is first divided into blocks and then filtering is performed over data blocks. To this end, the current estimates are obtained based on a block of current measurements and the previous block of optimal estimates. Here, the length of the data block is determined by the number of levels of the desired decomposition. An octave-band filter bank is employed as an effective vehicle for the multiresolutional decomposition.

11.2.1 Estimation and Decomposition of Random Signals

Now, consider a sequence of one-dimensional *random* signals, $\{x(N, k)\}$ at the highest resolution level (level N), governed by

$$x(N, k + 1) = A(N, k)x(N, k) + \xi(N, k), \qquad (11.17)$$

with measurement

$$v(N, k) = C(N, k)x(N, k) + \eta(N, k),\qquad (11.18)$$

where $\{\xi(N, k)\}$ and $\{\eta(N, k)\}$ are mutually independent Gaussian noise sequences with zero mean and variances $Q(N, k)$ and $R(N, k)$, respectively.

Given a sequence of measurements, $\{v(N, k)\}$, the conventional way of estimation and decomposition of the random signal $\{X(N, k)\}$ is performed in a sequential manner: first, find an estimation $\hat{x}(N, k)$ at the highest resolutional level; then apply a wavelet transform to decompose it into different resolutions.

In the following, an algorithm for simultaneous estimation and decomposition of the random signal is derived. For simplicity, only two-level decomposition and estimation will be discussed, i.e., from level N to levels $N - 1$ and $N - 2$. At all other levels, simultaneous estimation and decomposition procedures are exactly the same.

A data block with length $M = 2^2 = 4$ is chosen, where base 2 is used in order to design an octave filter bank, and power 2 is the same as the number of the levels in the decomposition. Up to instant k, we have

$$\underline{X}_k^N = \left[x(N, k - 3), x(N, k - 2), x(N, k - 1), x(N, k)\right]^\top,$$

for which the equivalent dynamical system in a data-block form is derived as follows. For notional convenience, the system and measurement equations, (11.17) and (11.18), are assumed to be time-invariant, and the level index N is dropped. The propagation is carried out over an interval of length M, yielding

$$x(N, k + 1) = A\,x(N, k) + \xi(N, k),\qquad (11.19)$$

or

$$x(N, k + 1) = A^2 x(N, k - 1) + A\xi(N, k - 1) + \xi(N, k),\qquad (11.20)$$

or

$$x(N, k + 1) = A^3 x(N, k - 2) + A^2 \xi(N, k - 2)$$
$$+ A\,\xi(N, k - 1) + \xi(N, k),\qquad (11.21)$$

or

$$x(N, k + 1) = A^4 x(N, k - 3) + A^3 \xi(N, k - 3) + A^2 \xi(N, k - 2)$$
$$+ A \xi(N, k - 1) + \xi(N, k). \tag{11.22}$$

Taking the average of (11.19), (11.20), (11.21) and (11.22), we obtain

$$x(N, k + 1) = \frac{1}{4} A^4 x(N, k - 3) + \frac{1}{4} A^3 x(N, k - 2)$$
$$+ \frac{1}{4} A^2 x(N, k - 1) + \frac{1}{4} A x(N, k) + \xi(1),$$

where

$$\xi(1) = \frac{1}{4} A^3 \xi(N, k - 3) + \frac{1}{2} A^2 \xi(N, k - 2)$$
$$+ \frac{3}{4} A \xi(N, k - 1) + \xi(N, k). \tag{11.23}$$

Also, taking into account all propagations, we finally have a dynamical system in a data-block form as follows (cf. Exercise 11.5):

$$
\underbrace{\begin{bmatrix} x(N, k + 1) \\ x(N, k + 2) \\ x(N, k + 3) \\ x(N, k + 4) \end{bmatrix}}_{\underline{X}^N_{k+1}} = \underbrace{\begin{bmatrix} \frac{1}{4} A^4 & \frac{1}{4} A^3 & \frac{1}{4} A^2 & \frac{1}{4} A \\ 0 & \frac{1}{3} A^4 & \frac{1}{3} A^3 & \frac{1}{3} A^2 \\ 0 & 0 & \frac{1}{2} A^4 & \frac{1}{2} A^3 \\ 0 & 0 & 0 & A^4 \end{bmatrix}}_{\overline{A}}
$$

$$
\times \underbrace{\begin{bmatrix} x(N, k - 3) \\ x(N, k - 2) \\ x(N, k - 1) \\ x(N, k) \end{bmatrix}}_{\underline{X}^N_k} + \underbrace{\begin{bmatrix} \xi(1) \\ \xi(2) \\ \xi(3) \\ \xi(4) \end{bmatrix}}_{\overline{W}^N_k}, \tag{11.24}
$$

where $\xi(i)$, $i = 2, 3, 4$, are similarly defiend, with

$$E\{\overline{W}^N_k\} = 0 \qquad and \qquad E\{\overline{W}^N_k (\overline{W}^N_k)^\top\} = \overline{Q},$$

and the elements of $\overline{Q}$ are given by

$$\bar{q}_{11} = \frac{1}{16} A^6 Q + \frac{1}{4} A^4 Q + \frac{9}{16} A^2 Q + Q, \qquad \bar{q}_{12} = \frac{1}{6} A^5 Q + \frac{1}{2} A^3 Q + AQ,$$

$$\bar{q}_{13} = \frac{3}{8} A^4 Q + A^2 Q, \qquad \bar{q}_{14} = A^3 Q, \qquad \bar{q}_{21} = \bar{q}_{12},$$

$$\bar{q}_{22} = \frac{1}{9} A^6 Q + \frac{4}{9} A^4 Q + A^2 Q + Q, \qquad \bar{q}_{23} = \frac{1}{3} A^5 Q + A^3 Q + AQ,$$

$$\bar{q}_{24} = A^4 Q + A^2 Q, \qquad \bar{q}_{31} = \bar{q}_{13}, \qquad \bar{q}_{32} = \bar{q}_{23},$$

$$\bar{q}_{33} = \frac{1}{4} A^6 Q + A^4 Q + A^2 Q + Q, \qquad \bar{q}_{34} = A^5 Q + A^3 Q + AQ,$$

$$\bar{q}_{41} = \bar{q}_{14}, \qquad \bar{q}_{42} = \bar{q}_{24}, \qquad \bar{q}_{43} = \bar{q}_{34}, \qquad \bar{q}_{44} = A^6 Q + A^4 Q + A^2 Q + Q.$$

The measurement equation associated with (11.24) can be easily constructed as

$$
\underbrace{\begin{bmatrix} v(N,k-3) \\ v(N,k-2) \\ v(N,k-1) \\ v(N,k) \end{bmatrix}}_{\underline{v}_k^N} = \underbrace{\begin{bmatrix} C & 0 & 0 & 0 \\ 0 & C & 0 & 0 \\ 0 & 0 & C & 0 \\ 0 & 0 & 0 & C \end{bmatrix}}_{\overline{\mathbf{C}}} \underbrace{\begin{bmatrix} x(N,k-3) \\ x(N,k-2) \\ x(N,k-1) \\ x(N,k) \end{bmatrix}}_{\underline{X}_k^N} + \underbrace{\begin{bmatrix} \eta(N,k-3) \\ \eta(N,k-2) \\ \eta(N,k-1) \\ \eta(N,k) \end{bmatrix}}_{\underline{\Pi}_k^N} ,
$$

where

$$
E\{\underline{\Pi}_k^N\} = 0 \qquad and \qquad E\{\underline{\Pi}_k^N (\underline{\Pi}_k^N)^\top\} = \mathbf{R} = diag\{R,R,R,R\} .
$$

Two-level decomposition is then performed, leading to

$$
\begin{bmatrix} \underline{X}_{k_L}^{N-2} \\ \underline{X}_{k_H}^{N-2} \\ \underline{X}_{k_H}^{N-1} \end{bmatrix} = \begin{bmatrix} \mathbf{H}^{N-2}\mathbf{H}^{N-1} \\ \mathbf{G}^{N-2}\mathbf{H}^{N-1} \\ \mathbf{G}^{N-1} \end{bmatrix} \underline{X}_k^N = \mathbf{T}^{N-2|N} \underline{X}_k^N . \tag{11.25}
$$

Substituting (11.25) into (11.24) results in

$$
\begin{bmatrix} \underline{X}_{k+1_L}^{N-2} \\ \underline{X}_{k+1_H}^{N-2} \\ \underline{X}_{k+1_H}^{N-1} \end{bmatrix} = \mathbf{A} \begin{bmatrix} \underline{X}_{k_L}^{N-2} \\ \underline{X}_{k_H}^{N-2} \\ \underline{X}_{k_H}^{N-1} \end{bmatrix} + \underline{W}_k^N , \tag{11.26}
$$

where

$$
\underline{W}_k^N = \mathbf{T}^{N-2|N}\overline{\underline{W}}_k^N , \qquad E\{\underline{W}_k^N\} = 0 , \qquad E\{\underline{W}_k^N (\underline{W}_k^N)^\top\} = \mathbf{Q} ,
$$
$$
\mathbf{A} = \mathbf{T}^{N-2|N}\overline{\mathbf{A}}(\mathbf{T}^{N-2|N})^\top , \qquad \mathbf{Q} = \mathbf{T}^{N-2|N}\overline{\mathbf{Q}}(\mathbf{T}^{N-2|N})^\top .
$$

Equation (11.26) describes a dynamical system for the decomposed quantities. Its associated measurement equation can also be derived by substituting (11.25) into (11.24), yielding

$$
\underline{Z}_k^N = \mathbf{C} \begin{bmatrix} \underline{X}_{k_L}^{N-2} \\ \underline{X}_{k_H}^{N-2} \\ \underline{X}_{k_H}^{N-1} \end{bmatrix} + \underline{\Pi}_k^N , \tag{11.27}
$$

where

$$
\mathbf{C} = \overline{\mathbf{C}}(\mathbf{T}^{N-2|N})^\top .
$$

Now, we have obtained the system and measurement equations, (11.26) and (11.27), for the decomposed quantities. The next task is to estimate these quantities using measurement data. The Kalman filter is readily applied to (11.26) and (11.27) to provide optimal estimates for these decomposed quantities.

11.2.2 An Example of Random Walk

A one-dimensional colored noise process (known also as the Brownian random walk) is studied in this section. This random process is governed by

$$x(N, k+1) = x(N, k) + \xi(N, k) \tag{11.28}$$

with the measurement

$$v(N, k) = x(N, k) + \eta(N, k), \tag{11.29}$$

where $\{\xi(N,k)\}$ and $\{\eta(N,k)\}$ are mutually independent zero mean Gaussian nose sequences with variances $Q(N,k) = 0.1$ and $R(N,k) = 1.0$, respectively.

The true $\{x(N,k)\}$ and the measurement $\{v(N,k)\}$ at the highest resolutional level are shown in Figures 11.2 (a) and (b). By applying the model of (11.26) and (11.27), Haar wavelets for the filter bank, and the standard Kalman filtering scheme through the process described in the last section, we obtain two-level estimates: $\widehat{\underline{X}}_{k_L}^{N-2}$, $\widehat{\underline{X}}_{k_H}^{N-2}$, and $\widehat{\underline{X}}_{k_H}^{N-1}$. The computation was performed on-line, and the results are shown in Figures 11.2 (c), (d), and 11.3 (a).

At the first glance, $\widehat{\underline{X}}_{k_H}^{N-2}$ and $\widehat{\underline{X}}_{k_H}^{N-1}$ both look like some kind of noise. Actually, $\widehat{\underline{X}}_{k_H}^{N-2}$ and $\widehat{\underline{X}}_{k_H}^{N-1}$ are estimates of the high frequency components of the true signal. We can use these high-frequency components to compose higher-level estimates by "adding" them to the estimates of the low-frequency components. For instance, $\widehat{\underline{X}}_{k_L}^{N-1}$ is derived by composing $\widehat{\underline{X}}_{k_L}^{N-2}$ and $\widehat{\underline{X}}_{k_H}^{N-2}$ as follows:

$$\widehat{\underline{X}}_{k_L}^{N-1} = \begin{bmatrix} (\mathbf{H}^{N-2})^\top & (\mathbf{G}^{N-2})^\top \end{bmatrix} \begin{bmatrix} \widehat{\underline{X}}_{k_L}^{N-2} \\ \widehat{\underline{X}}_{k_H}^{N-2} \end{bmatrix}, \tag{11.30}$$

which is depicted in Figure 11.3 (b). Similarly, $\widehat{\underline{X}}_{k_L}^{N}$ (at the highest resolutional level) can be obtained by composing $\widehat{\underline{X}}_{k_L}^{N-1}$ and $\widehat{\underline{X}}_{k_H}^{N-1}$, which is displayed in Figure 11.3 (c).

To compare the performance, the standard Kalman filter is applied directly to system (11.28) and (11.29), resulting in the curves shown in Figure 11.3 (d). We can see that both estimates at the highest resolutional level obtained by composing the estimates of decomposed quantities and by Kalman filtering along are very similar. To quantitatively compare these two approaches, a set of 200 Monte Carlo simulations were performed and the root-mean-square errors were found to be 0.2940 for the simultaneous estimation and decomposition algorithm but 0.3104 for the standard Kalman filtering scheme. This indicates that the simultaneous approach outperforms the direct Kalman filter, even for only two-level decomposition and estimation. The difference becomes more significant as the number of levels is allowed to increase.

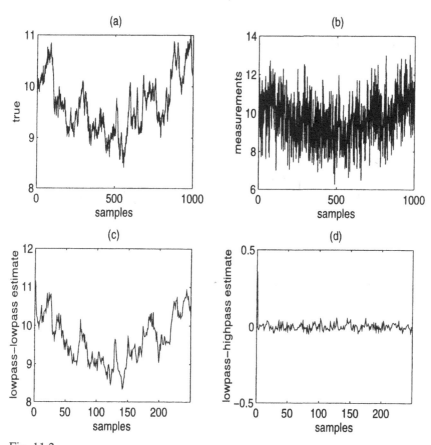

Fig. 11.2.

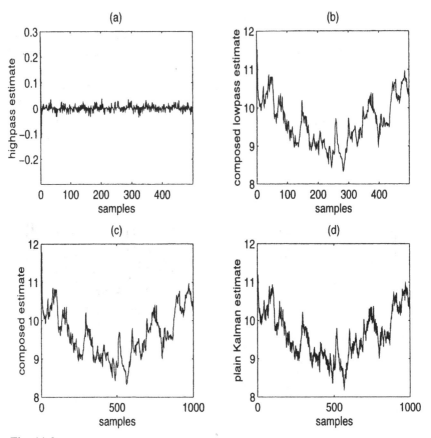

Fig. 11.3.

Exercises

11.1. The following Haar and triangle functions are typical window functions in the time domain:

$$\phi_H(t) = \begin{cases} 1 & 0 \leq t < 1 \\ 0 & otherwise; \end{cases}$$

$$\phi_T(t) = \begin{cases} t & 0 \leq t < 1, \\ 2 - t & 1 \leq t < 2, \\ 0 & otherwise. \end{cases}$$

Sketch these two functions, and verify that

$$\phi_T(t) = \int_{-\infty}^{\infty} \phi_H(\tau)\phi_H(t - \tau)d\tau.$$

Here, $\phi_H(t)$ and $\phi_T(t)$ are also called B-splines of degree zero and one, respectively. B-spline of degree n can be generated via

$$\phi_n(t) = \int_0^1 \phi_{n-1}(t - \tau)d\tau, \qquad n = 2, 3, \cdots.$$

Calculate and sketch $\phi_2(t)$ and $\phi_3(t)$. (Note that $\phi_0 = \phi_H$ and $\phi_1 = \phi_T$.)

11.2. Find the Fourier transforms of $\phi_H(t)$ and $\phi_T(t)$ defined above.

11.3. Based on the graphs of $\phi_H(t)$ and $\phi_T(t)$, sketch

$$\phi_{1,0}(t) = \phi_H(2t), \quad \phi_{1,1}(t) = \phi_H(2t-1), \; and \; \Phi(t) = \phi_{1,0}(t)+\phi_{1,1}(t).$$

Moreover, sketch the wavelets

$$\psi_H(t) = \phi_H(2t) - \phi_H(2t - 1)$$

and

$$\psi_T(t) = -\frac{1}{2}\phi_T(2t) - \frac{1}{2}\phi_T(2t - 2) + \phi_T(2t - 1).$$

11.4. Verify (11.16).

11.5. Verify (11.24) and (11.26).

12. Notes

The objective of this monograph has been to give a rigorous and yet elementary treatment of the Kalman filter theory and briefly introduce some of its real-time applications. No attempt was made to cover all the rudiment of the theory, and only a small sample of its applications has been included. There are many texts in the literature that were written for different purposes, including Anderson and Moore (1979), Balakrishnan (1984,87), Brammer and Sifflin (1989), Catlin (1989), Chen (1985), Chen, Chen and Hsu (1995), Goodwin and Sin (1984), Haykin (1986), Lewis (1986), Mendel (1987), Ruymgaart and Soong (1985,89), Sorenson (1985), Stengel (1986), and Young (1984). Unfortunately, in our detailed treatment we have had to omit many important topics; some of these will be introduced very briefly in this chapter. The interested reader is referred to the modest bibliography in this text for further study.

12.1 The Kalman Smoother

Suppose that data information is available over a discrete-time time-interval $\{1, 2, \cdots, N\}$. For any K with $1 \leq K < N$, the optimal estimate $\hat{x}_{K|N}$ of the state vector x_K using all the data information on this time-interval (i.e. past, present, and future information are being used) is called a *(digital) smoothing estimate* of x_K (cf. Definition 2.1 in Chapter 2). Although this smoothing problem is somewhat different from the real-time estimation which we considered in this text, it still has many useful applications. One such application is satellite orbit determination, where the satellite orbit estimation is allowed to be made after a certain period of time. More precisely, consider the following linear deterministic/stochastic system with a fixed terminal time N:

$$\begin{cases} \mathbf{x}_{k+1} = A_k\mathbf{x}_k + B_k\mathbf{u}_k + \Gamma_k\underline{\xi}_k \\ \mathbf{v}_k = C_k\mathbf{x}_k + D_k\mathbf{u}_k + \underline{\eta}_k, \end{cases}$$

where $1 \le k < N$ and in addition the variance matrices Q_k of $\underline{\xi}_k$ are assumed to be positive definite for all k (cf. Chapters 2 and 3 where Q_k were only assumed to be non-negative definite.)

The Kalman filtering process is to be applied to find the optimal smoothing estimate $\hat{\mathbf{x}}_{K|N}$ where the optimality is again in the sense of minimum error variance. First, let us denote by $\hat{\mathbf{x}}_K^f$ the usual Kalman filtering estimate $\hat{\mathbf{x}}_{K|K}$, using the data information $\{\mathbf{v}_1, \cdots, \mathbf{v}_K\}$, where the superscript f indicates that the estimation is a "forward" process. Similarly, denote by $\hat{\mathbf{x}}_K^b = \hat{\mathbf{x}}_{K|K+1}$ the "backward" optimal prediction obtained in estimating the state vector $\mathbf{x}_K$ by using the data information $\{\mathbf{v}_{K+1}, \cdots, \mathbf{v}_N\}$. Then the optimal smoothing estimate $\hat{\mathbf{x}}_{K|N}$ (usually called *Kalman smoothing estimate*) can be obtained by incorporating all the data information over the time-interval $\{1, 2, \cdots, N\}$ by the following recursive algorithm (see Lewis (1986) and also Balakrishnan (1984,87) for more details):

$$\begin{cases} G_K = P_K^f P_K^b (I + P_K^f P_K^b)^{-1} \\ P_K = (I - G_K)P_K^f \\ \hat{\mathbf{x}}_{K|N} = (I - G_K)\hat{\mathbf{x}}_K^f + P_K\hat{\mathbf{x}}_K^b, \end{cases}$$

where $P_K^f, \hat{\mathbf{x}}_K^f$ and $P_K^b, \hat{\mathbf{x}}_K^b$ are, respectively, computed recursively by using the following procedures:

$$\begin{cases} P_0^f = Var(\mathbf{x}_0) \\ P_{k,k-1}^f = A_{k-1}P_{k-1}^f A_{k-1}^\top + \Gamma_{k-1}Q_{k-1}\Gamma_{k-1}^\top \\ G_k^f = P_{k,k-1}^f C_k^\top (C_k P_{k,k-1}^f C_k^\top + R_k)^{-1} \\ P_k^f = (I - G_k^f C_k)P_{k,k-1}^f \\ \hat{\mathbf{x}}_0^f = E(\mathbf{x}_0) \\ \hat{\mathbf{x}}_k^f = A_{k-1}\hat{\mathbf{x}}_{k-1}^f + B_{k-1}\mathbf{u}_{k-1} \\ \qquad + G_k^f(\mathbf{v}_k - C_k A_{k-1}\hat{\mathbf{x}}_{k-1}^f - D_k\mathbf{u}_{k-1}) \\ k = 1, 2, \cdots, K, \end{cases}$$

and

$$\begin{cases} P_N^b = 0 \\ P_{k+1,N}^b = P_{k+1}^b + C_{k+1}^\top R_{k+1}^{-1} C_{k+1} \\ G_k^b = P_{k+1,N}^b \Gamma_{k+1}^\top (\Gamma_{k+1} P_{k+1,N}^b \Gamma_{k+1}^\top + Q_{k+1}^{-1})^{-1} \\ P_k^b = A_{k+1}^\top (I - G_k^b \Gamma_{k+1}^\top) P_{k,N}^b A_{k+1} \\ \hat{\mathbf{x}}_N^b = 0 \\ \hat{\mathbf{x}}_k^b = A_{k+1}^\top (I - G_k^b \Gamma_{k+1}^\top) \{ \hat{\mathbf{x}}_{k+1}^b + C_{k+1}^\top R_{k+1}^{-1} \mathbf{v}_{k+1} \\ \qquad\qquad - (C_{k+1}^\top R_{k+1}^{-1} D_{k+1} + P_{k+1,N}^b B_{k+1}) \mathbf{u}_{k+1} \} \\ k = N - 1, N - 2, \cdots, K. \end{cases}$$

12.2 The $\alpha - \beta - \gamma - \theta$ Tracker

Consider a time-invariant linear stochastic system with colored noise input described by the state-space description

$$\begin{cases} \mathbf{x}_{k+1} = A\mathbf{x}_k + \Gamma\underline{\xi}_k \\ \mathbf{v}_k = C\mathbf{x}_k + \underline{\eta}_k, \end{cases}$$

where A, C, Γ are $n \times n, q \times n$, and $n \times p$ constant matrices, $1 \leq p, q \leq n$, and

$$\begin{cases} \underline{\xi}_k = M\underline{\xi}_{k-1} + \underline{\beta}_k \\ \underline{\eta}_k = N\underline{\eta}_{k-1} + \underline{\gamma}_k \end{cases}$$

with M, N being constant matrices, and $\{\underline{\beta}_k\}$ and $\{\underline{\gamma}_k\}$ being independent Gaussian white noise sequences satisfying

$$E(\underline{\beta}_k \underline{\beta}_\ell^\top) = Q\delta_{k\ell}, \quad E(\underline{\gamma}_k \underline{\gamma}_\ell^\top) = R\delta_{k\ell}, \quad E(\underline{\beta}_k \underline{\gamma}_\ell^\top) = 0,$$

$$E(\mathbf{x}_0 \underline{\beta}_k^\top) = 0, \quad E(\mathbf{x}_0 \underline{\gamma}_k^\top) = 0, \quad \underline{\xi}_{-1} = \underline{\eta}_{-1} = 0.$$

The associated $\alpha - \beta - \gamma - \theta$ tracker for this system is defined to be the following algorithm:

$$\begin{cases} \check{X}_k = \begin{bmatrix} A & I & 0 \\ 0 & M & 0 \\ 0 & 0 & N \end{bmatrix} \check{X}_{k-1} \\ \qquad + \begin{bmatrix} \alpha \\ \beta \\ \gamma \\ \theta \end{bmatrix} [\mathbf{v}_k - [C\ 0\ I] \begin{bmatrix} A & I & 0 \\ 0 & M & 0 \\ 0 & 0 & N \end{bmatrix} \check{X}_{k-1}] \\ \check{X}_0 = \begin{bmatrix} E(\mathbf{x}_0) \\ 0 \\ 0 \end{bmatrix}, \end{cases}$$

where α, β, γ, and θ are certain constants, and $X_k := [\mathbf{x}_k^\top\ \underline{\xi}_k^\top\ \underline{\eta}_k^\top]^\top$.

For the real-time tracking system discussed in Section 9.2, with colored measurement noise input, namely:

$$A = \begin{bmatrix} 1 & h & h^2/2 \\ 0 & 1 & h \\ 0 & 0 & 1 \end{bmatrix}, \quad C = [1 \ 0 \ 0], \quad \Gamma = I_3,$$

$$Q = \begin{bmatrix} \sigma_p & 0 & 0 \\ 0 & \sigma_v & 0 \\ 0 & 0 & \sigma_a \end{bmatrix}, \quad R = [\sigma_m] > 0,$$

$$M = 0, \quad and \quad N = [r] > 0,$$

where $h > 0$ is the sampling time, this $\alpha - \beta - \gamma - \theta$ tracker becomes a (near-optimal) limiting Kalman filter if and only if the following conditions are satisfied: $\sigma_a > 0$, and

(1) $\gamma(2\alpha + 2\beta + \gamma) > 0$,

(2) $\alpha(r - 1)(r - 1 - r\theta) + r\beta\theta - r(r + 1)\gamma\theta/2(r - 1) > 0$,

(3) $(4\alpha + \beta)(\beta + \gamma) + 3(\beta^2 - 2\alpha\gamma) - 4\gamma(r - 1 - \gamma\theta)/(r - 1) \geq 0$,

(4) $\gamma(\beta + \gamma) \geq 0$,

(5) $\alpha(r + 1)(r - 1 + r\theta) + r(r + 1)\beta\theta/(r - 1)$
$\quad + r\gamma(r + 1)^2/2(r - 1)^2 - r^2 + 1 \geq 0$,

(6) $\sigma_v/\sigma_a = (\beta^2 - 2\alpha\gamma)/2h^2\gamma^2$,

(7) $\sigma_p/\sigma_a = \left[\alpha(2\alpha + 2\beta + \gamma) - 4\beta(r - 1 - r\theta)/(r - 1) - 4r\gamma\theta/(r - 1)^2\right]h^4/(2\gamma^2)$,

(8) $\sigma_m/\sigma_a = \left[\alpha(r + 1)(r - 1 + r\theta) + 1 - r^2 + r^2\theta^2\right.$
$\quad \left. + r(r + 1)\beta\theta/(r - 1) + r\gamma(r + 1)^2/2(r - 1)^2\right]h^4/\gamma^2$,

and

(9) the matrix $[p_{ij}]_{4\times4}$ is non-negative definite and symmetric, where

$$p_{11} = \frac{1}{r - 1}\left[\alpha(r - 1 - r\theta) + \frac{r\beta\theta}{r - 1} - \frac{r\gamma\theta(r + 1)}{2(r - 1)^2}\right],$$

$$p_{12} = \frac{1}{r - 1}\left[\beta(r - 1 - r\theta) + \frac{r\gamma\theta}{r - 1}\right], \quad p_{13} = \frac{1}{r - 1}\gamma(r - 1 - r\theta),$$

$$p_{14} = \frac{r\theta}{r - 1}\left[\alpha - \frac{\beta}{r - 1} + \frac{\gamma(r + 1)}{2(r - 1)^2}\right],$$

$$p_{22} = \frac{1}{4}(4\alpha + \beta)(\beta + \gamma) + \frac{3}{4}(\beta^2 - 2\alpha\gamma) - \frac{\gamma(r - 1 - r\theta)}{r - 1},$$

$$p_{23} = \gamma(\alpha + \beta/2), \quad p_{24} = \frac{r\theta}{r - 1}(\beta - \gamma/(r - 1)),$$

$$p_{33} = \gamma(\beta + \gamma), \quad p_{34} = r\gamma\theta/(r - 1), \quad and$$

$$p_{44} = \frac{1}{1 - r^2}\left[\alpha(r + 1)(r + r\theta - 1) + \frac{r\beta\theta(r + 1)}{r - 1} + \frac{r\gamma(r + 1)^2}{2(r - 1)^2} + 1 - r^2\right].$$

In particular, if, in addition, $N = 0$; that is, only the white noise input is considered, then the above conditions reduce to the ones obtained in Theorem 9.1.

Finally, by defining $\mathbf{x}_k := [x_k \quad \dot{x}_k \quad \ddot{x}_k \quad w_k]^\top$ and using the z-transform technique, the $\alpha - \beta - \gamma - \theta$ tracker can be decomposed as the following (decoupled) recursive equations:

$$(1) \quad x_k = a_1 x_{k-1} + a_2 x_{k-2} + a_3 x_{k-3} + a_4 x_{k-4} + \alpha v_k$$
$$+ (-2\alpha - r\alpha + \beta + \gamma/2)v_{k-1} + (\alpha - \beta + \gamma/2$$
$$+ r(2\alpha - \beta - \gamma/2)v_{k-2} - r(\alpha - \beta + \gamma/2)v_{k-3},$$

$$(2) \quad \dot{x}_k = a_1 \dot{x}_{k-1} + a_2 \dot{x}_{k-2} + a_3 \dot{x}_{k-3} + a_4 \dot{x}_{k-4}$$
$$+ (1/h)\{\beta v_k - [(2+r)\beta - \gamma]v_{k-1}$$
$$+ [\beta - \gamma + r(2\beta - \gamma)]v_{k-2} - r(\beta - \gamma)v_{k-3}\},$$

$$(3) \quad \ddot{x}_k = a_1 \ddot{x}_{k-1} + a_2 \ddot{x}_{k-2} + a_3 \ddot{x}_{k-3} + \ddot{x}_{k-4}$$
$$+ (\gamma/h^2)[v_k - (2+\gamma)v_{k-1} + (1+2r)v_{k-2} - rv_{k-3}],$$

$$(4) \quad w_k = a_1 w_{k-1} + a_2 w_{k-2} + a_3 w_{k-3} + a_4 w_{k-4}$$
$$+ \theta(v_k - 3v_{k-1} + 3v_{k-2} - v_{k-3}),$$

with initial conditions $x_{-1}, \dot{x}_{-1}, \ddot{x}_{-1}$, and w_{-1}, where

$$\begin{cases} a_1 = -\alpha - \beta - \gamma/2 + r(\theta - 1) + 3, \\ a_2 = 2\alpha + \beta - \gamma/2 + r(\alpha + \beta + \gamma/2 + 3\theta - 3) - 3, \\ a_3 = -\alpha + r(-2\alpha - \beta + \gamma/2 - 3\theta + 3) + 1, \\ a_4 = r(\alpha + \theta - 1). \end{cases}$$

We remark that the above decoupled formulas include the white noise input result obtained in Section 9.2, and this method works regardless of the near-optimality of the filter. For more details, see Chen and Chui (1986) and Chui (1984).

12.3 Adaptive Kalman Filtering

Consider the linear stochastic state-space description

$$\begin{cases} \mathbf{x}_{k+1} = A_k \mathbf{x}_k + \Gamma_k \underline{\xi}_k \\ \mathbf{v}_k = C_k \mathbf{x}_k + \underline{\eta}_k, \end{cases}$$

where $\{\underline{\xi}_k\}$ and $\{\underline{\eta}_k\}$ are uncorrelated zero-mean Gaussian white noise sequences with $Var(\underline{\xi}_k) = Q_k$ and $Var(\underline{\eta}_k) = R_k$. Then assuming that each R_k is positive definite and all the matrices

A_k, C_k, Γ_k, Q_k, and R_k, and initial conditions $E(\mathbf{x}_0)$ and $Var(\mathbf{x}_0)$ are known quantities, the Kalman filtering algorithm derived in Chapters 2 and 3 provides a very efficient optimal estimation process for the state vectors $\mathbf{x}_k$ in real-time. In fact, the estimation process is so effective that even for very poor choices of the initial conditions $E(\mathbf{x}_0)$ and/or $Var(\mathbf{x}_0)$, fairly desirable estimates are usually obtained in a relatively short period of time. However, if not all of the matrices $A_k, C_k, \Gamma_k, Q_k, R_k$ are known quantities, the filtering algorithm must be modified so that optimal estimations of the state as well as the unknown matrices are performed in real-time from the incoming data $\{\mathbf{v}_k\}$. Algorithms of this type may be called *adaptive algorithms*, and the corresponding filtering process *adaptive Kalman filtering*. If Q_k and R_k are known, then the adaptive Kalman filtering can be used to "identify" the system and/or measurement matrices. This problem has been discussed in Chapter 8 when partial information of A_k, Γ_k, and C_k is given. In general, the *identification problem* is a very difficult but important one [cf. Aström and Eykhoff (1971) and Mehra (1970,1972)].

Let us now discuss the situation where A_k, Γ_k, C_k are given. Then an adaptive Kalman filter for estimating the state as well as the noise variance matrices may be called a *noise-adaptive filter*. Although several algorithms are available for this purpose [cf. Aström and Eykhoff (1971), Chen and Chui (1991), Chin (1979), Jazwinski (1969), and Mehra (1970,1972) etc.], there is still no available algorithm that is derived from the truely optimality criterion. For simplicity, let us discuss the situation where A_k, Γ_k, C_k and Q_k are given. Hence, only R_k has to be estimated. The innovations approach [cf. Kailath (1968) and Mehra (1970)] seems to be very efficient for this situation. From the incoming data information $\mathbf{v}_k$ and the optimal prediction $\hat{\mathbf{x}}_{k|k-1}$ obtained in the previous step, the innovations sequence is defined to be

$$\mathbf{z}_k := \mathbf{v}_k - C_k \hat{\mathbf{x}}_{k|k-1}.$$

It is clear that

$$\mathbf{z}_k = C_k(\hat{\mathbf{x}}_k - \hat{\mathbf{x}}_{k|k-1}) + \underline{\eta}_k$$

which is, in fact, a zero-mean Gaussian white noise sequence. By taking variances on both sides, we have

$$S_k := Var(\mathbf{z}_k) = C_k P_{k,k-1} C_k^\mathsf{T} + R_k.$$

This yields an estimate of R_k; namely,

$$\hat{R}_k = \hat{S}_k - C_k P_{k,k-1} C_k^\mathsf{T},$$

where $\hat{S}_k$ is the statistical sample variance estimate of S_k given by

$$\hat{S}_k = \frac{1}{k-1} \sum_{i=1}^{k} (\mathbf{z}_i - \overline{\mathbf{z}}_i)(\mathbf{z}_i - \overline{\mathbf{z}}_i)^\top$$

with $\overline{\mathbf{z}}_i$ being the statistical sample mean defined by

$$\overline{\mathbf{z}}_i = \frac{1}{i} \sum_{j=1}^{i} \mathbf{z}_j \,,$$

(see, for example, Stengel (1986)).

12.4 Adaptive Kalman Filtering Approach to Wiener Filtering

In digital signal processing and system theory, an important problem is to determine the unit impulse responses of a digital filter or a linear system from the input/output information where the signal is contaminated with noise, see, for example, Chui and Chen (1992). More precisely, let $\{u_k\}$ and $\{v_k\}$ be known input and output signals, respectively, and $\{\eta_k\}$ be an unknown sequence of noise process. The problem is to "identify" the sequence $\{h_k\}$ from the relationship

$$v_k = \sum_{i=0}^{\infty} h_i u_{k-i} + \eta_k \,, \qquad k = 0, 1, \cdots .$$

The optimality criterion in the so-called *Wiener filter* is to determine a sequence $\{\hat{h}_k\}$ such that by setting

$$\hat{v}_k = \sum_{i=0}^{\infty} \hat{h}_i u_{k-i} \,,$$

it is required that

$$Var(v_k - \hat{v}_k) = \inf_{a_i} \left\{ Var(v_k - w_k) : \quad w_k = \sum_{i=0}^{\infty} a_i u_{k-i} \right\} .$$

Under the assumption that $h_i = 0$ for $i > M$; that is, when an *FIR* system is considered, the above problem can be recast in the state-space description as follows: Let

$$\mathbf{x} = [h_0 \quad h_1 \quad \cdots \quad h_M]^\top$$

be the state vector to be estimated. Since this is a constant vector, we may write

$$\mathbf{x}_{k+1} = \mathbf{x}_k = \mathbf{x}.$$

In addition, let C be the "observation matrix" defined by

$$C = [u_0 \quad u_1 \quad \cdots \quad u_M].$$

Then the input/output relationship can be written as

$$\begin{cases} \mathbf{x}_{k+1} = \mathbf{x}_k \\ \quad v_k = C\mathbf{x}_k + \eta_k, \end{cases} \qquad (a)$$

and we are required to give an optimal estimation $\hat{\mathbf{x}}_k$ of $\mathbf{x}_k$ from the data information $\{v_0, \cdots, v_k\}$. When $\{\eta_k\}$ is a zero-mean Gaussian white noise sequence with unknown variances $R_k = Var(\eta_k)$, the estimation can be done by applying the noise-adaptive Kalman filtering discussed in Section 12.3.

We remark that if an *IIR* system is considered, the adaptive Kalman filtering technique cannot be applied directly, since the corresponding linear system (a) becomes infinite-dimensional.

12.5 The Kalman-Bucy Filter

This book is devoted exclusively to the study of Kalman filtering for discrete-time models. The continuous-time analog, which we will briefly introduce in the following, is called the *Kalman-Bucy filter*.

Consider the continuous-time linear deterministic/stochastic system

$$\begin{cases} d\mathbf{x}(t) = A(t)\mathbf{x}(t)dt + B(t)\mathbf{u}(t)dt + \Gamma(t)\underline{\xi}(t)dt, \quad \mathbf{x}(0) = \mathbf{x}_0, \\ d\mathbf{v}(t) = C(t)\mathbf{x}(t)dt + \underline{\eta}(t)dt, \end{cases}$$

where $0 \le t \le T$ and the state vector $\mathbf{x}(t)$ is a random n- vector with initial position $\mathbf{x}(0) \sim N(0, \Sigma^2)$. Here, Σ^2, or at least an estimate of it, is given. The stochastic input or noise processes $\underline{\xi}(t)$ and $\underline{\eta}(t)$ are Wiener-Levy p- and q-vectors, respectively, with $1 \le p, q \le n$, the observation $\mathbf{v}(t)$ is a random q-vector, and $A(t), \Gamma(t)$, and $C(t)$ are $n \times n$, $n \times p$, and $q \times n$ deterministic matrix-valued continuous functions on the continuous-time interval $[0, T]$.

The Kalman-Bucy filter is given by the following recursive formula:

$$\hat{\mathbf{x}}(t) = \int_0^T P(\tau)C^\top(\tau)R^{-1}(\tau)d\mathbf{v}(\tau)$$

$$+ \int_0^T [A(\tau) - P(\tau)C^\top(\tau)R^{-1}(\tau)C(\tau)]\hat{\mathbf{x}}(\tau)d(\tau) + \int_0^T B(\tau)u(\tau)d\tau$$

where $R(t) = E\{\underline{\eta}(t)\underline{\eta}^\top(t)\}$, and $P(t)$ satisfies the matrix Riccati equation

$$\begin{cases} \dot{P}(t) = A(t)P(t) + P(t)A^\top(t) \\ \qquad - P(t)C^\top(t)R^{-1}(t)C(t)P(t) + \Gamma(t)Q(t)\Gamma^\top(t) \\ P(0) = Var(\mathbf{x}_0) = \Sigma^2\,, \end{cases}$$

with $Q(t) = E\{\underline{\xi}(t)\underline{\xi}^\top(t)\}$. For more details, the reader is referred to the original paper of Kalman and Bucy (1961), and the books by Ruymgaart and Soong (1985), and Fleming and Rishel (1975).

12.6 Stochastic Optimal Control

There is a vast literature on deterministic optimal control theory. For a brief discussion, the reader is referred to Chui and Chen (1989). The subject of stochastic optimal control deals with systems in which random disturbances are also taken into consideration. One of the typical stochastic optimal control problems is the so-called *linear regulator problem*. Since this model has continuously attracted most attention, we will devote this section to the discussion of this particular problem. The system and observation equations are given by the linear deterministic/stochastic differential equations:

$$\begin{cases} d\mathbf{x}(t) = A(t)\mathbf{x}(t)dt + B(t)\mathbf{u}(t)dt + \Gamma(t)\underline{\xi}(t)dt\,, \\ d\mathbf{v}(t) = C(t)\mathbf{x}(t)dt + \underline{\eta}(t)dt\,, \end{cases}$$

where $0 \leq t \leq T$ (cf. Section 12.5), and the cost functional to be minimized over a certain "admissible" class of control functions $\mathbf{u}(t)$ is given by

$$F(\mathbf{u}) = E\left\{ \int_0^T [\mathbf{x}^\top(t)W_x(t)\mathbf{x}(t) + \mathbf{u}^\top(t)W_u(t)\mathbf{u}(t)]dt \right\}.$$

Here, the initial state $\mathbf{x}(0)$ is assumed to be $\mathbf{x}_0 \sim N(0, \Sigma^2)$, Σ^2 being given, and $\underline{\xi}(t)$ and $\underline{\eta}(t)$ are uncorrelated zero-mean Gaussian white noise processes and are also independent of the initial state $\mathbf{x}_0$. In addition, the data item $\mathbf{v}(t)$ is known for $0 \leq t \leq T$, and $A(t), B(t), C(t), W_x(t)$, and $W_u(t)$ are known deterministic matrices of appropriate dimensions, with $W_x(t)$ being non-negative definite and symmetric, and $W_u(t)$ positive definite and symmetric. In general, the admissible class of control functions $\mathbf{u}(t)$ consists of vector-valued Borel measurable functions defined on $[0,T]$ with range in some closed subset of $\mathbf{R}^p$.

Suppose that the control function $\mathbf{u}(t)$ has partial knowledge of the system state via the observation data, in the sense that $\mathbf{u}(t)$ is a linear function of the data $\mathbf{v}(t)$ rather than the state vector $\mathbf{x}(t)$. For such a linear regulator problem, we may apply the so-called *separation principle*, which is one of the most useful results in stochastic optimal control theory. This principle essentially implies that the above "partially observed" linear regulator problem can be split into two parts: The first being an optimal estimation of the system state by means of the Kalman-Bucy filter discussed in Section 12.5, and the second a "completely observed" linear regulator problem whose solution is given by a linear feedback control function. More precisely, the optimal estimate $\hat{\mathbf{x}}(t)$ of the state vector $\mathbf{x}(t)$ satisfies the linear stochastic system

$$\begin{cases} d\hat{\mathbf{x}}(t) = A(t)\hat{\mathbf{x}}(t)dt + B(t)\mathbf{u}(t)dt \\ \qquad + P(t)C^\top(t)[d\mathbf{v}(t) - R^{-1}(t)C(t)\hat{\mathbf{x}}(t)dt] \\ \hat{\mathbf{x}}(0) = E(\mathbf{x}_0) , \end{cases}$$

where $R(t) = E\{\underline{\eta}(t)\underline{\eta}^\top(t)\}$, and $P(t)$ is the (unique) solution of the matrix Riccati equation

$$\begin{cases} \dot{P}(t) = A(t)P(t) + P(t)A^\top(t) - P(t)C^\top(t)R^{-1}(t)C(t)P(t) \\ \qquad + \Gamma(t)Q(t)\Gamma^\top(t) \\ P(0) = Var(\mathbf{x}_0) = \Sigma^2 , \end{cases}$$

with $Q(t) = E\{\underline{\xi}(t)\underline{\xi}^\top(t)\}$. On the other hand, an optimal control function $\mathbf{u}^*(t)$ is given by

$$\mathbf{u}^*(t) = -R^{-1}(t)B^\top(t)K(t)\hat{\mathbf{x}}(t) ,$$

where $K(t)$ is the (unique) solution of the matrix Riccati equation

$$\begin{cases} \dot{K}(t) = K(t)B(t)W_u^{-1}(t)B^\top(t)K(t) - K(t)A(t) - A^\top(t)K(t) - W_x(t) \\ K(T) = 0 , \end{cases}$$

with $0 \le t \le T$. For more details, the reader is referred to Wonham (1968), Kushner (1971), Fleming and Rishel (1975), Davis (1977), and more recently, Chen, Chen and Hsu (1995).

12.7 Square-Root Filtering and Systolic Array Implementation

The square-root filtering algorithm was first introduced by Potter (1963) and later improved by Carlson (1973) to give a fast computational scheme. Recall from Chapter 7 that this algorithm requires computation of the matrices $J_{k,k}$, $J_{k,k-1}$, and G_k, where

$$J_{k,k} = J_{k,k-1}[I - J_{k,k-1}^\top C_k^\top (H_k^\top)^{-1}(H_k + R_k^c)^{-1} C_k J_{k,k-1}], \qquad (a)$$

$J_{k,k-1}$ is a square-root of the matrix

$$[A_{k-1}J_{k-1,k-1} \quad \Gamma_{k-1}Q_{k-1}^{1/2}][A_{k-1}J_{k-1,k-1} \quad \Gamma_{k-1}Q_{k-1}^{1/2}]^\top,$$

and

$$G_k = J_{k,k-1}J_{k,k-1}^\top C_k^\top (H_k^\top)^{-1} H_k^{-1},$$

where $J_{k,k} = P_{k,k-1}^{1/2}$, $J_{k,k-1} = P_{k,k-1}^{1/2}$, $H_k = (C_k P_{k,k-1}C_k^\top + R_k)^c$ with M^c being the "square- root" of the matrix M in the form of a lower triangular matrix instead of being the positive definite square-root $M^{1/2}$ of M (cf. Lemma 7.1). It is clear that if we can compute $J_{k,k}$ directly from $J_{k,k-1}$ (or $P_{k,k}$ directly from $P_{k,k-1}$) without using formula (a), then the algorithm could be somewhat more efficient. From this point of view, Bierman (1973,1977) modified Carlson's method and made use of LU decomposition to give the following algorithm: First, consider the decompositions

$$P_{k,k-1} = U_1 D_1 U_1^\top \qquad and \qquad P_{k,k} = U_2 D_2 U_2^\top,$$

where U_i and D_i are upper triangular and diagonal matrices, respectively, $i = 1, 2$. The subscript k is omitted simply for convenience. Furthermore, define

$$D := D_1 - D_1 U_1^\top C_k^\top (H_k^\top)^{-1}(H_k)^{-1} C_k U_1 D_1$$

and decompose $D = U_3 D_3 U_3^\top$. Then it follows that

$$U_2 = U_1 U_3 \qquad and \qquad D_2 = D_3.$$

Bierman's algorithm requires $O(qn^2)$ arithmetical operations to obtain $\{U_2, D_2\}$ from $\{U_1, D_1\}$ where n and q are, respectively, the dimensions of the state and observation vectors. Andrews (1981) modified this algorithm by using parallel processing techniques and reduced the number of operations to $O(nq \, log \, n)$. More recently, Jover and Kailath (1986) made use of the Schur complement technique and applied systolic arrays [cf. Kung (1982), Mead and Conway (1980), and Kung (1985)] to further reduce the number of operations to $O(n)$ (or more precisely, approximately $4n$). In addition, the number of required arithmetic processors is reduced from $O(n^2)$ to $O(n)$. The basic idea of this approach can be briefly described as follows. Since $P_{k,k-1}$ is non-negative definite and symmetric, there is an orthogonal matrix M_1 such that

$$[A_{k-1}P_{k-1,k-1}^{1/2} \quad \Gamma_{k-1}Q_{k-1}^{1/2}]M_1 = [P_{k,k-1}^{1/2} \quad 0]. \qquad (b)$$

Consider the augmented matrix

$$A := \begin{bmatrix} H_k H_k^\top & C_k P_{k,k-1} \\ P_{k,k-1} C_k^\top & P_{k,k-1} \end{bmatrix},$$

which can be shown to have the following two decompositions:

$$A = \begin{bmatrix} I & C_k \\ 0 & I \end{bmatrix} \begin{bmatrix} R_k & 0 \\ 0 & P_{k,k} \end{bmatrix} \begin{bmatrix} I & 0 \\ C_k^\top & I \end{bmatrix}$$

and

$$A = \begin{bmatrix} I & 0 \\ P_{k,k-1} C_k^\top (H_k^\top)^{-1} H_k^{-1} & I \end{bmatrix} \begin{bmatrix} H_k H_k^\top & 0 \\ 0 & P_{k,k} \end{bmatrix}$$
$$\cdot \begin{bmatrix} I & (H_k^\top)^{-1} H_k^{-1} C_k P_{k,k-1} \\ 0 & I \end{bmatrix}.$$

Hence, by taking the upper block triangular "square-root" of the decomposition on the left, and the lower block triangular "square-root" of that on the right, there is an orthogonal matrix M_2 such that

$$\begin{bmatrix} R_k^{1/2} & C_k P_{k,k-1}^{1/2} \\ 0 & P_{k,k-1}^{1/2} \end{bmatrix} M_2 = \begin{bmatrix} H_k & 0 \\ P_{k,k-1} C_k^\top [(H_k^\top)^{-1/2} H_k^{-1/2}]^\top & P_{k,k}^{1/2} \end{bmatrix}. \qquad (c)$$

Now, using LU decompositions, where subscript k will be dropped for convenience, we have:

$$R_k = U_R D_R U_R^\top, \qquad H_k H_k^\top = U_H D_H U_H^\top,$$

$$P_{k,k-1} = U_1 D_1 U_1^\top, \qquad and \qquad P_{k,k} = U_2 D_2 U_2^\top.$$

It follows that the identity (c) may be written as

$$\begin{bmatrix} U_R & C_k U_1 \\ 0 & U_1 \end{bmatrix} \begin{bmatrix} D_R & 0 \\ 0 & D_1 \end{bmatrix}^{1/2} M_2$$

$$= \begin{bmatrix} U_H & 0 \\ P_{k,k-1} C_k^{\mathsf{T}} (H^{\mathsf{T}})^{-1} H^{-1} U_H & U_2 \end{bmatrix} \begin{bmatrix} D_H & 0 \\ 0 & D_2 \end{bmatrix}^{1/2},$$

so that by defining

$$M_3 = \begin{bmatrix} D_R & 0 \\ 0 & D_1 \end{bmatrix}^{1/2} M_2 \begin{bmatrix} D_H & 0 \\ 0 & D_2 \end{bmatrix}^{-1/2}$$

which is clearly an orthogonal matrix, we have

$$\begin{bmatrix} U_R & C_k U_1 \\ 0 & U_1 \end{bmatrix} M_3 = \begin{bmatrix} U_H & 0 \\ P_{k,k-1} C_k^{\mathsf{T}} (H^{\mathsf{T}})^{-1} H^{-1} U_H & U_2 \end{bmatrix}. \qquad (d)$$

By an algorithm posed by Kailath (1982), M_3 can be decomposed as a product of a finite number of elementary matrices without using the square-root operation. Hence, by an appropriate application of systolic arrays, $\{U_H, U_2\}$ can be computed from $\{U_R, U_1\}$ via (d) and $P_{k,k-1}^{1/2}$ from $P_{k-1,k-1}^{1/2}$ via (b) in approximately $4n$ arithmetical operations. Consequently, D_2 can be easily computed from D_1. For more details on this subject, see Jover and Kailath (1986) and Gaston and Irwin (1990).

References

Alfeld, G. and Herzberger, J. (1983): *Introduction to Interval Computations* (Academic, New York)

Anderson, B.D.O., Moore, J.B. (1979): *Optimal Filtering* (Prentice-Hall, Englewood Cliffs, NJ)

Andrews, A. (1981): "Parallel processing of the Kalman filter", IEEE Proc. Int. Conf. on Paral. Process., pp.216-220

Aoki, M. (1989): *Optimization of Stochastic Systems: Topics in Discrete-Time Dynamics* (Academic, New York)

Aström, K.J., Eykhoff, P. (1971): "System identification – a survey," Automatica, **7**, pp.123-162

Balakrishnan, A.V. (1984,87): *Kalman Filtering Theory* (Optimization Software, Inc., New York)

Bierman, G.J. (1973): "A comparison of discrete linear filtering algorithms," IEEE Trans. Aero. Elec. Systems, **9**, pp.28-37

Bierman, G.J. (1977): *Factorization Methods for Discrete Sequential Estimation* (Academic, New York)

Blahut, R.E. (1985): *Fast Algorithms for Digital Signal Processing* (Addison-Wesley, Reading, MA)

Bozic, S.M. (1979): *Digital and Kalman Filtering* (Wiley, New York)

Brammer, K., Sifflin, G. (1989): *Kalman-Bucy Filters* (Artech House, Boston)

Brown, R.G. and Hwang, P.Y.C. (1992,97): *Introduction to Random Signals and Applied Kalman Filtering* (Wiley, New York)

Bucy, R.S., Joseph, P.D. (1968): *Filtering for Stochastic Processes with Applications to Guidance* (Wiley, New York)

Burrus, C.S. , Gopinath, R.A. and Guo, H. (1998): *Introduction to Wavelets and Wavelet Transfroms: A Primer* (Prentice-Hall, Upper Saddle River, NJ)

Carlson, N.A. (1973): "Fast triangular formulation of the square root filter," J. ALAA, **11** pp.1259-1263

Catlin, D.E. (1989): *Estimation, Control, and the Discrete Kalman Filter* (Springer, New York)

Chen, G. (1992): "Convergence analysis for inexact mechanization of Kalman filtering," IEEE Trans. Aero. Elect. Syst., **28**, pp.612-621

Chen, G. (1993): *Approximate Kalman Filtering* (World Scientific, Singapore)

Chen, G., Chen, G. and Hsu, S.H. (1995): *Linear Stochastic Control Systems* (CRC, Boca Raton, FL)

Chen, G., Chui, C.K. (1986): "Design of near-optimal linear digital tracking filters with colored input," J. Comp. Appl. Math., **15**, pp.353-370

Chen, G., Wang, J. and Shieh, L.S. (1997): "Interval Kalman filtering," IEEE Trans. Aero. Elect. Syst., **33**, pp.250-259

Chen, H.F. (1985): *Recursive Estimation and Control for Stochastic Systems* (Wiley, New York)

Chui, C.K. (1984): "Design and analysis of linear prediction-correction digital filters," Linear and Multilinear Algebra, **15**, pp.47-69

Chui, C.K. (1997): *Wavelets: A Mathematical Tool for Signal Analysis*, (SIAM, Philadelphia)

Chui, C.K., Chen, G. (1989): *Linear Systems and Optimal Control*, Springer Ser. Inf. Sci., Vol. 18 (Springer, Berlin Heidelberg)

Chui, C.K., Chen, G. (1992,97): *Signal Processing and Systems Theory: Selected Topics*, Springer Ser. Inf. Sci., Vol. 26 (Springer, Berlin Heidelberg)

Chui, C.K., Chen, G. and Chui, H.C. (1990): "Modified extended Kalman filtering and a real-time parallel algorithm for system parameter identification," IEEE Trans. Auto. Control, **35**, pp.100-104

Davis, M.H.A. (1977): *Linear Estimation and Stochastic Control* (Wiley, New York)

Davis, M.H.A., Vinter, R.B. (1985): *Stochastic Modeling and Control* (Chapman and Hall, New York)

Fleming, W.H., Rishel, R.W. (1975): *Deterministic and Stochastic Optimal Control* (Springer, New York)

Gaston, F.M.F., Irwin, G.W. (1990): "Systolic Kalman filtering: An overview," IEE Proc.-D, **137**, pp.235-244

Goodwin, G.C., Sin, K.S. (1984): *Adaptive Filtering Prediction and Control* (Prentice-Hall, Englewood Cliffs, NJ)

Haykin, S. (1986): *Adaptive Filter Theory* (Prentice-Hall, Englewood Cliffs, NJ)

Hong, L., Chen, G. and Chui, C.K. (1998): "A filter-bank-based Kalman filtering technique for wavelet estimation and decomposition of random signals," IEEE Trans. Circ. Syst. (II), **45**, pp. 237-241.

Hong, L., Chen, G. and Chui, C.K. (1998): "Real-time simultaneous estimation and ecomposition of random signals," Multidim. Sys. Sign. Proc., **9**, pp. 273-289.

Jazwinski, A.H. (1969): "Adaptive filtering," Automatica, **5**, pp.475-485

Jazwinski, A.H. (1970): *Stochastic Processes and Filtering Theory* (Academic, New York)

Jover, J.M., Kailath, T. (1986): "A parallel architecture for Kalman filter measurement update and parameter estimation," Automatica, **22**, pp.43-57

Kailath, T. (1968): "An innovations approach to least-squares estimation, part I: linear filtering in additive white noise," IEEE Trans. Auto. Contr., **13**, pp.646-655

Kailath, T. (1982): *Course Notes on Linear Estimation* (Stanford University, CA)

Kalman, R.E. (1960): "A new approach to linear filtering and prediction problems," Trans. ASME, J. Basic Eng., **82**, pp.35-45

Kalman, R.E. (1963): "New method in Wiener filtering theory," Proc. Symp. Eng. Appl. Random Function Theory and Probability (Wiley, New York)

Kalman, R.E., Bucy, R.S. (1961): "New results in linear filtering and prediction theory," Trans. ASME J. Basic Eng., **83**, pp.95-108

Kumar, P.R., Varaiya, P. (1986): *Stochastic Systems: Estimation, Identification, and Adaptive Control* (Prentice-Hall, Englewood Cliffs, NJ)

Kung, H.T. (1982): "Why systolic architectures?" Computer, **15**, pp.37-46

Kung, S.Y. (1985): "VLSI arrays processors," IEEE ASSP Magazine, **2**, pp.4-22

Kushner, H. (1971): *Introduction to Stochastic Control* (Holt, Rinehart and Winston, Inc., New York)

Lewis, F.L. (1986): *Optimal Estimation* (Wiley, New York)

Lu, M., Qiao, X., Chen, G. (1992): "A parallel square-root algorithm for the modified extended Kalman filter," IEEE Trans. Aero. Elect. Syst., **28**, pp.153-163

Lu, M., Qiao, X., Chen, G. (1993): "Parallel computation of the modified extended Kalman filter," Int'l J. Comput. Math., **45**, pp.69-87

Maybeck, P.S. (1982): *Stochastic Models, Estimation, and Control*, Vol. 1,2,3 (Academic, New York)

Mead, C., Conway, L. (1980): *Introduction to VLSI systems* (Addison-Wesley, Reading, MA)

Mehra, R.K. (1970): "On the identification of variances and adaptive Kalman filtering," IEEE Trans. Auto. Contr., **15**, pp.175-184

Mehra, R.K. (1972): "Approaches to adaptive filtering," IEEE Trans. Auto. Contr., **17**, pp.693-698

Mendel, J.M. (1987): *Lessons in Digital Estimation Theory* (Prentice-Hall, Englewood Cliffs, New Jersey)

Potter, J.E. (1963): "New statistical formulas," Instrumentation Lab., MIT, Space Guidance Analysis Memo. # 40

Probability Group (1975), Institute of Mathematics, Academia Sinica, China (ed.): *Mathematical Methods of Filtering for Discrete-Time Systems* (in Chinese) (Beijing)

Ruymgaart, P.A., Soong, T.T. (1985,88): *Mathematics of Kalman-Bucy Filtering*, Springer Ser. Inf. Sci., Vol. 14 (Springer, Berlin Heidelberg)

Shiryayev, A.N. (1984): *Probability* (Springer-Verlag, New York)

Siouris, G., Chen, G. and Wang, J. (1997): "Tracking an incoming ballistic missile," IEEE Trans. Aero. Elect. Syst., **33**, pp.232-240

Sorenson, H.W., ed. (1985): *Kalman Filtering: Theory and Application* (IEEE, New York)

Stengel, R.F. (1986): *Stochastic Optimal Control: Theory and Application* (Wiley, New York)

Strobach, P. (1990): *Linear Prediction Theory: A Mathematical Basis for Adaptive Systems*, Springer Ser. Inf. Sci., Vol. 21 (Springer, Berlin Heidelberg)

Wang, E.P. (1972): "Optimal linear recursive filtering methods," J. Mathematics in Practice and Theory (in Chinese), **6**, pp.40-50

Wonham, W.M. (1968): "On the separation theorem of stochastic control," SIAM J. Control, **6**, pp.312-326

Xu, J.H., Bian, G.R., Ni, C.K., Tang, G.X. (1981): *State Estimation and System Identification* (in Chinese) (Beijing)

Young, P. (1984): *Recursive Estimation and Time-Series Analysis* (Springer, New York)

Answers and Hints to Exercises

Chapter 1

1.1. Since most of the properties can be verified directly by using the definition of the trace, we only consider $tr\,AB = tr\,BA$. Indeed,

$$tr\,AB = \sum_{i=1}^{n}\left(\sum_{j=1}^{m} a_{ij}b_{ji}\right) = \sum_{j=1}^{m}\left(\sum_{i=1}^{n} b_{ji}a_{ij}\right) = tr\,BA.$$

1.2.

$$(tr\,A)^2 = \left(\sum_{i=1}^{n} a_{ii}\right)^2 \leq n \sum_{i=1}^{n} a_{ii}^2 \leq n\,(tr\,AA^{\mathsf{T}}).$$

1.3.

$$A = \begin{bmatrix} 3 & 1 \\ 1 & 2 \end{bmatrix}, \qquad B = \begin{bmatrix} 2 & 0 \\ 0 & 1 \end{bmatrix}.$$

1.4. There exist unitary matrices P and Q such that

$$A = P \begin{bmatrix} \lambda_1 & & \\ & \ddots & \\ & & \lambda_n \end{bmatrix} P^{\mathsf{T}}, \quad B = Q \begin{bmatrix} \mu_1 & & \\ & \ddots & \\ & & \mu_n \end{bmatrix} Q^{\mathsf{T}},$$

and

$$\sum_{k=1}^{n} \lambda_k^2 \geq \sum_{k=1}^{n} \mu_k^2.$$

Let $P = [p_{ij}]_{n \times n}$ and $Q = [q_{ij}]_{n \times n}$. Then

$$p_{11}^2 + p_{21}^2 + \cdots + p_{n1}^2 = 1, p_{12}^2 + p_{22}^2 + \cdots + p_{n2}^2 = 1, \cdots,$$

$$p_{1n}^2 + p_{2n}^2 + \cdots + p_{nn}^2 = 1, \quad q_{11}^2 + q_{21}^2 + \cdots + q_{n1}^2 = 1,$$

$$q_{12}^2 + q_{22}^2 + \cdots + q_{n2}^2 = 1, \cdots, q_{1n}^2 + q_{2n}^2 + \cdots + q_{nn}^2 = 1,$$

and

$$tr AA^\top = tr\left\{ P \begin{bmatrix} \lambda_1^2 & & \\ & \ddots & \\ & & \lambda_n^2 \end{bmatrix} P^\top \right\}$$

$$= tr \begin{bmatrix} p_{11}^2\lambda_1^2 + p_{12}^2\lambda_2^2 & & & * \\ +\cdots+ p_{1n}^2\lambda_n^2 & & & \\ & p_{21}^2\lambda_1^2 + p_{22}^2\lambda_2^2 & & \\ & +\cdots+ p_{2n}^2\lambda_n^2 & & \\ & & p_{n1}^2\lambda_1^2 + p_{n2}^2\lambda_2^2 & \\ & & +\cdots+ p_{nn}^2\lambda_n^2 \\ * & & & \end{bmatrix}$$

$$= (p_{11}^2 + p_{21}^2 + \cdots + p_{n1}^2)\lambda_1^2 + \cdots + (p_{1n}^2 + p_{2n}^2 + \cdots + p_{nn}^2)\lambda_n^2$$

$$= \lambda_1^2 + \lambda_2^2 + \cdots + \lambda_n^2 .$$

Similarly, $tr BB^\top = \mu_1^2 + \mu_2^2 + \cdots + \mu_n^2$. Hence, $tr AA^\top \geq tr BB^\top$.

1.5.   Denote

$$I = \int_{-\infty}^{\infty} e^{-y^2} dy .$$

Then, using polar coordinates, we have

$$I^2 = \left(\int_{-\infty}^{\infty} e^{-y^2} dy \right) \left(\int_{-\infty}^{\infty} e^{-x^2} dx \right)$$

$$= \int_{-\infty}^{\infty} \int_{-\infty}^{\infty} e^{-(x^2+y^2)} dx dy$$

$$= \int_0^{2\pi} \int_0^{\infty} e^{-r^2} r dr d\theta = \pi .$$

1.6.   Denote

$$I(x) = \int_{-\infty}^{\infty} e^{-xy^2} dy .$$

Then, by Exercise 1.5,

$$I(x) = \frac{1}{\sqrt{x}} \int_{-\infty}^{\infty} e^{-(\sqrt{x}y)^2} d(\sqrt{x}y) = \sqrt{\pi/x} .$$

Hence,

$$\int_{-\infty}^{\infty} y^2 e^{-y^2} dy = -\frac{d}{dx} I(x) \Big|_{x=1}$$

$$= -\frac{1}{dx}\left(\sqrt{\pi/x} \right) \Big|_{x=1} = \frac{1}{2}\sqrt{\pi} .$$

1.7.   (a) Let P be a unitary matrix so that

$$R = P^\top diag[\lambda_1, \cdots, \lambda_n] P ,$$

and define

$$\mathbf{y} = \frac{1}{\sqrt{2}} \, diag[\sqrt{\lambda_1}, \cdots, \sqrt{\lambda_n}] P(\mathbf{x} - \underline{\mu}) \, .$$

Then

$$E(X) = \int_{-\infty}^{\infty} \mathbf{x} f(\mathbf{x}) d\mathbf{x}$$

$$= \int_{-\infty}^{\infty} (\underline{\mu} + \sqrt{2} P^{-1} \, diag[\, 1/\sqrt{\lambda_1}, \cdots, 1/\sqrt{\lambda_n} \,] \mathbf{y}) f(\mathbf{x}) d\mathbf{x}$$

$$= \underline{\mu} \int_{-\infty}^{\infty} f(\mathbf{x}) d\mathbf{x}$$

$$+ Const. \int_{-\infty}^{\infty} \cdots \int_{-\infty}^{\infty} \begin{bmatrix} y_1 \\ \vdots \\ y_n \end{bmatrix} e^{-y_1^2} \cdots e^{-y_n^2} dy_1 \cdots dy_n$$

$$= \underline{\mu} \cdot 1 + 0 = \underline{\mu} \, .$$

(b) Using the same substitution, we have

$$Var(X)$$

$$= \int_{-\infty}^{\infty} (\mathbf{x} - \underline{\mu})(\mathbf{x} - \underline{\mu})^{\mathsf{T}} f(\mathbf{x}) d\mathbf{x}$$

$$= \int_{-\infty}^{\infty} 2R^{1/2} \mathbf{y} \mathbf{y}^{\mathsf{T}} R^{1/2} f(\mathbf{x}) d\mathbf{x}$$

$$= \frac{2}{(\pi)^{n/2}} R^{1/2} \left\{ \int_{-\infty}^{\infty} \cdots \int_{-\infty}^{\infty} \begin{bmatrix} y_1^2 & \cdots & y_1 y_n \\ \vdots & & \vdots \\ y_n y_1 & \cdots & y_n^2 \end{bmatrix} \right.$$

$$\left. \cdot \, e^{-y_1^2} \cdots e^{-y_n^2} dy_1 \cdots dy_n \right\} R^{1/2}$$

$$= R^{1/2} I R^{1/2} = R \, .$$

1.8. All the properties can be easily verified from the definitions.

1.9. We have already proved that if X_1 and X_2 are independent then $Cov(X_1, X_2) = 0$. Suppose now that $R_{12} = Cov(X_1, X_2) = 0$. Then $R_{21} = Cov(X_2, X_1) = 0$ so that

$$f(X_1, X_2) = \frac{1}{(2\pi)^{n/2} det R_{11} det R_{22}}$$

$$\cdot \, e^{-\frac{1}{2}(X_1 - \underline{\mu}_1)^{\mathsf{T}} R_{11}(X_1 - \underline{\mu}_1)} e^{-\frac{1}{2}(X_2 - \underline{\mu}_2)^{\mathsf{T}} R_{22}(X_2 - \underline{\mu}_2)}$$

$$= f_1(X_1) \cdot f_2(X_2) \, .$$

Hence, X_1 and X_2 are independent.

1.10. (1.35) can be verified by a direct computation. First, the following formula may be easily obtained:

$$\begin{bmatrix} I & -R_{xy}R_{yy}^{-1} \\ 0 & I \end{bmatrix} \begin{bmatrix} R_{xx} & R_{xy} \\ R_{yx} & R_{yy} \end{bmatrix} \begin{bmatrix} I & 0 \\ -R_{yy}^{-1}R_{xy}^{\mathsf{T}} & I \end{bmatrix}$$
$$= \begin{bmatrix} R_{xx} - R_{xy}R_{yy}^{-1}R_{yx} & 0 \\ 0 & R_{yy} \end{bmatrix}.$$

This yields, by taking determinants,

$$det \begin{bmatrix} R_{xx} & R_{xy} \\ R_{yx} & R_{yy} \end{bmatrix} = det[R_{xx} - R_{xy}R_{yy}^{-1}R_{yx}] \cdot det R_{yy}$$

and

$$\left(\begin{bmatrix} \mathbf{x} \\ \mathbf{y} \end{bmatrix} - \begin{bmatrix} \mu_x \\ \mu_y \end{bmatrix} \right)^{\mathsf{T}} \begin{bmatrix} R_{xx} & R_{xy} \\ R_{yx} & R_{yy} \end{bmatrix}^{-1} \left(\begin{bmatrix} \mathbf{x} \\ \mathbf{y} \end{bmatrix} - \begin{bmatrix} \mu_x \\ \mu_y \end{bmatrix} \right)$$
$$= (\mathbf{x} - \tilde{\mu})^{\mathsf{T}} [R_{xx} - R_{xy}R_{yy}^{-1}R_{yx}]^{-1}(\mathbf{x} - \tilde{\mu}) + (\mathbf{y} - \mu_y)^{\mathsf{T}} R_{yy}^{-1}(\mathbf{y} - \mu_y),$$

where

$$\tilde{\mu} = \mu_x + R_{xy}R_{yy}^{-1}(\mathbf{y} - \mu_y).$$

The remaining computational steps are straightforward.

1.11. Let $\mathbf{p}_k = C_k^{\mathsf{T}} W_k \mathbf{z}_k$ and $\sigma^2 = E[\mathbf{p}_k^{\mathsf{T}}(C_k^{\mathsf{T}} W_k C_k)^{-1}\mathbf{p}_k]$. Then it can be easily verified that

$$F(\mathbf{y}_k) = \mathbf{y}_k^{\mathsf{T}}(C_k^{\mathsf{T}} W_k C_k)\mathbf{y}_k - \mathbf{p}_k^{\mathsf{T}}\mathbf{y}_k - \mathbf{y}_k^{\mathsf{T}}\mathbf{p}_k + \sigma^2.$$

From

$$\frac{dF(\mathbf{y}_k)}{d\mathbf{y}_k} = 2(C_k^{\mathsf{T}} W_k C_k)\mathbf{y}_k - 2\mathbf{p}_k = 0,$$

and the assumption that the matrix $(C_k^{\mathsf{T}} W_k C_k)$ is nonsingular, we have

$$\hat{\mathbf{y}}_k = (C_k^{\mathsf{T}} W_k C_k)^{-1}\mathbf{p}_k = (C_k^{\mathsf{T}} W_k C_k)^{-1}C_k^{\mathsf{T}} W_k \mathbf{z}_k.$$

1.12.

$$E\hat{\mathbf{x}}_k = (C_k^{\mathsf{T}} R_k^{-1}C_k)^{-1}C_k^{\mathsf{T}} R_k^{-1}E(\mathbf{v}_k - D_k\mathbf{u}_k)$$
$$= (C_k^{\mathsf{T}} R_k^{-1}C_k)^{-1}C_k^{\mathsf{T}} R_k^{-1}E(C_k\mathbf{x}_k + \underline{\eta}_k)$$
$$= E\mathbf{x}_k.$$

Chapter 2

2.1.

$$W_{k,k-1}^{-1} = Var(\underline{\varepsilon}_{k,k-1}) = E(\underline{\varepsilon}_{k,k-1}\underline{\varepsilon}_{k,k-1}^{\top})$$

$$= E(\overline{\mathbf{v}}_{k-1} - H_{k,k-1}\mathbf{x}_k)(\overline{\mathbf{v}}_{k-1} - H_{k,k-1}\mathbf{x}_k)^{\top}$$

$$= \begin{bmatrix} R_0 & & \\ & \ddots & \\ & & R_{k-1} \end{bmatrix} + Var \begin{bmatrix} C_0 \sum_{i=1}^{k} \Phi_{0i}\Gamma_{i-1}\underline{\xi}_{i-1} \\ \vdots \\ C_{k-1}\Phi_{k-1,k}\Gamma_{k-1}\underline{\xi}_{k-1} \end{bmatrix}.$$

2.2.   For any nonzero vector $\mathbf{x}$, we have $\mathbf{x}^{\top}A\mathbf{x} > 0$ and $\mathbf{x}^{\top}B\mathbf{x} \geq 0$ so that

$$\mathbf{x}^{\top}(A+B)\mathbf{x} = \mathbf{x}^{\top}A\mathbf{x} + \mathbf{x}^{\top}B\mathbf{x} > 0.$$

Hence, $A + B$ is positive definite.

2.3.

$$W_{k,k-1}^{-1}$$
$$= E(\underline{\varepsilon}_{k,k-1}\underline{\varepsilon}_{k,k-1}^{\top})$$
$$= E(\underline{\varepsilon}_{k-1,k-1} - H_{k,k-1}\Gamma_{k-1}\underline{\xi}_{k-1})(\underline{\varepsilon}_{k-1,k-1} - H_{k,k-1}\Gamma_{k-1}\underline{\xi}_{k-1})^{\top}$$
$$= E(\underline{\varepsilon}_{k-1,k-1}\underline{\varepsilon}_{k-1,k-1}^{\top}) + H_{k,k-1}\Gamma_{k-1}E(\underline{\xi}_{k-1}\underline{\xi}_{k-1}^{\top})\Gamma_{k-1}^{\top}H_{k,k-1}^{\top}$$
$$= W_{k-1,k-1}^{-1} + H_{k-1,k-1}\Phi_{k-1,k}\Gamma_{k-1}Q_{k-1}\Gamma_{k-1}^{\top}\Phi_{k-1,k}^{\top}H_{k-1,k-1}^{\top}.$$

2.4.   Apply Lemma 1.2 to $A_{11} = W_{k-1,k-1}^{-1}, A_{22} = Q_{k-1}^{-1}$ and

$$A_{12} = A_{21}^{\top} = H_{k-1,k-1}\Phi_{k-1,k}\Gamma_{k-1}.$$

2.5.   Using Exercise 2.4, or (2.9), we have

$$H_{k,k-1}^{\top}W_{k,k-1}$$
$$= \Phi_{k-1,k}^{\top}H_{k-1,k-1}^{\top}W_{k-1,k-1}$$
$$\quad - \Phi_{k-1,k}^{\top}H_{k-1,k-1}^{\top}W_{k-1,k-1}H_{k,k-1}\Phi_{k-1,k}\Gamma_{k-1}$$
$$\quad \cdot (Q_{k-1}^{-1} + \Gamma_{k-1}^{\top}\Phi_{k-1,k}^{\top}H_{k-1,k-1}^{\top}W_{k-1,k-1}H_{k-1,k-1}\Phi_{k-1,k}\Gamma_{k-1})^{-1}$$
$$\quad \cdot \Gamma_{k-1}^{\top}\Phi_{k-1,k}^{\top}H_{k-1,k-1}^{\top}W_{k-1,k-1}$$
$$= \Phi_{k-1,k}^{\top}\{I - H_{k-1,k-1}^{\top}W_{k-1,k-1}H_{k-1,k-1}\Phi_{k-1,k}\Gamma_{k-1}$$
$$\quad \cdot (Q_{k-1}^{-1} + \Gamma_{k-1}^{\top}\Phi_{k-1,k}^{\top}H_{k-1,k-1}^{\top}W_{k-1,k-1}H_{k-1,k-1}\Phi_{k-1,k}\Gamma_{k-1})^{-1}$$
$$\quad \cdot \Gamma_{k-1}^{\top}\Phi_{k-1,k}^{\top}\}H_{k-1,k-1}^{\top}W_{k-1,k-1}.$$

2.6. Using Exercise 2.5, or (2.10), and the identity $H_{k,k-1} = H_{k-1,k-1}\Phi_{k-1,k}$, we have

$$(H_{k,k-1}^\top W_{k,k-1} H_{k,k-1})\Phi_{k,k-1}$$
$$\cdot (H_{k-1,k-1}^\top W_{k-1,k-1} H_{k-1,k-1})^{-1} H_{k-1,k-1}^\top W_{k-1,k-1}$$
$$= \Phi_{k-1,k}^\top \{I - H_{k-1,k-1}^\top W_{k-1,k-1} H_{k-1,k-1}\Phi_{k-1,k}\Gamma_{k-1}$$
$$\cdot (Q_{k-1}^{-1} + \Gamma_{k-1}^\top \Phi_{k-1,k}^\top H_{k-1,k-1}^\top W_{k-1,k-1} H_{k-1,k-1}\Phi_{k-1,k}\Gamma_{k-1})^{-1}$$
$$\cdot \Gamma_{k-1}^\top \Phi_{k-1,k}^\top\} H_{k-1,k-1}^\top W_{k-1,k-1}$$
$$= H_{k,k-1}^\top W_{k,k-1} \,.$$

2.7.

$$P_{k,k-1} C_k^\top (C_k P_{k,k-1} C_k^\top + R_k)^{-1}$$
$$= P_{k,k-1} C_k^\top (R_k^{-1} - R_k^{-1} C_k (P_{k,k-1}^{-1} + C_k^\top R_k^{-1} C_k)^{-1} C_k^\top R_k^{-1})$$
$$= (P_{k,k-1} - P_{k,k-1} C_k^\top R_k^{-1} C_k (P_{k,k-1}^{-1} + C_k^\top R_k^{-1} C_k)^{-1}) C_k^\top R_k^{-1}$$
$$= (P_{k,k-1} - P_{k,k-1} C_k^\top (C_k P_{k,k-1} C_k^\top + R_k)^{-1}$$
$$\cdot (C_k P_{k,k-1} C_k^\top + R_k) R_k^{-1} C_k (P_{k,k-1}^{-1} + C_k^\top R_k^{-1} C_k)^{-1}) C_k^\top R_k^{-1}$$
$$= (P_{k,k-1} - P_{k,k-1} C_k^\top (C_k P_{k,k-1} C_k^\top + R_k)^{-1}$$
$$\cdot (C_k P_{k,k-1} C_k^\top R_k^{-1} C_k + C_k)(P_{k,k-1}^{-1} + C_k^\top R_k^{-1} C_k)^{-1}) C_k^\top R_k^{-1}$$
$$= (P_{k,k-1} - P_{k,k-1} C_k^\top (C_k P_{k,k-1} C_k^\top + R_k)^{-1} C_k P_{k,k-1}$$
$$\cdot (C_k^\top R_k^{-1} C_k + P_{k,k-1}^{-1})(P_{k,k-1}^{-1} + C_k^\top R_k^{-1} C_k)^{-1}) C_k^\top R_k^{-1}$$
$$= (P_{k,k-1} - P_{k,k-1} C_k^\top (C_k P_{k,k-1} C_k^\top + R_k)^{-1} C_k P_{k,k-1}) C_k^\top R_k^{-1}$$
$$= P_{k,k} C_k^\top R_k^{-1}$$
$$= G_k \,.$$

2.8.

$$P_{k,k-1}$$
$$= (H_{k,k-1}^\top W_{k,k-1} H_{k,k-1})^{-1}$$
$$= (\Phi_{k-1,k}^\top (H_{k-1,k-1}^\top W_{k-1,k-1} H_{k-1,k-1}$$
$$\quad - H_{k-1,k-1}^\top W_{k-1,k-1} H_{k-1,k-1}\Phi_{k-1,k}\Gamma_{k-1}$$
$$\cdot (Q_{k-1}^{-1} + \Gamma_{k-1}^\top \Phi_{k-1,k}^\top H_{k-1,k-1}^\top W_{k-1,k-1} H_{k-1,k-1}\Phi_{k-1,k}\Gamma_{k-1})^{-1}$$
$$\cdot \Gamma_{k-1}^\top \Phi_{k-1,k}^\top H_{k-1,k-1}^\top W_{k-1,k-1} H_{k-1,k-1})\Phi_{k-1,k})^{-1}$$
$$= (\Phi_{k-1,k}^\top P_{k-1,k-1}^{-1}\Phi_{k-1,k} - \Phi_{k-1,k}^\top P_{k-1,k-1}^{-1}\Phi_{k-1,k}\Gamma_{k-1}$$
$$\cdot (Q_{k-1}^{-1} + \Gamma_{k-1}^\top \Phi_{k-1,k}^\top P_{k-1,k-1}^{-1}\Phi_{k-1,k}\Gamma_{k-1})^{-1}$$
$$\cdot \Gamma_{k-1}^\top \Phi_{k-1,k}^\top P_{k-1,k-1}^{-1}\Phi_{k-1,k})^{-1}$$
$$= (\Phi_{k-1,k}^\top P_{k-1,k-1}^{-1}\Phi_{k-1,k})^{-1} + \Gamma_{k-1} Q_{k-1}\Gamma_{k-1}^\top$$
$$= A_{k-1} P_{k-1,k-1} A_{k-1}^\top + \Gamma_{k-1} Q_{k-1}\Gamma_{k-1}^\top \,.$$

2.9.

$$E(\mathbf{x}_k - \hat{\mathbf{x}}_{k|k-1})(\mathbf{x}_k - \hat{\mathbf{x}}_{k|k-1})^{\top}$$
$$=E(\mathbf{x}_k - (H_{k,k-1}^{\top}W_{k,k-1}H_{k,k-1})^{-1}H_{k,k-1}^{\top}W_{k,k-1}\bar{\mathbf{v}}_{k-1})$$
$$\cdot (\mathbf{x}_k - (H_{k,k-1}^{\top}W_{k,k-1}H_{k,k-1})^{-1}H_{k,k-1}^{\top}W_{k,k-1}\bar{\mathbf{v}}_{k-1})^{\top}$$
$$=E(\mathbf{x}_k - (H_{k,k-1}^{\top}W_{k,k-1}H_{k,k-1})^{-1}H_{k,k-1}^{\top}W_{k,k-1}$$
$$\cdot (H_{k,k-1}\mathbf{x}_k + \bar{\underline{\xi}}_{k,k-1}))(\mathbf{x}_k - (H_{k,k-1}^{\top}W_{k,k-1}H_{k,k-1})^{-1}$$
$$\cdot H_{k,k-1}^{\top}W_{k,k-1}(H_{k,k-1}\mathbf{x}_k + \bar{\underline{\xi}}_{k,k-1}))^{\top}$$
$$=(H_{k,k-1}^{\top}W_{k,k-1}H_{k,k-1})^{-1}H_{k,k-1}^{\top}W_{k,k-1}E(\bar{\underline{\xi}}_{k,k-1}\bar{\underline{\xi}}_{k,k-1}^{\top})W_{k,k-1}$$
$$\cdot H_{k,k-1}(H_{k,k-1}^{\top}W_{k,k-1}H_{k,k-1})^{-1}$$
$$=(H_{k,k-1}^{\top}W_{k,k-1}H_{k,k-1})^{-1}$$
$$=P_{k,k-1}.$$

The derivation of the second identity is similar.

2.10. Since

$$\sigma^2 = Var(x_k) = E(ax_{k-1} + \xi_{k-1})^2$$
$$= a^2 Var(x_{k-1}) + 2aE(x_{k-1}\xi_{k-1}) + E(\xi_{k-1}^2)$$
$$= a^2\sigma^2 + \mu^2,$$

we have

$$\sigma^2 = \mu^2/(1 - a^2).$$

For $j = 1$, we have

$$E(x_k x_{k+1}) = E(x_k(ax_k + \xi_k))$$
$$= aVar(x_k) + E(x_k\xi_k)$$
$$= a\sigma^2.$$

For $j = 2$, we have

$$E(x_k x_{k+2}) = E(x_k(ax_{k+1} + \xi_{k+1}))$$
$$= aE(x_k x_{k+1}) + E(x_k + \xi_{k+1})$$
$$= aE(x_k x_{k+1})$$
$$= a^2\sigma^2,$$

etc. If j is negative, then a similar result can be obtained. By induction, we may conclude that $E(x_k x_{k+j}) = a^{|j|}\sigma^2$ for all integers j.

2.11. Using the Kalman filtering equations (2.17), we have

$$P_{0,0} = Var(x_0) = \mu^2,$$

$$P_{k,k-1} = P_{k-1,k-1},$$

$$G_k = P_{k,k-1}(P_{k,k-1} + R_k)^{-1} = \frac{P_{k-1,k-1}}{P_{k-1,k-1} + \sigma^2},$$

and

$$P_{k,k} = (1 - G_k)P_{k,k-1} = \frac{\sigma^2 P_{k-1,k-1}}{\sigma^2 + P_{k-1,k-1}}.$$

Observe that

$$P_{1,1} = \frac{\sigma^2 \mu^2}{\mu^2 + \sigma^2},$$

$$P_{2,2} = \frac{\sigma^2 P_{1,1}}{P_{1,1} + \sigma^2} = \frac{\sigma^2 \mu^2}{2\mu^2 + \sigma^2}.$$

$$\cdots$$

$$P_{k,k} = \frac{\sigma^2 \mu^2}{k\mu^2 + \sigma^2}.$$

Hence,

$$G_k = \frac{P_{k-1,k-1}}{P_{k-1,k-1} + \sigma^2} = \frac{\mu^2}{k\mu^2 + \sigma^2}$$

so that

$$\hat{x}_{k|k} = \hat{x}_{k|k-1} + G_k(v_k - \hat{x}_{k|k-1})$$

$$= \hat{x}_{k-1|k-1} + \frac{\mu^2}{\sigma^2 + k\mu^2}(v_k - \hat{x}_{k-1|k-1})$$

with $\hat{x}_{0|0} = E(x_0) = 0$. It follows that

$$\hat{x}_{k|k} = \hat{x}_{k-1|k-1}$$

for large values of k.

2.12.

$$\hat{Q}_N = \frac{1}{N} \sum_{k=1}^{N} (\mathbf{v}_k \mathbf{v}_k^{\mathsf{T}})$$

$$= \frac{1}{N}(\mathbf{v}_N \mathbf{v}_N^{\mathsf{T}}) + \frac{1}{N} \sum_{k=1}^{N-1} (\mathbf{v}_k \mathbf{v}_k^{\mathsf{T}})$$

$$= \frac{1}{N}(\mathbf{v}_N \mathbf{v}_N^{\mathsf{T}}) + \frac{N-1}{N}\hat{Q}_{N-1}$$

$$= \hat{Q}_{N-1} + \frac{1}{N}[(\mathbf{v}_N \mathbf{v}_N^{\mathsf{T}}) - \hat{Q}_{N-1}]$$

with the initial estimation $\hat{Q}_1 = \mathbf{v}_1 \mathbf{v}_1^{\mathsf{T}}$.

2.13. Use superimposition.

2.14. Set $\mathbf{x}_k = [(\mathbf{x}_k^1)^\top \cdots (\mathbf{x}_k^N)^\top]^\top$ for each $k, k = 0, 1, \cdots$, with $\mathbf{x}_j = 0$ (and $\mathbf{u}_j = 0$) for $j < 0$, and define

$$\mathbf{x}_k^1 = B_1 \mathbf{x}_{k-1}^1 + \mathbf{x}_{k-1}^2 + (A_1 + B_1 A_0)\mathbf{u}_{k-1},$$

$$\cdots\cdots$$

$$\mathbf{x}_k^M = B_M \mathbf{x}_{k-1}^1 + \mathbf{x}_{k-1}^{M+1} + (A_M + B_M A_0)\mathbf{u}_{k-1},$$

$$\mathbf{x}_k^{M+1} = B_{M+1}\mathbf{x}_{k-1}^1 + \mathbf{x}_{k-1}^{M+2} + B_{M+1} A_0 \mathbf{u}_{k-1},$$

$$\cdots\cdots$$

$$\mathbf{x}_k^{N-1} = B_{N-1}\mathbf{x}_{k-1}^1 + \mathbf{x}_{k-1}^N + B_{N-1} A_0 \mathbf{u}_{k-1},$$

$$\mathbf{x}_k^N = B_N \mathbf{x}_{k-1}^1 + B_N A_0 \mathbf{u}_{k-1}.$$

Then, substituting these equations into

$$\mathbf{v}_k = C\mathbf{x}_k + D\mathbf{u}_k = \mathbf{x}_k^1 + A_0 \mathbf{u}_k$$

yields the required result. Since $\mathbf{x}_j = 0$ and $\mathbf{u}_j = 0$ for $j < 0$, it is also clear that $\mathbf{x}_0 = 0$.

Chapter 3

3.1. Let $A = BB^\top$ where $B = [b_{ij}] \neq 0$. Then $\mathrm{tr}A = \mathrm{tr}BB^\top = \sum_{i,j} b_{ij}^2 > 0$.

3.2. By Assumption 2.1, $\underline{\eta}_\ell$ is independent of $\mathbf{x}_0, \underline{\xi}_0, \cdots, \underline{\xi}_{j-1}, \underline{\eta}_0, \cdots, \underline{\eta}_{j-1}$, since $\ell \geq j$. On the other hand,

$$\hat{\mathbf{e}}_j = C_j(\mathbf{x}_j - \hat{\mathbf{y}}_{j-1})$$

$$= C_j\left(A_{j-1}\mathbf{x}_{j-1} + \Gamma_{j-1}\underline{\xi}_{j-1} - \sum_{i=0}^{j-1} \hat{P}_{j-1,i}(C_i\mathbf{x}_i + \underline{\eta}_i)\right)$$

$$\cdots\cdots$$

$$= B_0 \mathbf{x}_0 + \sum_{i=0}^{j-1} B_{1i}\underline{\xi}_i + \sum_{i=0}^{j-1} B_{2i}\underline{\eta}_i$$

for some constant matrices B_0, B_{1i} and B_{2i}. Hence, $\langle \underline{\eta}_\ell, \hat{\mathbf{e}}_j \rangle = O_{q \times q}$ for all $\ell \geq j$.

3.3. Combining (3.8) and (3.4), we have

$$\mathbf{e}_j = \|\mathbf{z}_j\|_q^{-1}\mathbf{z}_j = \|\mathbf{z}_j\|_q^{-1}\mathbf{v}_j - \sum_{i=0}^{j-1}\left(\|\mathbf{z}_j\|_q^{-1} C_j \hat{P}_{j-1,i}\right)\mathbf{v}_i;$$

that is, e_j can be expressed in terms of $v_0, v_1, \cdots, v_j$. Conversely, we have

$$
\begin{aligned}
v_0 &= z_0 = \|z_0\|_q e_0 , \\
v_1 &= z_1 + C_1 \hat{y}_0 = z_1 + C_1 \hat{P}_{0,0} v_0 \\
&= \|z_1\|_q e_1 + C_1 \hat{P}_{0,0} \|z_0\|_q e_0 , \\
&\quad \cdots\cdots
\end{aligned}
$$

that is, v_j can also be expressed in terms of $e_0, e_1, \cdots, e_j$. Hence, we have

$$
Y(e_0, \cdots, e_k) = Y(v_0, \cdots, v_k) .
$$

3.4. By Exercise 3.3, we have

$$
v_i = \sum_{\ell=0}^{i} L_\ell e_\ell
$$

for some $q \times q$ constant matrices L_ℓ, $\ell = 0, 1, \cdots, i$, so that

$$
\langle v_i,\ z_k \rangle = \sum_{\ell=0}^{i} L_\ell \langle e_\ell,\ e_k \rangle \|z_k\|_q^\top = O_{q \times q} ,
$$

$i = 0, 1, \cdots, k-1$. Hence, for $j = 0, 1, \cdots, k-1$,

$$
\begin{aligned}
\langle \hat{y}_j,\ z_k \rangle &= \left\langle \sum_{i=0}^{j} \hat{P}_{j,i} v_i,\ z_k \right\rangle \\
&= \sum_{i=0}^{j} \hat{P}_{j,i} \langle v_i,\ z_k \rangle \\
&= O_{n \times q} .
\end{aligned}
$$

3.5. Since

$$
\begin{aligned}
x_k &= A_{k-1} x_{k-1} + \Gamma_{k-1} \underline{\xi}_{k-1} \\
&= A_{k-1}(A_{k-2} x_{k-2} + \Gamma_{k-2} \underline{\xi}_{k-2}) + \Gamma_{k-1} \underline{\xi}_{k-1} \\
&= \cdots\cdots \\
&= B_0 x_0 + \sum_{i=0}^{k-1} B_{1i} \underline{\xi}_i
\end{aligned}
$$

for some constant matrices B_0 and B_{1i} and $\underline{\xi}_k$ is independent of x_0 and $\underline{\xi}_i$ ($0 \le i \le k-1$), we have $\langle x_k,\ \underline{\xi}_k \rangle = 0$. The rest can be shown in a similar manner.

3.6.   Use superimposition.

3.7.   Using the formula obtained in Exercise 3.6, we have

$$\begin{cases} \hat{d}_{k|k} = \hat{d}_{k-1|k-1} + hw_{k-1} + G_k(v_k - \Delta d_k - \hat{d}_{k-1|k-1} - hw_{k-1}) \\ \hat{d}_{0|0} = E(d_0), \end{cases}$$

where G_k is obtained by using the standard algorithm (3.25) with $A_k = C_k = \Gamma_k = 1$.

3.8.   Let

$$\mathbf{x}_k = \begin{bmatrix} \mathbf{x}_k^1 \\ \mathbf{x}_k^2 \\ \mathbf{x}_k^3 \end{bmatrix}, \quad \mathbf{x}_k^1 = \begin{bmatrix} \Sigma_k \\ \dot{\Sigma}_k \\ \ddot{\Sigma}_k \end{bmatrix}, \quad \mathbf{x}_k^2 = \begin{bmatrix} \Delta A_k \\ \Delta \dot{A}_k \\ \Delta \ddot{A}_k \end{bmatrix}, \quad \mathbf{x}_k^3 = \begin{bmatrix} \Delta E_k \\ \Delta \dot{E}_k \\ \Delta \ddot{E}_k \end{bmatrix},$$

$$\underline{\xi}_k = \begin{bmatrix} \xi_k^1 \\ \xi_k^2 \\ \xi_k^3 \end{bmatrix}, \quad \underline{\eta}_k = \begin{bmatrix} \eta_k^1 \\ \eta_k^2 \\ \eta_k^3 \end{bmatrix}, \quad \mathbf{v}_k = \begin{bmatrix} v_k^1 \\ v_k^2 \\ v_k^3 \end{bmatrix},$$

$$A = \begin{bmatrix} 1 & h & h^2/2 \\ 0 & 1 & h \\ 0 & 0 & 1 \end{bmatrix}, \quad and \quad C = [\, 1 \ 0 \ 0 \,].$$

Then the system described in Exercise 3.8 can be decomposed into three subsystems:

$$\begin{cases} \mathbf{x}_{k+1}^i = A\mathbf{x}_k^i + \Gamma_k^i \underline{\xi}_k^i \\ v_k^i = C\mathbf{x}_k^i + \eta_k^i, \end{cases}$$

$i = 1, 2, 3$, where for each k, $\mathbf{x}_k$ and $\underline{\xi}_k$ are 3-vectors, v_k and η_k are scalars, Q_k a 3×3 non-negative definite symmetric matrix, and $R_k > 0$ a scalar.

Chapter 4

4.1.   Using (4.6), we have

$$\begin{aligned} L(A&\mathbf{x} + B\mathbf{y}, \ \mathbf{v}) \\ &= E(A\mathbf{x} + B\mathbf{y}) + \langle A\mathbf{x} + B\mathbf{y}, \ \mathbf{v} \rangle [Var(\mathbf{v})]^{-1}(\mathbf{v} - E(\mathbf{v})) \\ &= A\{E(\mathbf{x}) + \langle \mathbf{x}, \ \mathbf{v} \rangle [Var(\mathbf{v})]^{-1}(\mathbf{v} - E(\mathbf{v}))\} \\ &\quad + B\{E(\mathbf{y}) + \langle \mathbf{y}, \ \mathbf{v} \rangle [Var(\mathbf{v})]^{-1}(\mathbf{v} - E(\mathbf{v}))\} \\ &= AL(\mathbf{x}, \ \mathbf{v}) + BL(\mathbf{y}, \ \mathbf{v}). \end{aligned}$$

4.2.   Using (4.6) and the fact that $E(\mathbf{a}) = \mathbf{a}$ so that

$$\langle \mathbf{a}, \ \mathbf{v} \rangle = E(\mathbf{a} - E(\mathbf{a})) \ (\mathbf{v} - E(\mathbf{v})) = 0 \,,$$

we have

$$L(\mathbf{a}, \ \mathbf{v}) = E(\mathbf{a}) + \langle \mathbf{a}, \ \mathbf{v} \rangle \, [Var(\mathbf{v})]^{-1} (\mathbf{v} - E(\mathbf{v})) = \mathbf{a} \,.$$

4.3.   By definition, for a real-valued function f and a matrix $A = [a_{ij}]$, $df/dA = [\partial f/\partial a_{ji}]$. Hence,

$$
\begin{aligned}
0 &= \frac{\partial}{\partial H} \left(tr \| \mathbf{x} - \mathbf{y} \|_n^2 \right) \\
&= \frac{\partial}{\partial H} E((\mathbf{x} - E(\mathbf{x})) - H(\mathbf{v} - E(\mathbf{v})))^{\mathsf{T}} ((\mathbf{x} - E(\mathbf{x})) - H(\mathbf{v} - E(\mathbf{v}))) \\
&= E \frac{\partial}{\partial H} ((\mathbf{x} - E(\mathbf{x})) - H(\mathbf{v} - E(\mathbf{v})))^{\mathsf{T}} ((\mathbf{x} - E(\mathbf{x})) - H(\mathbf{v} - E(\mathbf{v}))) \\
&= E(-2(\mathbf{x} - E(\mathbf{x})) - H(\mathbf{v} - E(\mathbf{v}))) (\mathbf{v} - E(\mathbf{v}))^{\mathsf{T}} \\
&= 2 \big(H \ E(\mathbf{v} - E(\mathbf{v})) \ (\mathbf{v} - E(\mathbf{v}))^{\mathsf{T}} - E(\mathbf{x} - E(\mathbf{x})) \ (\mathbf{v} - E(\mathbf{v}))^{\mathsf{T}} \big) \\
&= 2 \big(H \| \mathbf{v} \|_q^2 - \langle \mathbf{x}, \ \mathbf{v} \rangle \big) \,.
\end{aligned}
$$

This gives

$$H^* = \langle \mathbf{x}, \ \mathbf{v} \rangle \Big[\| \mathbf{v} \|_q^2 \Big]^{-1}$$

so that

$$\mathbf{x}^* = E(\mathbf{x}) - \langle \mathbf{x}, \ \mathbf{v} \rangle \Big[\| \mathbf{v} \|_q^2 \Big]^{-1} (E(\mathbf{v}) - \mathbf{v}) \,.$$

4.4.   Since $\mathbf{v}^{k-2}$ is a linear combination (with constant matrix coefficients) of

$$\mathbf{x}_0, \ \underline{\xi}_0, \ \cdots, \ \underline{\xi}_{k-3}, \ \underline{\eta}_0, \ \cdots, \ \underline{\eta}_{k-2}$$

which are all uncorrelated with $\underline{\xi}_{k-1}$ and $\underline{\eta}_{k-1}$, we have

$$\langle \underline{\xi}_{k-1}, \ \mathbf{v}^{k-2} \rangle = 0 \qquad and \qquad \langle \underline{\eta}_{k-1}, \ \mathbf{v}^{k-2} \rangle = 0 \,.$$

Similarly, we can verify the other formulas [where (4.6) may be used].

4.5.   The first identity follows from the Kalman gain equation (cf. Theorem 4.1(c) or (4.19)), namely:

$$G_k (C_k P_{k,k-1} C_k^{\mathsf{T}} + R_k) = P_{k,k-1} C_k^{\mathsf{T}} \,,$$

so that

$$G_k R_k = P_{k,k-1} C_k^\top - G_k C_k P_{k,k-1} C_k^\top$$
$$= (I - G_k C_k) P_{k,k-1} C_k^\top.$$

To prove the second equality, we apply (4.18) and (4.17) to obtain

$$\langle \mathbf{x}_{k-1} - \hat{\mathbf{x}}_{k-1|k-1}, \ \Gamma_{k-1}\underline{\xi}_{k-1} - K_{k-1}\underline{\eta}_{k-1} \rangle$$

$$= \langle \mathbf{x}_{k-1} - \hat{\mathbf{x}}_{k-1|k-2} - \langle \mathbf{x}^\#{}_{k-1}, \ \mathbf{v}^\#{}_{k-1} \rangle \big[\|\mathbf{v}^\#{}_{k-1}\|^2 \big]^{-1} \mathbf{v}^\#{}_{k-1},$$
$$\Gamma_{k-1}\underline{\xi}_{k-1} - K_{k-1}\underline{\eta}_{k-1} \rangle$$

$$= \langle \mathbf{x}^\#{}_{k-1} - \langle \mathbf{x}^\#{}_{k-1}, \ \mathbf{v}^\#{}_{k-1} \rangle \big[\|\mathbf{v}^\#{}_{k-1}\|^2 \big]^{-1} (C_{k-1}\mathbf{x}^\#{}_{k-1} + \underline{\eta}_{k-1}),$$
$$\Gamma_{k-1}\underline{\xi}_{k-1} - K_{k-1}\underline{\eta}_{k-1} \rangle$$

$$= -\langle \mathbf{x}^\#{}_{k-1}, \ \mathbf{v}^\#{}_{k-1} \rangle \big[\|\mathbf{v}^\#{}_{k-1}\|^2 \big]^{-1} (S_{k-1}^\top \Gamma_{k-1}^\top - R_{k-1} K_{k-1}^\top)$$

$$= O_{n \times n},$$

in which since $K_{k-1} = \Gamma_{k-1} S_{k-1} R_{k-1}^{-1}$, we have

$$S_{k-1}^\top \Gamma_{k-1}^\top - R_{k-1} K_{k-1}^\top = O_{n \times n}.$$

4.6. Follow the same procedure in the derivation of Theorem 4.1 with the term $\mathbf{v}_k$ replaced by $\mathbf{v}_k - D_k \mathbf{u}_k$, and with

$$\hat{\mathbf{x}}_{k|k-1} = L(A_{k-1}\mathbf{x}_{k-1} + B_{k-1}\mathbf{u}_{k-1} + \Gamma_{k-1}\underline{\xi}_{k-1}, \ \mathbf{v}^{k-1})$$

instead of

$$\hat{\mathbf{x}}_{k|k-1} = L(\mathbf{x}_k, \ \mathbf{v}^{k-1}) = L(A_{k-1}\mathbf{x}_{k-1} + \Gamma_{k-1}\underline{\xi}_{k-1}, \ \mathbf{v}^{k-1}).$$

4.7. Let

$$w_k = -a_1 v_{k-1} + b_1 u_{k-1} + c_1 e_{k-1} + w_{k-1},$$
$$w_{k-1} = -a_2 v_{k-2} + b_2 u_{k-2} + w_{k-2},$$
$$w_{k-2} = -a_3 v_{k-3},$$

and define $\mathbf{x}_k = [\ w_k \quad w_{k-1} \quad w_{k-2}\ ]^\top$. Then,

$$\begin{cases} \mathbf{x}_{k+1} = A\mathbf{x}_k + B u_k + \Gamma e_k \\ v_k = C\mathbf{x}_k + D u_k + \Delta e_k, \end{cases}$$

where

$$A = \begin{bmatrix} -a_1 & 1 & 0 \\ -a_2 & 0 & 1 \\ -a_3 & 0 & 0 \end{bmatrix}, \qquad B = \begin{bmatrix} b_1 - a_1 b_0 \\ b_2 - a_2 b_0 \\ -a_3 b_0 \end{bmatrix}, \qquad \Gamma = \begin{bmatrix} c_1 - a_1 c_0 \\ -a_2 c_0 \\ -a_3 b_0 \end{bmatrix},$$

$$C = [\ 1 \quad 0 \quad 0\ ], \qquad D = [b_0] \qquad and \qquad \Delta = [c_0].$$

4.8. Let

$$w_k = -a_1 v_{k-1} + b_1 u_{k-1} + c_1 e_{k-1} + w_{k-1},$$

$$w_{k-1} = -a_2 v_{k-2} + b_2 u_{k-2} + c_2 e_{k-2} + w_{k-2},$$

$$\cdots\cdots$$

$$w_{k-n+1} = -a_n v_{k-n} + b_n u_{k-n} + c_n e_{k-n},$$

where $b_j = 0$ for $j > m$ and $c_j = 0$ for $j > \ell$, and define

$$\mathbf{x}_k = [\; w_k \quad w_{k-1} \quad \cdots \quad w_{k-n+1}\;]^{\mathsf{T}}.$$

Then

$$\begin{cases} \mathbf{x}_{k+1} = A\mathbf{x}_k + Bu_k + \Gamma e_k \\ v_k = C\mathbf{x}_k + Du_k + \Delta e_k, \end{cases}$$

where

$$A = \begin{bmatrix} -a_1 & 1 & 0 & \cdots & 0 \\ -a_2 & 0 & 1 & \cdots & 0 \\ \vdots & \vdots & \vdots & & \vdots \\ -a_{n-1} & 0 & 0 & \cdots & 1 \\ -a_n & 0 & 0 & \cdots & 0 \end{bmatrix},$$

$$B = \begin{bmatrix} b_1 - a_1 b_0 \\ \vdots \\ b_m - a_m b_0 \\ -a_{m+1} b_0 \\ \vdots \\ -a_n b_0 \end{bmatrix}, \quad \Gamma = \begin{bmatrix} c_1 - a_1 c_0 \\ \vdots \\ c_\ell - a_\ell c_0 \\ -a_{\ell+1} \\ \vdots \\ -a_n c_0 \end{bmatrix},$$

$$C = [\,1 \quad 0 \quad \cdots\cdots \quad 0\,], \qquad D = [b_0], \qquad and \qquad \Delta = [c_0].$$

Chapter 5

5.1. Since $\mathbf{v}^k$ is a linear combination (with constant matrices as coefficients) of

$$\mathbf{x}_0, \; \underline{\eta}_0, \; \underline{\gamma}_0, \; \cdots \; , \; \underline{\gamma}_k, \; \underline{\xi}_0, \; \underline{\beta}_0, \; \cdots \; , \; \underline{\beta}_{k-1}$$

which are all independent of $\underline{\beta}_k$, we have

$$\langle \underline{\beta}_k, \mathbf{v}^k \rangle = 0.$$

On the other hand, $\underline{\beta}_k$ has zero-mean, so that by (4.6) we have

$$L(\tilde{\underline{\beta}}_k, \mathbf{v}^k) = E(\tilde{\underline{\beta}}_k) - \langle \underline{\beta}_k, \mathbf{v}^k \rangle \left[\|\mathbf{v}^k\|^2 \right]^{-1} \left(E(\mathbf{v}^k) - \mathbf{v}^k \right) = 0.$$

5.2. Using Lemma 4.2 with $\mathbf{v} = \mathbf{v}^{k-1}$, $\mathbf{v}^1 = \mathbf{v}^{k-2}$, $\mathbf{v}^2 = \mathbf{v}_{k-1}$ and

$$\mathbf{v}^{\#}_{k-1} = \mathbf{v}_{k-1} - L(\mathbf{v}_{k-1}, \mathbf{v}^{k-2}),$$

we have, for $\mathbf{x} = \mathbf{v}_{k-1}$,

$$L(\mathbf{v}_{k-1}, \mathbf{v}^{k-1})$$
$$= L(\mathbf{v}_{k-1}, \mathbf{v}^{k-2}) + \langle \mathbf{v}^{\#}_{k-1}, \mathbf{v}^{\#}_{k-1} \rangle \left[\|\mathbf{v}^{\#}_{k-1}\|^2 \right]^{-1} \mathbf{v}^{\#}_{k-1}$$
$$= L(\mathbf{v}_{k-1}, \mathbf{v}^{k-2}) + \mathbf{v}_{k-1} - L(\mathbf{v}_{k-1}, \mathbf{v}^{k-2})$$
$$= \mathbf{v}_{k-1}.$$

The equality $L(\underline{\gamma}_k, \mathbf{v}^{k-1}) = 0$ can be shown by imitating the proof in Exercise 5.1.

5.3. It follows from Lemma 4.2 that

$$\mathbf{z}_{k-1} - \hat{\mathbf{z}}_{k-1}$$
$$= \mathbf{z}_{k-1} - L(\mathbf{z}_{k-1}, \mathbf{v}^{k-1})$$
$$= \mathbf{z}_{k-1} - E(\mathbf{z}_{k-1}) + \langle \mathbf{z}_{k-1}, \mathbf{v}^{k-1} \rangle \left[\|\mathbf{v}^{k-1}\|^2 \right]^{-1} \left(E(\mathbf{v}^{k-1}) - \mathbf{v}^{k-1} \right)$$
$$= \begin{bmatrix} \mathbf{x}_{k-1} \\ \underline{\xi}_{k-1} \end{bmatrix} - \begin{bmatrix} E(\mathbf{x}_{k-1}) \\ E(\underline{\xi}_{k-1}) \end{bmatrix}$$
$$+ \begin{bmatrix} \langle \mathbf{x}_{k-1}, \mathbf{v}^{k-1} \rangle \\ \langle \underline{\xi}_{k-1}, \mathbf{v}^{k-1} \rangle \end{bmatrix} \left[\|\mathbf{v}^{k-1}\|^2 \right]^{-1} \left(E(\mathbf{v}^{k-1}) - \mathbf{v}^{k-1} \right)$$

whose first n-subvector and last p-subvector are, respectively, linear combinations (with constant matrices as coefficients) of

$$\mathbf{x}_0, \ \underline{\xi}_0, \ \underline{\beta}_0, \ \cdots \ , \ \underline{\beta}_{k-2}, \ \underline{\eta}_0, \ \underline{\gamma}_0, \ \cdots \ , \ \underline{\gamma}_{k-1},$$

which are all independent of $\underline{\gamma}_k$. Hence, we have

$$B \langle \mathbf{z}_{k-1} - \hat{\mathbf{z}}_{k-1}, \ \underline{\gamma}_k \rangle = 0.$$

5.4. The proof is similar to that of Exercise 5.3.

5.5. For simplicity, denote

$$B = [C_0 Var(\mathbf{x}_0) C_0^\top + R_0]^{-1}.$$

It follows from (5.16) that

$$Var(\mathbf{x}_0 - \hat{\mathbf{x}}_0)$$
$$= Var(\mathbf{x}_0 - E(\mathbf{x}_0)$$
$$\quad - [Var(\mathbf{x}_0)]C_0^\top [C_0 Var(\mathbf{x}_0)C_0^\top + R_0]^{-1}(\mathbf{v}_0 - C_0 E(\mathbf{x}_0)))$$
$$= Var(\mathbf{x}_0 - E(\mathbf{x}_0) - [Var(\mathbf{x}_0)]C_0^\top B(C_0(\mathbf{x}_0 - E(\mathbf{x}_0)) + \underline{\eta}_0))$$
$$= Var((I - [Var(\mathbf{x}_0)]C_0^\top BC_0)(\mathbf{x}_0 - E(\mathbf{x}_0)) - [Var(\mathbf{x}_0)]C_0^\top B\underline{\eta}_0)$$
$$= (I - [Var(\mathbf{x}_0)]C_0^\top BC_0)Var(\mathbf{x}_0)\ (I - C_0^\top BC_0[Var(\mathbf{x}_0)])$$
$$\quad + [Var(\mathbf{x}_0)]C_0^\top BR_0 BC_0[Var(\mathbf{x}_0)]$$
$$= Var(\mathbf{x}_0) - [Var(\mathbf{x}_0)]C_0^\top BC_0[Var(\mathbf{x}_0)]$$
$$\quad - [Var(\mathbf{x}_0)]C_0^\top BC_0[Var(\mathbf{x}_0)]$$
$$\quad + [Var(\mathbf{x}_0)]C_0^\top BC_0[Var(\mathbf{x}_0)]C_0^\top BC_0[Var(\mathbf{x}_0)]$$
$$\quad + [Var(\mathbf{x}_0)]C_0^\top BR_0 BC_0[Var(\mathbf{x}_0)]$$
$$= Var(\mathbf{x}_0) - [Var(\mathbf{x}_0)]C_0^\top BC_0[Var(\mathbf{x}_0)]$$
$$\quad - [Var(\mathbf{x}_0)]C_0^\top BC_0[Var(\mathbf{x}_0)] + [Var(\mathbf{x}_0)]C_0^\top BC_0[Var(\mathbf{x}_0)]$$
$$= Var(\mathbf{x}_0) - [Var(\mathbf{x}_0)]C_0^\top BC_0[Var(\mathbf{x}_0)]\,.$$

5.6.   From $\hat{\underline{\xi}}_0 = 0$, we have

$$\hat{\mathbf{x}}_1 = A_0 \hat{\mathbf{x}}_0 + G_1(\mathbf{v}_1 - C_1 A_0 \hat{\mathbf{x}}_0)$$

and $\hat{\underline{\xi}}_1 = 0$, so that

$$\hat{\mathbf{x}}_2 = A_1 \hat{\mathbf{x}}_1 + G_2(\mathbf{v}_2 - C_2 A_1 \hat{\mathbf{x}}_1)\,,$$

etc. In general, we have

$$\hat{\mathbf{x}}_k = A_{k-1}\hat{\mathbf{x}}_{k-1} + G_k(\mathbf{v}_k - C_k A_{k-1}\hat{\mathbf{x}}_{k-1})$$
$$= \hat{\mathbf{x}}_{k|k-1} + G_k(\mathbf{v}_k - C_k \hat{\mathbf{x}}_{k|k-1})\,.$$

Denote

$$P_{0,0} = \left[[Var(\mathbf{x}_0)]^{-1} + C_0^\top R_0^{-1} C_0 \right]^{-1}$$

and

$$P_{k,k-1} = A_{k-1}P_{k-1,k-1}A_{k-1}^\top + \Gamma_{k-1}Q_{k-1}\Gamma_{k-1}^\top\,.$$

Then

$$G_1 = \begin{bmatrix} A_0 & \Gamma_0 \\ 0 & 0 \end{bmatrix} \begin{bmatrix} P_{0,0} & 0 \\ 0 & Q_0 \end{bmatrix} \begin{bmatrix} A_0^\top & C_1^\top \\ \Gamma_0^\top & C_1^\top \end{bmatrix}$$
$$\cdot \left([\, C_1 A_0 \;\; C_1 \Gamma_0\,] \begin{bmatrix} P_{0,0} & 0 \\ 0 & Q_0 \end{bmatrix} \begin{bmatrix} A_0^\top & C_1^\top \\ \Gamma_0^\top & C_1^\top \end{bmatrix} + R_1 \right)^{-1}$$
$$= \begin{bmatrix} P_{1,0}C_1^\top (C_1 P_{1,0}C_1^\top + R_1)^{-1} \\ 0 \end{bmatrix}\,,$$

$$P_1 = \left(\begin{bmatrix} A_0 & \Gamma_0 \\ 0 & 0 \end{bmatrix} - G_1 [\, C_1 A_0 \;\; C_1 \Gamma_0 \,] \right) \begin{bmatrix} P_{0,0} & 0 \\ 0 & Q_0 \end{bmatrix} \begin{bmatrix} A_0^\top & 0 \\ \Gamma_0^\top & 0 \end{bmatrix}$$

$$+ \begin{bmatrix} 0 & 0 \\ 0 & Q_1 \end{bmatrix}$$

$$= \begin{bmatrix} [\, I_n - P_{1,0} C_1^\top (C_1 P_{1,0} C_1^\top + R_1)^{-1} C_1 \,] P_{1,0} & 0 \\ 0 & Q_1 \end{bmatrix},$$

and, in general,

$$G_k = \begin{bmatrix} P_{k,k-1} C_k^\top (C_k P_{k,k-1} C_k^\top + R_k)^{-1} \\ 0 \end{bmatrix},$$

$$P_k = \begin{bmatrix} [\, I_n - P_{k,k-1} C_k^\top (C_k P_{k,k-1} C_k^\top + R_k)^{-1} C_k \,] P_{k,k-1} & 0 \\ 0 & Q_k \end{bmatrix}.$$

Finally, if we use the unbiased estimate $\hat{\mathbf{x}}_0 = E(\mathbf{x}_0)$ of $\mathbf{x}_0$ instead of the somewhat more superior initial state estimate

$$\hat{\mathbf{x}}_0 = E(\mathbf{x}_0) - [Var(\mathbf{x}_0)] C_0^\top [C_0 Var(\mathbf{x}_0) C_0^\top + R_0]^{-1} [C_0 E(\mathbf{x}_0) - \mathbf{v}_0],$$

and consequently set

$$P_0 = E \left(\begin{bmatrix} \mathbf{x}_0 \\ \underline{\xi}_0 \end{bmatrix} - \begin{bmatrix} E(\mathbf{x}_0) \\ E(\underline{\xi}_0) \end{bmatrix} \right) \left(\begin{bmatrix} \mathbf{x}_0 \\ \underline{\xi}_0 \end{bmatrix} - \begin{bmatrix} E(\mathbf{x}_0) \\ E(\underline{\xi}_0) \end{bmatrix} \right)^\top$$

$$= \begin{bmatrix} Var(\mathbf{x}_0) & 0 \\ 0 & Q_0 \end{bmatrix},$$

then we obtain the Kalman filtering algorithm derived in Chapters 2 and 3.

5.7. Let

$$\overline{P}_0 = [\, [Var(\mathbf{x}_0)]^{-1} + C_0^\top R_0^{-1} C_0 \,]^{-1}$$

and

$$\overline{H}_{k-1} = [\, C_k A_{k-1} - N_{k-1} C_{k-1} \,].$$

Starting with (5.17b), namely:

$$P_0 = \begin{bmatrix} (\, [Var(\mathbf{x}_0)]^{-1} + C_0 R_0^{-1} C_0)^{-1} & 0 \\ 0 & Q_0 \end{bmatrix} = \begin{bmatrix} \overline{P}_0 & 0 \\ 0 & Q_0 \end{bmatrix},$$

we have

$$G_1 = \begin{bmatrix} A_0 & \Gamma_0 \\ 0 & 0 \end{bmatrix} \begin{bmatrix} \overline{P}_0 & 0 \\ 0 & Q_0 \end{bmatrix} \begin{bmatrix} \overline{H}_0^\top \\ \Gamma_0^\top C_1^\top \end{bmatrix}$$

$$\cdot \left([\, H_0 \;\; C_1 \Gamma_0 \,] \begin{bmatrix} \overline{P}_0 & 0 \\ 0 & Q_0 \end{bmatrix} \begin{bmatrix} \overline{H}_0^\top \\ \Gamma_0^\top C_1^\top \end{bmatrix} + R_1 \right)^{-1}$$

$$= \begin{bmatrix} (A_0 \overline{P}_0 \overline{H}_0^\top + \Gamma_0 Q_0 \Gamma_0^\top C_1^\top)(\overline{H}_0 \overline{P}_0 \overline{H}_0^\top + C_1 \Gamma_0 Q_0 \Gamma_0^\top C_1^\top + R_1)^{-1} \\ 0 \end{bmatrix}$$

$$:= \begin{bmatrix} \overline{G}_1 \\ 0 \end{bmatrix}$$

and

$$P_1 = \left(\begin{bmatrix} A_0 & \Gamma_0 \\ 0 & 0 \end{bmatrix} - \begin{bmatrix} \overline{G}_1 \\ 0 \end{bmatrix} [\ \overline{H}_0 \ \ C_1\Gamma_0\ ] \right) \begin{bmatrix} \overline{P}_0 & 0 \\ 0 & Q_0 \end{bmatrix} \begin{bmatrix} A_0^\top & 0 \\ \Gamma_0^\top & 0 \end{bmatrix}$$
$$+ \begin{bmatrix} 0 & 0 \\ 0 & Q_1 \end{bmatrix}$$
$$= \begin{bmatrix} (A_0 - \overline{G}_1\overline{H}_0)\overline{P}_0 A_0^\top + (I - \overline{G}_1 C_1)\Gamma_0 Q_0 \Gamma_0^\top & 0 \\ 0 & Q_1 \end{bmatrix}$$
$$:= \begin{bmatrix} \overline{P}_1 & 0 \\ 0 & Q_1 \end{bmatrix}.$$

In general, we obtain

$$\begin{cases} \hat{\mathbf{x}}_k = A_{k-1}\hat{\mathbf{x}}_{k-1} + \bar{G}_k(\mathbf{v}_k - N_{k-1}\mathbf{v}_{k-1} - \bar{H}_{k-1}\hat{\mathbf{x}}_{k-1}) \\ \hat{\mathbf{x}}_0 = E(\mathbf{x}_0) - [Var(\mathbf{x}_0)]C_0^\top [C_0 Var(\mathbf{x}_0)C_0^\top + R_0]^{-1}[C_0 E(\mathbf{x}_0) - \mathbf{v}_0] \\ \bar{H}_{k-1} = [\ C_k A_{k-1} - N_{k-1}C_{k-1}\ ] \\ \bar{P}_k = (A_{k-1} - \bar{G}_k\bar{H}_{k-1})\bar{P}_{k-1}A_{k-1}^\top + (I - \bar{G}_k C_k)\Gamma_{k-1}Q_{k-1}\Gamma_{k-1}^\top \\ \bar{G}_k = (A_{k-1}\bar{P}_{k-1}\bar{H}_{k-1}^\top + \Gamma_{k-1}Q_{k-1}\Gamma_{k-1}^\top C_k^\top) \cdot \\ \qquad (\bar{H}_{k-1}\bar{P}_{k-1}\bar{H}_{k-1}^\top + C_k\Gamma_{k-1}Q_{k-1}\Gamma_{k-1}^\top C_k^\top + R_{k-1})^{-1} \\ \bar{P}_0 = [\ [Var(\mathbf{x}_0)]^{-1} + C_0^\top R_0^{-1}C_0]^{-1} \\ k = 1, 2, \cdots. \end{cases}$$

By omitting the "bar" on $\bar{H}_k, \bar{G}_k$, and $\bar{P}_k$, we have (5.21).

5.8.   (a)

$$\begin{cases} \underline{X}_{k+1} = A_c\underline{X}_k + \underline{\zeta}_k \\ v_k = C_c\underline{X}_k. \end{cases}$$

(b)

$$P_{0,0} = \begin{bmatrix} Var(\mathbf{x}_0) & 0 & 0 \\ 0 & Var(\underline{\xi}_0) & 0 \\ 0 & 0 & Var(\eta_0) \end{bmatrix},$$

$$P_{k,k-1} = A_c P_{k-1,k-1} A_c^\top + \begin{bmatrix} 0 & & \\ & 0 & \\ & & 0 \\ & & Q_{k-1} \\ & & & r_{k-1} \end{bmatrix},$$

$$G_k = P_{k,k-1}C_c^\top \left(C_c^\top P_{k,k-1}C_c \right)^{-1},$$
$$P_{k,k} = (I - G_k C_c)P_{k,k-1},$$
$$\hat{\underline{X}}_0 = \begin{bmatrix} E(\mathbf{x}_0) \\ 0 \\ 0 \end{bmatrix},$$
$$\hat{\underline{X}}_k = A_c\hat{\underline{X}}_{k-1} + G_k(v_k - C_c A_c\hat{\underline{X}}_{k-1}).$$

(c) The matrix $C_c^\top P_{k,k-1} C_c$ may not be invertible, and the extra estimates $\hat{\underline{\xi}}_k$ and $\hat{\eta}_k$ in $\hat{\underline{X}}_k$ are needed.

Chapter 6

6.1. Since

$$\mathbf{x}_{k-1} = A\mathbf{x}_{k-2} + \Gamma\underline{\xi}_{k-2} = \cdots = A^n\mathbf{x}_{k-n-1} + noise$$

and

$$
\begin{aligned}
\tilde{\mathbf{x}}_{k-1} &= A^n[N_{CA}^\top N_{CA}]^{-1}(C^\top \mathbf{v}_{k-n-1} + A^\top C^\top \mathbf{v}_{k-n} \\
&\quad + \cdots + (A^\top)^{n-1}C^\top \mathbf{v}_{k-2}) \\
&= A^n[N_{CA}^\top N_{CA}]^{-1}(C^\top C\mathbf{x}_{k-n-1} + A^\top C^\top CA\mathbf{x}_{k-n-1} \\
&\quad + \cdots + (A^\top)^{n-1}C^\top CA^{n-1}\mathbf{x}_{k-n-1} + noise) \\
&= A^n[N_{CA}^\top N_{CA}]^{-1}[N_{CA}^\top N_{CA}]\mathbf{x}_{k-n-1} + noise \\
&= A^n\mathbf{x}_{k-n-1} + noise,
\end{aligned}
$$

we have $E(\tilde{\mathbf{x}}_{k-1}) = E(A^n\mathbf{x}_{k-n-1}) = E(\mathbf{x}_{k-1})$.

6.2. Since

$$\frac{d}{ds}\left[A^{-1}(s)A(s)\right] = \frac{d}{ds}I = 0\,,$$

we have

$$A^{-1}(s)\left[\frac{d}{ds}A(s)\right] + \left[\frac{d}{ds}A^{-1}(s)\right]A(s) = 0\,.$$

Hence,

$$\frac{d}{ds}A^{-1}(s) = -A^{-1}(s)\left[\frac{d}{ds}A(s)\right]A^{-1}(s)\,.$$

6.3. Let $P = U diag[\ \lambda_1, \cdots, \lambda_n\ ]U^{-1}$. Then

$$P - \lambda_{min}I = U diag[\ \lambda_1 - \lambda_{min}, \cdots, \lambda_n - \lambda_{min}\ ]U^{-1} \geq 0\,.$$

6.4. Let $\lambda_1, \cdots, \lambda_n$ be the eigenvalues of F and J be its Jordan canonical form. Then there exists a nonsingular matrix U such that

$$U^{-1}FU = J = \begin{bmatrix} \lambda_1 & * & & & \\ & \lambda_2 & * & & \\ & & \ddots & \ddots & \\ & & & \ddots & * \\ & & & & \lambda_n \end{bmatrix}$$

with each $*$ being 1 or 0. Hence,

$$F^k = UJ^kU^{-1} = U \begin{bmatrix} \lambda_1^k & * & \cdots & \cdots & * \\ & \lambda_2^k & * & \cdots & * \\ & & \ddots & & \vdots \\ & & & \ddots & * \\ & & & & \lambda_n^k \end{bmatrix},$$

where each $*$ denotes a term whose magnitude is bounded by

$$p(k)|\lambda_{max}|^k$$

with $p(k)$ being a polynomial of k and $|\lambda_{max}| = max(|\lambda_1|, \cdots, |\lambda_n|)$. Since $|\lambda_{max}| < 1$, $F^k \to 0$ as $k \to \infty$.

6.5. Since

$$0 \leq (A - B)(A - B)^\top = AA^\top - AB^\top - BA^\top + BB^\top,$$

we have

$$AB^\top + BA^\top \leq AA^\top + BB^\top.$$

Hence,

$$(A + B)(A + B)^\top = AA^\top + AB^\top + BA^\top + BB^\top$$
$$\leq 2(AA^\top + BB^\top).$$

6.6. Since $\mathbf{x}_{k-1} = A\mathbf{x}_{k-2} + \Gamma\underline{\xi}_{k-2}$ is a linear combination (with constant matrices as coefficients) of $\mathbf{x}_0, \underline{\xi}_0, \cdots, \underline{\xi}_{k-2}$ and

$$\vec{\mathbf{x}}_{k-1} = A\vec{\mathbf{x}}_{k-2} + G(\mathbf{v}_{k-1} - CA\vec{\mathbf{x}}_{k-2})$$
$$= A\vec{\mathbf{x}}_{k-2} + G(CA\mathbf{x}_{k-2} + CT\underline{\xi}_{k-2} + \underline{\eta}_{k-1}) - GCA\vec{\mathbf{x}}_{k-2}$$

is an analogous linear combination of $\mathbf{x}_0, \underline{\xi}_0, \cdots, \underline{\xi}_{k-2}$ and $\underline{\eta}_{k-1}$, which are uncorrelated with $\underline{\xi}_{k-1}$ and $\underline{\eta}_k$, the two identities follow immediately.

6.7. Since

$$P_{k,k-1}C_k^\top G_k^\top - G_k C_k P_{k,k-1} C_k^\top G_k^\top$$
$$= G_k C_k P_{k,k-1} C_k^\top G_k^\top + G_k R_k G_k^\top - G_k C_k P_{k,k-1} C_k^\top G_k^\top$$
$$= G_k R_k G_k^\top,$$

we have

$$-(I - G_k C)P_{k,k-1}C^\top G_k^\top + G_k R G_k^\top = 0.$$

Hence,

$$
\begin{aligned}
P_{k,k} &= (I - G_k C) P_{k,k-1} \\
&= (I - G_k C) P_{k,k-1} (I - G_k C)^{\top} + G_k R G_k^{\top} \\
&= (I - G_k C)\ (A P_{k-1,k-1} A^{\top} + \Gamma Q \Gamma^{\top})\ (I - G_k C)^{\top} + G_k R G_k^{\top} \\
&= (I - G_k C) A P_{k-1,k-1} A^{\top} (I - G_k C)^{\top} \\
&\quad + (I - G_k C)\Gamma Q \Gamma^{\top}(I - G_k C)^{\top} + G_k R G_k^{\top}\ .
\end{aligned}
$$

6.8. Imitating the proof of Lemma 6.8 and assuming that $|\lambda| \geq 1$, where λ is an eigenvalue of $(I - GC)A$, we arrive at a contradiction to the controllability condition.

6.9. The proof is similar to that of Exercise 6.6.

6.10. From

$$
\begin{aligned}
0 &\leq \langle \underline{\epsilon}_j - \underline{\delta}_j, \underline{\epsilon}_j - \underline{\delta}_j \rangle \\
&= \langle \underline{\epsilon}_j, \underline{\epsilon}_j \rangle - \langle \underline{\epsilon}_j, \underline{\delta}_j \rangle - \langle \underline{\delta}_j, \underline{\epsilon}_j \rangle + \langle \underline{\delta}_j, \underline{\delta}_j \rangle
\end{aligned}
$$

and Theorem 6.2, we have

$$
\begin{aligned}
&\langle \underline{\epsilon}_j, \underline{\delta}_j \rangle + \langle \underline{\delta}_j, \underline{\epsilon}_j \rangle \\
&\leq \langle \underline{\epsilon}_j, \underline{\epsilon}_j \rangle + \langle \underline{\delta}_j, \underline{\delta}_j \rangle \\
&= \langle \hat{\mathbf{x}}_j - \mathbf{x}_j + \mathbf{x}_j - \vec{\mathbf{x}}_j, \hat{\mathbf{x}}_j - \mathbf{x}_j + \mathbf{x}_j - \vec{\mathbf{x}}_j \rangle + \|\mathbf{x}_j - \vec{\mathbf{x}}_j\|_n^2 \\
&= \|\mathbf{x}_j - \hat{\mathbf{x}}_j\|_n^2 + \langle \mathbf{x}_j - \vec{\mathbf{x}}_j, \hat{\mathbf{x}}_j - \mathbf{x}_j \rangle \\
&\quad + \langle \hat{\mathbf{x}}_j - \mathbf{x}_j, \mathbf{x}_j - \vec{\mathbf{x}}_j \rangle + 2\|\mathbf{x}_j - \vec{\mathbf{x}}_j\|_n^2 \\
&\leq 2\|\mathbf{x}_j - \hat{\mathbf{x}}_j\|_n^2 + 3\|\mathbf{x}_j - \hat{\mathbf{x}}_j\|_n^2 \\
&\to 5(P^{-1} + C^{\top} R^{-1} C)^{-1}
\end{aligned}
$$

as $j \to \infty$. Hence, $B_j = \langle \underline{\epsilon}_j, \underline{\delta}_j \rangle A^{\top} C^{\top}$ are componentwise uniformly bounded.

6.11. Using Lemmas 1.4, 1.6, 1.7 and 1.10 and Theorem 6.1, and applying Exercise 6.10, we have

$$
\begin{aligned}
&tr[F B_{k-1-i}(G_{k-i} - G)^{\top} + (G_{k-i} - G)B_{k-1-i}^{\top} F^{\top}] \\
&\leq (n\ tr F B_{k-1-i}(G_{k-i} - G)^{\top}(G_{k-i} - G)B_{k-1-i}^{\top} F^{\top})^{1/2} \\
&\quad + (n\ tr(G_{k-i} - G)B_{k-1-i}^{\top} F^{\top} F B_{k-1-i}(G_{k-i} - G)^{\top})^{1/2} \\
&\leq (n\ tr F F^{\top} \cdot tr B_{k-1-i} B_{k-1-i}^{\top} \cdot tr(G_{k-i} - G)^{\top}(G_{k-i} - G))^{1/2} \\
&\quad + (n\ tr(G_{k-i} - G)\ (G_{k-i} - G)^{\top} \cdot tr B_{k-1-i}^{\top} B_{k-1-i} \cdot tr F^{\top} F)^{1/2} \\
&= 2(n\ tr(G_{k-i} - G)\ (G_{k-i} - G)^{\top} \cdot tr B_{k-1-i}^{\top} B_{k-1-i} \cdot tr F^{\top} F)^{1/2} \\
&\leq C_1 r_1^{k+1-i}
\end{aligned}
$$

for some real number r_1, $0 < r_1 < 1$, and some positive constant C independent of i and k.

6.12. First, solving the Riccati equation (6.6); that is,

$$c^2 p^2 + [(1 - a^2)r - c^2\gamma^2 q]p - rq\gamma^2 = 0,$$

we obtain

$$p = \frac{1}{2c^2}\{c^2\gamma^2 q + (a^2 - 1)r + \sqrt{[(1 - a^2)r - c^2\gamma^2 q]^2 + 4c^2\gamma^2 qr}\,\}.$$

Then, the Kalman gain is given by

$$g = pc/(c^2 p + r).$$

Chapter 7

7.1.   The proof of Lemma 7.1 is constructive. Let $A = [a_{ij}]_{n \times n}$ and $A^c = [\ell_{ij}]_{n \times n}$. It follows from $A = A^c(A^c)^\top$ that

$$a_{ii} = \sum_{k=1}^{i} \ell_{ik}^2, \qquad i = 1, 2, \cdots, n,$$

and

$$a_{ij} = \sum_{k=1}^{j} \ell_{ik}\ell_{jk}, \quad j \neq i; \qquad i, j = 1, 2, \cdots, n.$$

Hence, it can be easily verified that

$$\ell_{ii} = \left(a_{ii} - \sum_{k=1}^{i-1} \ell_{ik}^2\right)^{1/2}, \qquad i = 1, 2, \cdots, n,$$

$$\ell_{ij} = \left(a_{ij} - \sum_{k=1}^{j-1} \ell_{ik}\ell_{jk}\right)/\ell_{jj}, \qquad j = 1, 2, \cdots, i - 1; \quad i = 2, 3, \cdots, n,$$

and

$$\ell_{ij} = 0, \qquad j = i + 1, i + 2, \cdots, n; \quad i = 1, 2, \cdots, n.$$

This gives the lower triangular matrix A^c. This algorithm is called the *Cholesky decomposition*. For the general case, we can use a (standard) singular value decomposition (SVD) algorithm to find an orthogonal matrix U such that

$$U \ diag[s_1, \cdots, s_r, 0, \cdots, 0]U^\top = AA^\top,$$

where $1 \le r \le n, s_1, \cdots, s_r$ are singular values (which are positive numbers) of the non-negative definite and symmetric matrix $AA^\top$, and then set

$$\tilde{A} = U \ diag[\sqrt{s_1}, \cdots, \sqrt{s_r}, 0, \cdots, 0].$$

7.2.

(a) $L = \begin{bmatrix} 1 & 0 & 0 \\ 2 & 2 & 0 \\ 3 & -2 & 1 \end{bmatrix}$. (b) $L = \begin{bmatrix} \sqrt{2} & 0 & 0 \\ \sqrt{2}/2 & \sqrt{2.5} & 0 \\ \sqrt{2}/2 & 1.5/\sqrt{2.5} & \sqrt{2.6} \end{bmatrix}$.

7.3.

(a)

$$L^{-1} = \begin{bmatrix} 1/\ell_{11} & 0 & 0 \\ -\ell_{21}/\ell_{11}\ell_{22} & 1/\ell_{22} & 0 \\ -\ell_{31}/\ell_{11}\ell_{33} + \ell_{32}\ell_{21}/\ell_{11}\ell_{22}\ell_{33} & -\ell_{32}/\ell_{22}\ell_{33} & 1/\ell_{33} \end{bmatrix}.$$

(b)

$$L^{-1} = \begin{bmatrix} b_{11} & 0 & 0 & \cdots & 0 \\ b_{21} & b_{22} & 0 & \cdots & 0 \\ \vdots & \vdots & \vdots & & 0 \\ b_{n1} & b_{n2} & b_{n3} & \cdots & b_{nn} \end{bmatrix},$$

where

$$\begin{cases} b_{ii} = \ell_{ii}^{-1}, & i = 1, 2, \cdots, n; \\ b_{ij} = -\ell_{jj}^{-1} \sum_{k=j+1}^{i} b_{ik}\ell_{kj}, \\ j = i-1, i-2, \cdots, 1; & i = 2, 3, \cdots, n. \end{cases}$$

7.4. In the standard Kalman filtering process,

$$P_{k,k} \simeq \begin{bmatrix} 0 & 0 \\ 0 & 1 \end{bmatrix},$$

which is a singular matrix. However, its "square-root" is

$$P_{k,k}^{1/2} = \begin{bmatrix} \epsilon/\sqrt{1-\epsilon^2} & 0 \\ 0 & 1 \end{bmatrix} \simeq \begin{bmatrix} \epsilon & 0 \\ 0 & 1 \end{bmatrix}$$

which is a nonsingular matrix.

7.5. Analogous to Exercise 7.1, let $A = [a_{ij}]_{n \times n}$ and $A^u = [\ell_{ij}]_{n \times n}$. It follows from $A = A^u(A^u)^\mathsf{T}$ that

$$a_{ii} = \sum_{k=i}^{n} \ell_{ik}^2 , \qquad i = 1, 2. \cdots, n,$$

and

$$a_{ij} = \sum_{k=j}^{n} \ell_{ik}\ell_{jk} , \qquad j \neq i; \quad i, j = 1, 2, \cdots, n.$$

Hence, it can be easily verified that

$$\ell_{ii} = \left(a_{ii} - \sum_{k=i+1}^{n} \ell_{ik}^2 \right)^{1/2} , \qquad i = 1, 2, \cdots, n,$$

$$\ell_{ij} = \left(a_{ij} - \sum_{k=j+1}^{n} \ell_{ik}\ell_{jk} \right) / \ell_{jj} ,$$

$$j = i+1, \cdots, n; \quad i = 1, 2, \cdots, n.$$

and

$$\ell_{ij} = 0 , \qquad j = 1, 2, \cdots, i-1; \quad i = 2, 3, \cdots, n.$$

This gives the upper-triangular matrix A^u.

7.6. The new formulation is the same as that studied in this chapter except that every lower triangular matrix with superscript c must be replaced by the corresponding upper triangular matrix with superscript u.

7.7. The new formulation is the same as that given in Section 7.3 except that all lower triangular matrix with superscript c must be replaced by the corresponding upper triangular matrix with superscript u.

Chapter 8

8.1. (a) Since $r^2 = x^2 + y^2$, we have

$$\dot{r} = \frac{x}{r}\dot{x} + \frac{y}{r}\dot{y} ,$$

so that $\dot{r} = v \, sin\theta$ and

$$\ddot{r} = \dot{v} \, sin\theta + v\dot{\theta} \, cos\theta .$$

On the other hand, since $tan\theta = y/x$, we have $\dot{\theta}sec^2\theta = (x\dot{y} - \dot{x}y)/x^2$ or

$$\dot{\theta} = \frac{x\dot{y} - \dot{x}y}{x^2\ sec^2\theta} = \frac{x\dot{y} - \dot{x}y}{r^2} = \frac{v}{r}cos\theta,$$

so that

$$\ddot{r} = a\ sin\theta + \frac{v^2}{r}cos^2\theta$$

and

$$\ddot{\theta} = \left(\frac{\dot{v}r - v\dot{r}}{r^2}\right)cos\theta - \frac{v}{r}\dot{\theta}sin\theta$$

$$= \left(\frac{ar - v^2sin\theta}{r^2}\right)cos\theta - \frac{v^2}{r^2}sin\theta cos\theta.$$

(b)

$$\dot{\mathbf{x}} = \mathbf{f}(\mathbf{x}) := \begin{bmatrix} v\ sin\theta \\ a\ sin\theta + \frac{v^2}{r}cos^2\theta \\ (ar - v^2sin\theta)cos\theta/r^2 - v^2sin\theta\ cos\theta/r^2 \end{bmatrix}.$$

(c)

$$\mathbf{x}_{k+1} = \begin{bmatrix} x_k[1] + hv\ sin(x_k[3]) \\ x_k[2] + ha\ sin(x_k[3]) + v^2cos^2(x_k[3])/x_k[1] \\ vcos(x_k[3])/x_k[1] \\ (ax_k[1] - v^2sin(x_k[3]))cos(x_k[3])/x_k[1]^2 \\ -v^2sin(x_k[3])cos(x_k[3])/x_k[1]^2 \end{bmatrix} + \underline{\xi}_k$$

and

$$v_k = [\,1\ \ 0\ \ 0\ \ 0\,]\mathbf{x}_k + \eta_k\,,$$

where $\mathbf{x}_k := [\ x_k[1]\ \ x_k[2]\ \ x_k[3]\ \ x_k[4]\ ]^\mathsf{T}$.

(d) Use the formulas in (8.8).

8.2.   The proof is straightforward.

8.3.   The proof is straightforward. It can be verified that

$$\hat{\mathbf{x}}_{k|k-1} = A_{k-1}\hat{\mathbf{x}}_{k-1} + B_{k-1}\mathbf{u}_{k-1} = \mathbf{f}_{k-1}(\hat{\mathbf{x}}_{k-1})\,.$$

8.4.   Taking the variances of both sides of the modified "observation equation"

$$\mathbf{v}_0 - C_0(\theta)E(\mathbf{x}_0) = C_0(\theta)\mathbf{x}_0 - C_0(\theta)E(\mathbf{x}_0) + \underline{\eta}_0\,,$$

and using the estimate $(\mathbf{v}_0 - C_0(\theta)E(\mathbf{x}_0))(\mathbf{v}_0 - C_0(\theta)E(\mathbf{x}_0))^\top$
for $Var(\mathbf{v}_0 - C_0(\theta)E(\mathbf{x}_0))$ on the left-hand side, we have

$$(\mathbf{v}_0 - C_0(\theta)E(\mathbf{x}_0))(\mathbf{v}_0 - C_0(\theta)E(\mathbf{x}_0))^\top$$
$$= C_0(\theta)Var(\mathbf{x}_0)C_0(\theta)^\top + R_0 .$$

Hence, (8.13) follows immediately.

8.5. Since
$$E(\mathbf{v}_1) = C_1(\theta)A_0(\theta)E(\mathbf{x}_0) ,$$

taking the variances of both sides of the modified "observation equation"

$$\mathbf{v}_1 - C_1(\theta)A_0(\theta)E(\mathbf{x}_0)$$
$$= C_1(\theta)(A_0(\theta)\mathbf{x}_0 - C_1(\theta)A_0(\theta)E(\mathbf{x}_0) + \Gamma(\theta)\underline{\xi}_0) + \underline{\eta}_1 ,$$

and using the estimate $(\mathbf{v}_1 - C_1(\theta)A_0(\theta)E(\mathbf{x}_0))(\mathbf{v}_1 - C_1(\theta)A_0(\theta)$
$\cdot E(\mathbf{x}_0))^\top$ for the variance $Var(\mathbf{v}_1 - C_1(\theta)A_0(\theta)E(\mathbf{x}_0))$ on the
left-hand side, we have

$$(\mathbf{v}_1 - C_1(\theta)A_0(\theta)E(\mathbf{x}_0))(\mathbf{v}_1 - C_1(\theta)A_0(\theta)E(\mathbf{x}_0))^\top$$
$$= C_1(\theta)A_0(\theta)Var(\mathbf{x}_0)A_0^\top(\theta)C_1^\top(\theta) + C_1(\theta)\Gamma_0(\theta)Q_0\Gamma_0^\top(\theta)C_1^\top(\theta) + R_1 .$$

Then (8.14) follows immediately.

8.6. Use the formulas in (8.8) directly.

8.7. Since $\underline{\theta}$ is a constant vector, we have $S_k := Var(\underline{\theta}) = 0$, so
that
$$P_{0,0} = Var\binom{\mathbf{x}}{\underline{\theta}} = \begin{bmatrix} Var(\mathbf{x}_0) & 0 \\ 0 & 0 \end{bmatrix} .$$

It follows from simple algebra that

$$P_{k,k-1} = \begin{bmatrix} * & 0 \\ 0 & 0 \end{bmatrix} \quad and \quad G_k = \begin{bmatrix} * \\ 0 \end{bmatrix}$$

where $*$ indicates a constant block in the matrix. Hence,
the last equation of (8.15) yields $\hat{\underline{\theta}}_{k|k} \equiv \hat{\underline{\theta}}_{k-1|k-1}$.

8.8.
$$\begin{cases} \begin{bmatrix} \hat{x}_0 \\ \hat{c}_0 \end{bmatrix} = \begin{bmatrix} x^0 \\ c^0 \end{bmatrix} , \qquad P_{0,0} = \begin{bmatrix} p_0 & 0 \\ 0 & s_0 \end{bmatrix} \\[2ex] For \quad k = 1, 2, \cdots , \\[1ex] P_{k,k-1} = P_{k-1,k-1} + \begin{bmatrix} q_{k-1} & 0 \\ 0 & s_{k-1} \end{bmatrix} \\[2ex] G_k = P_{k,k-1} \begin{bmatrix} \hat{c}_{k-1} \\ 0 \end{bmatrix} [[\hat{c}_{k-1}\ 0]P_{k,k-1} \begin{bmatrix} \hat{c}_{k-1} \\ 0 \end{bmatrix} + r_k]^{-1} \\[2ex] P_{k,k} = [I - G_k[\hat{c}_{k-1}\ 0]]P_{k,k-1} \\[2ex] \begin{bmatrix} \hat{x}_k \\ \hat{c}_k \end{bmatrix} = \begin{bmatrix} \hat{x}_{k-1} \\ \hat{c}_{k-1} \end{bmatrix} + G_k(v_k - \hat{c}_{k-1}\hat{x}_{k-1}) , \end{cases}$$

where c^0 is an estimate of $\hat{c}_0$ given by (8.13); that is,

$$v_0^2 - 2v_0 x^0 c^0 + [(x^0)^2 - p_0](c^0)^2 - r_0 = 0.$$

Chapter 9

9.1. (a) Let $\bar{\mathbf{x}}_k = [x_k \ \dot{x}_k]^\mathsf{T}$. Then

$$\begin{cases} x_k = -(\alpha + \beta - 2)x_{k-1} - (1-\alpha)x_{k-2} + \alpha v_k + (-\alpha + \beta)v_{k-1} \\ \dot{x}_k = -(\alpha + \beta - 2)\dot{x}_{k-1} - (1-\alpha)\dot{x}_{k-2} + \dfrac{\beta}{h}v_k - \dfrac{\beta}{h}v_{k-1}. \end{cases}$$

(b) $0 < \alpha < 1$ and $0 < \beta < \frac{\alpha^2}{1-\alpha}$.

9.2. System (9.11) follows from direct algebraic manipulation.

9.3. (a)

$$\Phi = \begin{bmatrix} 1-\alpha & (1-\alpha)h & (1-\alpha)h^2/2 & -s\alpha \\ -\beta/h & 1-\beta & h - \beta h/2 & -s\beta/h \\ -\gamma/h^2 & 1-\gamma/h & 1-\gamma/2 & -s\gamma/h^2 \\ -\theta & -\theta/h & -\theta h^2/2 & s(1-\theta) \end{bmatrix}$$

(b)

$$det[zI - \Phi] =$$
$$z^4 + [(\alpha - 3) + \beta + \gamma/2 - (\theta - 1)s]z^3$$
$$+ [(3 - 2\alpha) - \beta + \gamma/2 + (3 - \alpha - \beta - \gamma/2 - 3\theta)s]z^2$$
$$+ [(\alpha - 1) - (3 - 2\alpha - \beta + \gamma/2 - 3\theta)s]z + (1 - \alpha - \theta)s.$$

$$\tilde{X}_1 = \frac{zV(z-s)}{det[zI - \Phi]}\{\alpha z^2 + (\gamma/2 + \beta - 2\alpha)z + (\gamma/2 - \beta + \alpha)\},$$

$$\tilde{X}_2 = \frac{zV(z-1)(z-s)}{det[zI - \Phi]}\{\beta z - \beta + \gamma\}/h,$$

$$\tilde{X}_3 = \frac{zV(z-1)^2(z-s)}{det[zI - \Phi]}\gamma/h^2,$$

and

$$W = \frac{zV(z-1)^3}{det[zI - \Phi]}\theta.$$

(c) Let $\check{X}_k = [\; x_k \;\; \dot{x}_k \;\; \ddot{x}_k \;\; w_k \;]^{\mathsf{T}}$. Then

$$x_k = a_1 x_{k-1} + a_2 x_{k-2} + a_3 x_{k-3} + a_4 x_{k-4} + \alpha v_k$$
$$+ (-2\alpha - s\alpha + \beta + \gamma/2)v_{k-1} + [\alpha - \beta + \gamma/2$$
$$+ (2\alpha - \beta - \gamma/2)s]v_{k-2} - (\alpha - \beta + \gamma/2)sv_{k-3}\,,$$

$$\dot{x}_k = a_1 \dot{x}_{k-1} + a_2 \dot{x}_{k-2} + a_3 \dot{x}_{k-3} a_4 \dot{x}_{k-4} + (\beta/h)v_k$$
$$- [(2+s)\beta/h - \gamma/h]v_{k-1} + [\beta/h - \gamma/h$$
$$+ (2\beta - \gamma)s/h]v_{k-2} - [(\beta - \gamma)s/h]v_{k-3}\,,$$

$$\ddot{x}_k = a_1 \ddot{x}_{k-1} + a_2 \ddot{x}_{k-2} + a_3 \ddot{x}_{k-3} + a_4 \ddot{x}_{k-4} + (\gamma/h)v_k$$
$$- [(2+\gamma)\gamma/h^2]v_{k-1} + (1+2s)v_{k-2} - sv_{k-3}\,,$$

$$w_k = a_1 w_{k-1} + a_2 w_{k-2} + a_3 w_{k-3} + a_4 w_{k-4}$$
$$+ (\gamma/h^2)(v_k - 3v_{k-1} + 3v_{k-2} - v_{k-3})\,,$$

with the initial conditions $x_{-1} = \dot{x}_{-1} = \ddot{x}_{-1} = w_0 = 0$, where

$$a_1 = -\alpha - \beta - \gamma/2 + (\theta - 1)s + 3\,,$$
$$a_2 = 2\alpha + \beta - \gamma/2 + (\alpha + \beta h + \gamma/2 + 3\theta - 3)s - 3\,,$$
$$a_3 = -\alpha + (-2\alpha - \beta + \gamma/2 - 3\theta + 3)s + 1\,,$$

and

$$a_4 = (\alpha + \theta - 1)s\,.$$

(d) The verification is straightforward.

9.4. The verifications are tedious but elementary.

9.5. Study (9.19) and (9.20). We must have $\sigma_p, \sigma_v, \sigma_a \geq 0$, $\sigma_m > 0$, and $P > 0$.

9.6. The equations can be obtained by elementary algebraic manipulation.

9.7. Only algebraic manipulation is required.

Chapter 10

10.1. For (1) and (4), let $* \in \{+, -, \cdot, /\}$. Then

$$X * Y = \{x * y | x \in X, y \in Y\}$$
$$= \{y * x | y \in Y, x \in X\}$$
$$= Y * X\,.$$

The others can be verified in a similar manner. As to part (c) of (7), without loss of generality, we may only consider the situation where both $x \geq 0$ and $y \geq 0$ in $X = [\underline{x}, \overline{x}]$ and $Y = [\underline{y}, \overline{y}]$, and then discuss different cases of $\underline{z} \geq 0$, $\overline{z} \leq 0$, and $\underline{z}\overline{z} < 0$.

10.2. It is straightforward to verify all the formulas by definition. For instance, for part (j.1), we have

$$
A^I(BC) = \left[\sum_{j=1}^{n} A^I(i,j) \left[\sum_{\ell=1}^{n} B_{j\ell}C_{\ell k} \right] \right]
$$

$$
\subseteq \left[\sum_{j=1}^{n} \sum_{\ell=1}^{n} A^I(i,j)B_{j\ell}C_{\ell k} \right]
$$

$$
= \left[\sum_{\ell=1}^{n} \left[\sum_{j=1}^{n} A^I(i,j)B_{j\ell} \right] C_{\ell k} \right]
$$

$$
= (A^I B)C .
$$

10.3. See: Alefeld, G. and Herzberger, J. (1983).

10.4. Similar to Exercise 1.10.

10.5. Observe that the filtering results for a boundary system and any of its neighboring system will be inter-crossing from time to time.

10.6. See: Siouris, G., Chen, G. and Wang, J. (1997).

Chapter 11

11.1.

$$
\phi_2(t) = \begin{cases} \dfrac{1}{2}t^2 & 0 \leq t < 1 \\[2mm] -t^2 + 3t - \dfrac{3}{2} & 1 \leq t < 2 \\[2mm] \dfrac{1}{2}t^2 - 3t + \dfrac{9}{2} & 2 \leq t < 3 \\[2mm] 0 & otherwise. \end{cases}
$$

$$
\phi_3(t) = \begin{cases} \dfrac{1}{6}t^3 & 0 \leq t < 1 \\[2mm] -\dfrac{1}{2}t^3 + 2t^2 - 2t + \dfrac{2}{3} & 1 \leq t < 2 \\[2mm] \dfrac{1}{2}t^3 - 4t^2 + 10t - \dfrac{22}{3} & 2 \leq t < 3 \\[2mm] -\dfrac{1}{6}t^3 + 2t^2 - 8t + \dfrac{32}{3} & 3 \leq t < 4 \\[2mm] 0 & otherwise. \end{cases}
$$

11.2.
$$\widehat{\phi_n}(\omega) = \left(\frac{1 - e^{-i\omega}}{i\omega}\right)^n = e^{-in\omega/2}\left(\frac{sin(\omega/2)}{\omega/2}\right)^n.$$

11.3. Simple graphs.

11.4. Straightforward algebraic operations.

11.5. Straightforward algebraic operations.

Subject Index